The Biosaline Concept

An Approach to the Utilization of Underexploited Resources

Environmental Science Research

Editorial Board

Alexander Hollaender
Associated Universities, Inc.
Washington, D.C.

Ronald F. Probstein
Massachusetts Institute of Technology
Cambridge, Massachusetts

Bruce L. Welch
Environmental Biomedicine Research, Inc.
and
The Johns Hopkins University School of Medicine
Baltimore, Maryland

Recent Volumes in this Series

The Biosaline Concept

An Approach to the Utilization of Underexploited Resources

Edited by

Alexander Hollaender
Associated Universities, Inc.
Washington, D.C.

and

James C. Aller National Science Foundation
Emanuel Epstein University of California, Davis
Anthony San Pietro Indiana University
Oskar R. Zaborsky National Science Foundation

PLENUM PRESS / NEW YORK AND LONDON

Library of Congress Cataloging in Publication Data

Main entry under title:

The Biosaline concept.

(Environmental science research)
Includes index.
1. Biosaline resources—Congresses. 2. Mariculture—Congresses. 3. Salt-tolerant
crops—Congresses. I. Hollaender, Alexander, 1898-
SH138.B56   631   79-18804
ISBN 0-306-40295-5

© 1979 Plenum Press, New York
A Division of Plenum Publishing Corporation
227 West 17th Street, New York, N.Y. 10011

Printed in the United States of America

Preface

There are many areas on this world which might lend themselves
to agricultural development and which are, at the present, not used
for this purpose. Two of the most obvious are desert areas where
the salt concentration is very high, both land and water areas. With
the development of new approaches and careful research, considerably
more productive capability could be developed in these. This volume
points out some of the possible approaches as well as results obtained by a combination of creative research, practical understanding
of the problems involved and inventive ways to overcome some of the
handicaps of utilizing biosaline areas.

This volume grew out of the "International Workshop on Biosaline
Research" organized by Mr. Gilbert Devey of the Division of International programs of the National Science Foundation and directed by
Dr. Anthony San Pietro of the Department of Biology of Indiana University. Since the proceedings of the workshop appeared somewhat
limited, it was thought to broaden the spectra of chapters and include several topics briefly discussed at the Kiawah workshop.

As in any quickly developing field, it was not possible to obtain all the contributions which we thought should be included in
this volume. However, we have presented enough material to stimulate further investigation of this interesting field. When asked by
my colleagues to edit this volume, I was very glad to do this because
of my interest in mutagenesis, which seems to be one way to make
plants produce under biosaline conditions. Since no one is an expert
in the biosaline field, I couldn't have completed this book without
the cooperation and support of my co-editors and contributors. All
of us look forward to the day when there are biosaline experts.

Alexander Hollaender

v

Contents

THE BIOSALINE CONCEPT

James C. Aller and Oskar R. Zaborsky

National Science Foundation

Washington, DC   20550

"The real and legitimate goal of the sciences is the endowment
of human life with new inventions and riches."
                              -Sir Francis Bacon,
                                  Novum Organum, 1620.

INTRODUCTION

For centuries man has known that the earth is a sphere and that
there are physical limitations to its size and resources. However,
only in the last few decades have we really seen our planet as an island in the void of space. From this dramatic vantage point, we have
also gotten the harsh message that, indeed, our earth has only finite
resources and that we need to manage them more effectively. At the
very same time, the world's population has increased to over four
billion, and severe political, social and economic instabilities are
prevalent in many parts of the world.

We should also be keenly aware that obtaining more resources
from fragile ecosystems or marginal lands can, at times, be counter
productive. Detrimental effects can manifest themselves in desertification or the collapse of fisheries. While some increases in production can be obtained through the use of well-known methods, as in
the recovery of land from the sea in the Netherlands or new irrigation projects in many parts of the world, these attempts are by them-

----

The views, conclusions and recommendations expressed herein are those
of the authors and do not necessarily reflect the views and policies
of the National Science Foundation.

selves insufficient. The world faces a problem of potentially cata-
strophic dimensions unless more resources can be discovered and managed properly. Yet, there is an opportunity. It is predicated on
using past scientific discoveries and on developing new knowledge and
technologies poised to tap the presently underutilized biological and
physical resources. The opportunity is based on the "biosaline concept."

THE BIOSALINE CONCEPT

The biosaline concept envisions the harmonious interplay of
biological systems with saline environments for ultimate benefit to
man. In particular, the desire is to provide alternative options to
critical material needs from renewable resources in an environmentally acceptable manner. The essential elements of the concept are biological systems and a saline or marine environment. Usually, the environment of primary interest is at high solar flux and elevated temperatures. Biological systems represent animals, plants or microorganisms or their individual cellular constituents; in particular,
the critical catalysts of concern--enzymes--are of prime importance.
Thus, the boundary conditions for biosaline research are biological
organisms, or their essential constituents, living in or being able
to adapt to and tolerate saline or marine environments. Arid lands
are an intrinsic part of the biosaline concept because they comprise
a resource that is currently underutilized and exists at the interface of many oceans. Figure 1 presents the essential elements of
the concept in diagrammatic form. The objective of the biosaline
concept is to provide us with material needs -- foods, feeds, energy,
fuels, chemicals, fertilizers, structural materials, fibers, and
medicinal compounds in an environmentally acceptable fashion.

The simplicity of the biosaline concept is that most of man's
activities to convert the enormous resources offered by desert areas
to productive uses are based on modifying the environment to be suitable for plants and animals that are already understood and used.
One example of such an environmental change is the removal of salt
and minerals from brackish water and the use of the desalinated water
on plants which require fresh water. In fact, nature does not levy
this fresh water requirement. Throughout the world--in the seas,
brackish waters, and the salt ponds and lakes of the deserts--many
plants and organisms exist which not only tolerate the presence of
dissolved minerals, including salt, but actually thrive on them.

Before expanding on this concept, we also need to recognize that
a problem of the magnitude posed by an expanding population and dimi-
nishing resources is not alleviated by one unique or magical solution
but needs rational development of alternatives suitable to individual
countries and regions. A number of possibilities exist. Although we

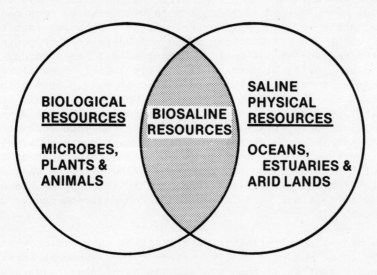

FIGURE 1

The essential elements of the Biosaline concept.

feel that the biosaline concept is a promising candidate, years or
decades may be required to bring certain elements to maturity. Consequently, other measures need to be pursued while the necessary research is underway.

One alternative to the present pattern of development is the
concept of conservation. Clearly, with a finite limit of available
resources, we ultimately need to utilize our materials much more
efficiently. Conservation is a sound principle and institutional
incentives and rewards are needed for effective management of our
presently known resources. For example, with food, it does not make
sense in the long term to grow more and more and at the same time
waste more and more. With the present food management system, the
production of more food means that more will also be consumed by
pests and subjected to other losses. It is a sobering fact to realize that about one-third of the world's food is lost to pests and
that for rice, one of the most important of all human foods, this
loss is a staggering 46 percent (1). This loss in food represents
the potential to increase our food supply by nearly 50 percent if
pest losses could indeed be eliminated. Underscoring this is the
fact that losses are highest in those regions of the world where the
human population is greatest in relation to the food resource. It
is clear that in some regions, for some crops, pests reap a harvest
much larger than does man. Pursuit of such alternatives as conservation will provide immediate gains and allow time for the careful
investigations and evaluations necessary if such approaches as the
biosaline concept are to be exploited in an environmentally acceptable manner.

The biosaline alternative is predicated on the fact that we have
not appreciated nor used wisely certain rather abundant physical and
biological resources of this planet. In particular, vast oceans and
seas of the world have not been utilized to any extent and certainly
not in any managed, systematic approach. The oceans cover 71 percent of the earth's surface and contain many valuable minerals and
living resources. Oceans and seas also account for about 97 percent
of all water on this planet (2). Of the remaining three percent,
about three quarters is stored in the polar ice caps and in glaciers.
Water in lakes, swamps, streams and the atmosphere accounts for less
than one percent of the total available fresh water. Ground water
and soil moisture account for approximately 22 percent of all fresh
water that is available. Likewise, the marginal lands at the interface of the oceans have been neglected both in scientific investigations as well as for possibly deriving essential materials. The
coastal deserts alone are an extensive domain covering the earth
for approximately 30,000 kilometers (3)..Also, the semi-arid lands
of the world have not been fully recognized as having merit for potential applications. These lands account for 36 percent of the
land availability or the equivalent of a land area of 51,970,000

square miles (4). As with these physical resources, except perhaps
for oceans, the biological resources associated with these domains
have also been considered detrimental or undesirable. Yet, there
exists a great diversity of microorganisms, plants and animal life
in these seemingly unfit environments which are rich for intellec-
tual endeavors leading toward applications to serve man's increased
needs.

In particular, the microbes living in these habitats of high
salinity and hot climates are an immense reservoir of clues on how
life began on this planet and on how we can better utilize the remaining portions of our finite world. Even with our land-based
plants we are currently using only about 200-300 species for commercial purposes and only about 80-90 to produce crops in the U.S. which
are valued at more than one billion dollars (5). With the great
diversity of biological life in the oceans (6), the estuarine environment and the arid regions of the world, there exists a great potential for developing new systems for our present and emerging needs.

An obvious element of the biosaline concept is to grow conventional crops with saline water. This approach has been described by
many investigators and a report was issued by the National Academy
of Sciences (7). Irrigation with saline waters of such crops as
cotton, barley, wheat, sugar beets, rye grass, date palm, and pis-
tachio can produce usable biomass and products. Mixed strategies of
using fresh or low salinity water for seed germination combined with
highly saline waters at other stages is also possible. Boyko is
generally credited with demonstrating that under suitable soil conditions (i.e., sandy), saline waters and even sea water can be used to
obtain respectable plant growth (8). However, this is but one of
two separate paths in reaching the goal of useful biosaline-derived
products. We can start with a plant such as barley or rice which
has a known useful product and is grown on fresh water. Then,
through research, we can capitalize on the plant's genetic diversity
to select those cultivars exhibiting salt tolerance which can be
grown on saline lands or with sea water. This approach is shown in
Figure 2 (lower part) and characterizes much of the work of Boyko (8)
and others (9,10).

However, there is another, perhaps more powerful method and this
is shown as the alternate (upper) path in Figure 2. In this approach, we start with biological sources which, over the ages, have
been selected by nature itself to adapt to and survive in saline
water. These biological resources are precious and not yet fully
identified. Then, through appropriate research we can create even
superior cultivars. As with the other approach, the use of naturally
salt-tolerant biological resources has been described but only to a
limited extent (3). Figure 2 also stresses that the products may not
be just the cereals of present agriculture but can well be chemicals

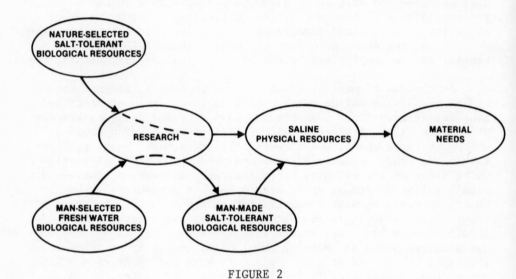

FIGURE 2

Alternatives for producing biosaline resources.

or other essential products.

The biosaline concept envisions the use and further development of advanced technologies to produce and process biosaline-derived resources. Two examples may illustrate the importance of advanced technologies in developing new industries: one is based on enzyme technology (11) and the other on the unique abilities of salt-tolerant microbes.

Sucrose or table sugar is an important food ingredient consumed in vast quantities throughout the world. Current total world production is estimated to be 85 million metric tons. Most sucrose is used in the solid crystalline form, but a large amount is handled as a liquid. Furthermore, most liquid sugar is partially or wholly inverted, i.e., the sucrose is hydrolyzed into its two component monosaccharide sugars, glucose and fructose. For this reason, liquid sugar markets have been an enticing target for the corn syrup and glucose manufacturers for some time. A deficiency of corn syrup is that glucose is only 0.7 times as sweet as sucrose. Hence, if a product as sweet as sucrose could be made, then a greater market penetration into sucrose could be achieved with corn syrups. The isomerization of glucose to fructose is the most obvious method for achieving this goal because fructose is about twice as sweet as glucose.

Although isomerization of glucose to fructose can be achieved with alkaline catalysts at high pH, the product made by this route was not a commercial success because excessive amounts of by-products were formed that imparted dark color and undesirable flavors. The answer to the long-sought goal of isomerizing glucose to fructose was realized through enzyme technology.

A most significant discovery was the enzyme "xylose isomerase" which converts not only xylose to xylulose but also D-glucose to D-fructose. After the discovery of this conversion in 1957, industrial development was pursued and the first commercial high-fructose corn syrup was introduced ten years later. The product contained 15% fructose and was made by a batch process using a soluble glucose isomerase. In 1968, the Clinton Corn Processing Company introduced a 42% fructose solution made with an immobilized enzyme. A continuous system using the immobilized enzyme was put into operation in 1972 which produces a high-fructose corn syrup containing 42% fructose, 50% glucose and 8% other saccharides on a dry basis. Since that time, a host of enzyme producers and corn wet milling companies have provided significant improvements to the overall process.

The impact of this enzyme technology innovation--an alternative sweetener source--has been dramatic. It has been estimated that high-fructose corn syrup will penetrate up to 30% of the market by 1985. This year over one million tons will be produced in the U.S. alone.

The second example to illustrate the potential of the biosaline concept deals with the use of salt-tolerant microorganisms able to produce a high concentration of a desirable material--glycerol. Microorganisms, of course, can be used in a variety of ways and have contributed to our quality of life and material well-being since antiquity. However, these elementary forms of life can be applied to provide us with foods, fuels, feeds, fertilizers, medicinals and a host of other products through ingenuity and modern science. As with higher plants and other living resources, an integrated and multi-product approach is the most desirable and economic path to achieve commercial realities. Such an approach may be especially attractive for developing countries (12).

Glycerol is a polyhydroxy, three-carbon compound useful as a solvent, plasticizer, sweetener, antifreeze, lubricant, fermentation nutrient, etc. Of course, it is used in the manufacture of dynamite. Glycerol can be obtained as a by-product in the production of soap and fatty acids from fats or from propylene obtained from petroleum. Glycerol can also be produced by fermentation using appropriate microbes. Interestingly, glycerol is produced within the cells of a number of algae and, in particular, high concentrations of this compound have been observed in the halophilic green alga, Dunaliella (13). The particular species producing large quantities of glycerol was isolated from the highly saline Dead Sea and the commercialization of this process is being developed in Israel (14). Koor, a large industrial complex, plans to grow these algae from which glycerol can be extracted. However, the entire process will also involve using the residual algae as an animal feed or as a source of desirable pigments (15). Preliminary estimates for producing glycerol by this route look encouraging, but much more time and effort are needed to verify and establish this process as a commercial reality. Yet, this development points toward the future by showing that relatively quick applications may be expected.

When considering the biosaline concept, we do not advocate increasing the salinity of land or bodies of water nor the indiscriminate use of salt water for irrigating semi-arid lands. On the contrary, we stress the need for constant evaluation of any developmental effort of fragile ecosystems in order to not irreversibly destroy them. Any conversion of biosaline research into full-scale use should be done with caution. Leopold has commented on the deterioration of good range lands by the invasion of downy chess or cheat grass (16). Other examples of detrimental effects are given in the National Academy of Sciences report on aquatic weeds (17). A major concern has been the invasion of waterways by water hyacinths.

Still, there is another equally important point. Many barren sea
coasts are in their present state as a result of man's activities
in disturbing the ecological balance. Conversion of these barren
areas into productive resources can restore the ecological balance.
The long-term effects to soils, aquafirs, and existing plant-animal
systems need to be analyzed before large-scale applications are
undertaken. Nevertheless, we should observe that sound use of biosaline research offers the opportunity for environmental improvements.
These improvements may involve mixed strategies. It is quite possible that presently barren coast lines can be returned to productivity by relatively conventional management. Tolba made this point
in his keynote address at an International Workshop on the Application of Science and Technology for Desert Development (18). He
quoted historians of 2000 years ago on the fertility of land between
Alexandria and the Libyan border which is now a barren coast. The
area was kept productive until the 11th Century through careful
methods of soil and water conservation and ecologically sound systems. It may well be that carefully controlled biosaline projects
can be added to enhance even more the potential of this and other
coastlines.

                               RESEARCH NEEDS

     In order to attain the goals envisioned in the biosaline con-
cept, it is necessary to establish an effective information base of
the relevant fields of science and to translate basic facts into
applied and developmental efforts. Ultimately, new major industries
would be started and old ones augmented by new, sophisticated technologies.

     With regard to the needed research areas encompassing the bio-
saline concept, the identification process started with the generation of the concept (19). However, it was also recognized immediately that an immense number of disciplines and fields of science
and engineering are intimately connected to this broad concept and
that it would be best to develop it initially on a limited basis.
Yet, this does not imply that the biosaline concept excludes areas
not mentioned nor highlighted in this introductory paper.

     The genesis of the biosaline concept was founded on the exciting
developments of biotechnology and especially in enzyme technology
and genetic engineering (20). As was discussed previously, enzyme
technology is dramatically revolutionizing the sweetener industry
and represents an excellent opportunity for industrial innovation.
Numerous informal discussion were held with a number of scientists
in 1976 which ultimately led Dr. San Pietro to propose a workshop
on biosaline research areas. San Pietro identified many topics for
possible discussion within the biosaline area--from biophotolysis

TABLE 1. SPECIFIC BIOSALINE RESEARCH AREAS

LAND PLANTS AND CONTROLLED ENVIRONMENT
AGRICULTURE

o Identify and characterize arid land plants
   for useful constituents

o Establish effective cultivation and harvesting
   practices

o Examine biochemical pathways and physiological
   control points

o Devise more effective and inexpensive controlled
   environment agriculture (CEA) systems

o Improve plants for CEA compatability

PHOTOSYNTHETIC MARINE ORGANISMS

o Identify and characterize photosynthetic marine
   microorganisms and higher plants

o Establish effective large-scale production practices for promising candidates

o Investigate relevant biochemical, physiological,
   genetic, and environmental factors

o Examine general toxicity and nutritional characteristics of marine organisms

o Establish integrated use of marine resources
   (by-product development)

SALT-TOLERANT PLANTS AND MICROORGANISMS

o Determine the structure/function relationship of
   membranes from salt-tolerant organisms

o Select salt-tolerant cereals, legumes and forage
   crops for agricultural production

o Transfer salt-tolerant genetic traits to salt-
   sensitive lines

o Select halo-tolerant mycorrhizae for enhanced
   mineral nutrition of plants

o Enhance or introduce N-fixing capability in saline
   species

BIOLOGICAL WASTE TREATMENT

o Determine the influence of salinity and temperature on algal growth and bacterial oxidation of wastes

o Devise new procedures utilizing local materials
   and easily harvested algae

o Develop methods which optimize solar energy
   utilization, reclaim nitrogen and other resources
   and minimize loss of water

o Investigate the processing of waste-grown algae

o Develop effective detection and control systems
   for infectious agents and refractory residuals

ENZYME TECHNOLOGY

o Identify and characterize thermo- and halotolerant enzymes

o Develop effective procedures for the stabilization and immobilization of biosaline-derived
   enzymes and intact cells

o Examine the utility of biosaline microbes for
   processing advantages (sterilization, energy
   effectiveness)

o Assess the feasibility of using microbes and/
   or enzymes to produce industrial chemicals
   (hydrogen, alcohols) and other materials

o Explore the use of enzyme to upgrade low-
   quality biosaline-derived foods and feeds

to the use of salt-tolerant crops--but a more limited basis had to be
adopted in order to arrive at a tangible and realistic research program. Consequently, a steering committee was established to focus
on a limited number of high-priority research topics consistent with
the embryonic concept. Ultimately, the number of topics to be discussed was reduced to five general areas:

o Land Plants and Controlled Environment Agriculture

o Photosynthetic Marine Organisms

o Salt-Tolerant Plants and Microorganisms

o Biological Waste Treatment

o Enzyme Technology

A number of relevant topics were excluded from open discussion
because of a severe time constraint as well as the need to focus on
new topics, especially as they relate to the interface of various
key disciplines. For example, energy and climatology were not included in the research list because a number of conferences, workshops and articles had been devoted to these topics. Of course,
this omission from the conference did not imply that they are not
important or that they do not form an integral element of any coherent research plan. Also, the emphasis of the research initially
envisioned was not to focus on conventional agricultural systems.
The five research areas were examined in the following manner at the
International Workshop on Biosaline Research held at Kiawah Island
in September, 1977. A position paper was first prepared by an
acknowledged expert who then delivered a summary of his paper and
held general discussions with participants in order to arrive at an
assessment of research needs, priorities and associated research
costs. Finally, a research document was prepared for all five areas
and a summary report was issued (21). A diverse array of research
topics was identified at the Kiawah Island Conference and in the
various documents generated in the further evolution of the concept.
The topics spanned from basic to applied research to demonstration
or field projects, and many disciplines and fields of science and
engineering are involved (See Table 1).

Additionally, a number of other conferences have been held.
These have augmented the research needs identified initially at the
Kiawah Island conference as well as highlighted other important
areas. A most important meeting was held in Kuwait in 1977 on Micro-
bial Conversion Systems for Food and Fodder Production and Waste
Management. The meeting was organized by the Kuwait Institute for
Scientific Research and Kuwait University and was sponsored by
UNESCO, ICRO and IFIAS. The meeting focused on the regional develop-

ment of the Kuwait area, as well as other countries of the Middle
East, in terms of single-cell protein, marine algae and plants as
food components, arid zone and saline plant productivity, enzyme
technology, and biological waste treatments. A more recent conference, devoted in part to the biosaline concept, was the Cairo International Workshop on the Application of Science and Technology for
Desert Development held in September of 1978. At this meeting, a
prototype directory of current biosaline research projects was an-
nounced and further refinements were solicited. The identification
and further development of research areas and a coherent program
have not stopped. In fact, the various chapters in this book represent an additional refinement of the process begun several years
ago.

Although specific research areas may be highlighted in this
chapter and other documents available on biosaline research, it is
important to remember the concept in its entirety. This means that
the research should not be too narrowly focused on very specific
objectives or end products. The aim is to utilize the entire available resource for a multitude of products and needs.

CONCLUSION

The implications of the biosaline concept are simple but dramatic. With diligence and ingenuity, it can provide for a viable alternative to the present strategy of economic development and may
be especially suited to those areas that have the physical and biological resources now viewed with much disdain. In fact, the oceans,
arid lands, estuaries and saline soils should be viewed as a valuable
resource in both physical and biological terms. The timely application of presently known science and technology, coupled with new in-
sights and scientific facts that can be derived from indigenous biological organisms, make it an extremely fruitful endeavor for all
nations concerned. For those countries that have a biosaline environment, it provides an unequaled opportunity to develop their
physical, biological and human resources for their own benefit without necessarily duplicating the patterns of the presently developed
countries. At the same time, the more developed, affluent countries
have an opportunity to establish for themselves and for other less-
developed nations an industrial base for the future when their non-
renewable fossil resources will be depleted. At the very least, the
biosaline concept should serve to unify diverse elements of science
toward a common objective.

REFERENCES

1. F. L. McEwen (1978) BioScience 28, 773.

2. H. L. Penman (1970) in The Biosphere, pp. 39-45, W.H. Freeman and Company, San Francisco.

3. P. J. Mudie (1974) in Ecology of Halophytes (R.J. Reinold and W. H. Queens, eds.) pp. 565-597, Academic Press, New York.

4. H. L. Shantz (1956) in The Future of Arid Lands (G.G. White, ed.), pp. 3-25, Am. Assoc. Adv. Science, Washington, DC.

5. D. L. Miller (1975) in Proceedings of the Eighth Cellulose Conference I. Wood Chemicals - A Future Challenge, Applied Polymer Symposia No. 28, (T.E. Timell, ed.), pp. 21-28, John Wiley, New York.

6. J. D. Costlow, Jr., ed (1971) Fertility of the Sea, Proceedings of an International Symposium, Sao Paulo, Brazil, 1969, Vol. 1 and II, Gordon and Breach Science Publishers, New York.

7. Report of an Ad Hoc Panel of the Advisory Committee on Technology Innovation Board on Science and Technology for International Development (1974) More Water for Arid Lands, Promising Technologies and Research Opportunities, National Academy of Sciences, Washington, DC.

8. H. Boyko (1967), Sci. Am., 215,89.

9. E. Epstein, (1977) in Plant Adaptation to Mineral Stress in Problem Soils (M.J. Wright, ed.), pp. 73-82, Cornell University Agricultural Experiment Station, New York.

10. M. Akbar, A. Shakoor and M.S. Sajjad (1977) in Genetic Diversity in Plants (A. Muhammed, R. Aksel and R.C. von Borstel, eds.) pp. 291-299, Plenum Press, New York.

11. O. R. Zaborsky, (1977) in Proceedings of the Regional Seminar on Microbial Conversion Systems in Food and Fodder Production and Waste Management (T. Overmire, ed.) pp. 193-209, Kuwait Institute for Scientific Research, Kuwait.

12. E. J. DaSilva, R. Olembo and A. Burgers (1978) Impact of Science on Society, 28, 159.

13. A. Ben-Amotz and M. Avron (1973) Plant Physiol. 51, 875.

14. Anon (1978) Inside R&D 7, No. 17, April 26.

15. A. San Pietro in Proceedings of Cairo International Workshop on Application of Science and Technology for Desert Development, Sept. 1978, in press.

16. A. Leopold (1966) A Sand County Almanac, pp. 164-168, Ballantine, New York.

17. Report of Ad Hoc Panel of the Advisory Committee on Technology Innovation Board on Science and Technology for International Development (1976) Making Aquatic Weeds Useful: Some Perspectives for Developing Countries, National Academy of Sciences, Washington, D.C.

18. M.K. Tolba in Proceedings of Cairo International Workshop on Application of Science and Technology for Desert Development, September, 1978, in press.

19. J. Aller, L. Mayfield and O. Zaborsky (1976), Oct. 29 Memorandum.

20. J. K. Setlow and A. Hollaender, eds. (1979) Genetic Engineering, Principles and Methods, Vol. 1, Plenum Press, New York.

21. A. San Pietro (1977) Summary Report of International Workshop on Biosaline Research, Indiana University, Bloomington.

FEED AND FOOD FROM DESERT ENVIRONMENTS

James A. Bassham

Lawrence Berkeley Laboratory
University of California
Berkeley, CA 94720

INTRODUCTION

The Bio-Saline Concept. The term "Bio-Saline Research" was originally designated by the National Science Foundation scientists to
describe a concept "that in time will lead to sustainable industries
based specifically on plants selected to grow in salt water",[1] particularly in arid land areas of the Earth, where there is an abundance
of solar radiation, sea water and dry climate.[2] A broader consideration of the possibilities of such areas indicates that it is also
worthwhile to consider the potential of utilizing plants which are
not salt-tolerant. The bio-saline concept has been extended to em-
compass "the elements of research, development, demonstration, and
utilization to apply modern biological sciences and technologies for
deriving essential resource materials in an environmentally harmonious
manner from marine and arid land mass systems."[3] Within this latter
framework are the two main topics of this article: the use of plants
adapted to arid lands, and modification of the environment through
Controlled Environmental Agricultural Technology (CEA)[2] to allow a
broad range of conventional and unconventional crops to be grown with
a very limited supply of fresh or brackish water. This water could
be derived from the sea, from saline lakes, or from waste water treatment.

The Limitation Due to Lack of Water. Availability of water is,
in fact, the most serious limitation to crop productivity in large
areas of the world with high solar radiation and long growing seasons.
This limitation is felt not only in clearly arid and semi-arid regions
but also in areas where the rainfall is highly variable over a period
of time. Such variability may occur over a short span of a few years

17

or over a much longer span of centuries. What makes this kind of
variation so dangerous at the present time is the fact that the population of the Earth is now so much greater than during those past
periods of history when there was low rainfall in areas presently
used for agriculture. The margin of productivity over need is now
precarious. Also according to some meteorological forecasts, an increasing CO_2 level in the atmosphere is leading to warming of the
average temperature of the earth which could result in decreased rainfall over some extensive areas that are presently productive. It is
true that such predictions also forecast other areas where rainfall
will increase, but there may not be an equalizing of effects on global
productivity, given the location of land masses, mountains and plains,
etc. In any event, severe local consequences might be experienced.

Desertification. Another cause of problems in areas of variable
rainfall is over-grazing or in some cases over-cultivation of the land
which accelerates the process of desertification--the conversion of
land with plant cover to barren rocky or sandy desert. When subnormal
rainfall is combined with such overuse, the results are particularly
disastrous. Clearly, there is a pressing need to develop technologies
to help those nations in arid and semi-arid regions whose principal
energy resource is the sun.

BENEFITS OF INCREASED PRODUCTIVITY

The economic and sociological implications of increased crop
productivity in the dry-land countries are enormous. There is great
diversity among the countries with arid lands and high solar energy
input. The need for more feed and food, as well as for chemicals and
materials from plants varies in both kind and amount from one country
to another. Most or all would welcome increased crop production, even
though the reasons may vary. The technology to be employed in achieving such improvements ultimately must be tailored to the specific opportunities and needs of each country.

Among the rewards of a successful program are the following:

1. The development of new and improved agricultural, agrochemi-
cal, and agromaterial industries, which in the case of countries rich
in petroleum resources can augment industry based on fossil fuels as
these become less abundant.

2. The improvement of nutrition and health in all dry land countries, but especially those with presently inadequate access to agricultural products, or subject to threat of famine from drought, desertification, attack by insects, and plant diseases.

3. The alleviation of costly imports of food and chemicals

through improved agriculture.

    4. The possibility of environmental benefits to land and water-
sheds by growing of more plants tolerant to local conditions, and
by relieving fragile grasslands of excessive grazing pressure by
growing more forage crops in small irrigated areas.

    Research and Costs. These important benefits must be kept in
mind when assessing the cost of the research and development program
needed to bring them about. This program will involve the disciplines
of not only plant physiology and agronomy, but also of genetics, chemistry, physics, and several kinds of engineering, applied to a very
diversified group of objectives. Some demonstration projects may
of necessity have to be rather large in scale to be meaningful, and
economic feasibility may be achievable only when highly complex systems are fully integrated. Some simpler concepts may also prove to
be feasible, even if the benefits may be more limited.

    Mankind has reached an extremely critical stage in history,
where population, technology, education, science, and political systems all are changing at an accelerating rate. It is imperative that
we make full use of the advanced state of science to design systems
for the future when present day economic parameters will no longer
be valid. We must ask the economists not what is feasible now but
what will be feasible in 10 to 50 years, and we must supply them with
the best possible projections of advances in science and technology
to use with their projections of supply and demand. We should not
consider this to be an impossible task, but rather should do the best
we can now, and constantly update the projections as new data are
gained from an intensive program of research and development.

DRY LANDS CHARACTERISTICS

    Geography of Dry Lands and Coastlines. The arid and semi-arid
lands of the world constitute about 36 percent of the land area.[4]
Although some dry lands are found in the arctic regions, the arid
and semi-arid lands with high annual solar energy lie mostly in the
regions between 15° and 40° north and south of the equator.

    Those areas with the most abundant solar energy average more
than 200 Kcal/cm^2 · yr., and there are extensive additional areas
averaging over 160Kcal/cm^2 · yr (Figure 1).[5]

    There are very extensive lengths of coastlines with seas bor-
dering arid lands with high annual incidence of solar energy. Such
lands border the South Coast of the Mediterranean Sea, the Red Sea,
the Arabian Sea and the Gulf of Omar and the Persian Gulf, the Atlantic cost of Africa north of about 15°N and south of 10°S latitude,

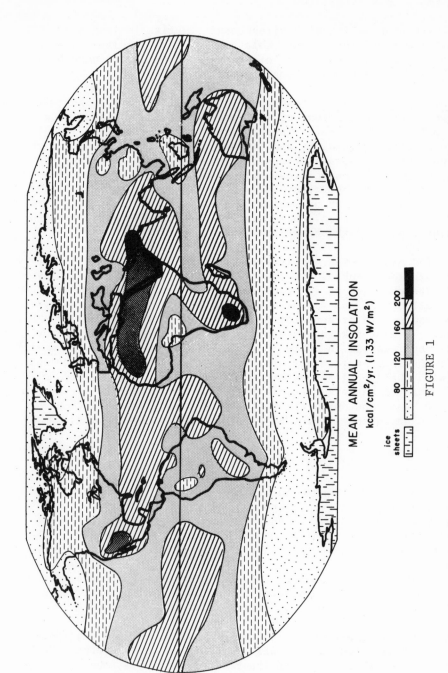

MEAN ANNUAL INSOLATION
kcal/cm²/yr. (1.33 W/m²)

ice
sheets

80 120 160 200

FIGURE 1

Mean Annual Insolation—Worldwide. Smaller area local variations are
omitted in this global map in order to provide a general view of incidence of
solar energy at the earth's surface. The figures are for total insolation over
one year.

and the Gulf of California and Pacific Ocean coast of Baja California
in Mexico. This proximity makes it possible to obtain water from the
sea either for use with salt-tolerant plants, or following desalination, for use with plants requiring fresh water. Also, temperature
control of land installations such as greenhouses may be possible by
using the lower temperature of the sea in the summer.

In a few cases, saline to brackish water lakes are found in desert areas, for example, Lake Chad in Africa. Moreover, there have
been proposals to admit sea water to certain below sea level areas in
the deserts, such as the 20,000 Km^2 Qattara Depression in Egypt. A
few other such projects are possible if deemed advisable. Conceiv-
ably, salt water canals could be employed to bring sea water to interior areas, if an agricultural industry based on salt water were to
evolve.

Productivity of Dry Lands. Without irrigation, the photosynthetic productivity of dry lands is naturally extremely low. Desert
Scrub Dry Desert, and Chaparral lands constitute 19% of the continental area (Table I),[6] but only 2.3% of the primary photosynthetic
productivity is found there. Most of the continental productivity is
in the forests, grasslands, woodlands, wet-lands and lakes, and in
cultivated fields. Very little of the lands in arid regions are cultivated, but where irrigation is possible, as in the Nile Valley of
Egypt, or the Imperial Valley of the United States, crop productivity
is very high, even though the problem of salt-buildup can become
serious in time. In such areas, further extension of agriculture is
usually limited by the availability of good land, water, or both.
Some lands, initially unsuited to conventional agriculture because
they are too sandy, salty, or lack humus, can be improved with appropriate treatment.

Efficient Use of Water. Efficient utilization of water can be
accomplished in several ways. Plants capable of growing in dry lands
can be exploited, application of water to the plants can be made more
economical, as by trickle irrigation, and waste water from municipal
and industrial uses can be reclaimed, purified and reused for agriculture.[7] Evaporative losses can be greatly reduced by utilization
of CEA, provided the system is sealed to prevent water loss, or sea
water evaporation is employed to saturate the air over the plants
with water vapor.

Desert Science. As indicated earlier, agriculture in the desert
without abundant sources of fresh water for irrigation or the use of
salt tolerant plants would appear to be limited to two possibilities.
The first is exploitation and perhaps cultivation of plants native
to arid environments. The second is to use CEA systems in which a
small amount of fresh water obtained from salt water or limited
sources such as springs, wells, or waste water reclamation can be

TABLE I

PRIMARY PHOTOSYNTHETIC PRODUCTIVITY OF THE EARTH

Area (total = 510 million Km^2)			Net Productivity (total = 155.2 billion tons dry wt/yr)	
Total Earth	100		100	
Continents	29.2		64.6	
Forests	9.8		41.6	
Tropical Rain		3.3		21.9
Raingreen		1.5		7.3
Summer Green		1.4		4.5
Chaparral		0.3		0.7
Warm Temperate				
Mixed		1.0		3.2
Boreal (Northern)		2.4		3.9
Woodland	1.4		2.7	
Dwarf and Scrub	5.1		1.5	
Tundra		1.6		0.7
Desert Scrub		3.5		0.8
Grassland	4.7		9.7	
Tropical		2.9		6.8
Temperate		1.8		2.9
Desert (Extreme)	4.7		0	
Dry		1.7		0
Ice		3.0		0
Cultivated Land	2.7		5.9	
Freshwater	0.8		3.2	
Swamp & Marsh		0.4		2.6
Lake & Stream		0.4		0.6
Oceans	70.8		35.4	
Reefs & Estuaries		0.4		2.6
Continental Shelf		5.1		6.0
Open Ocean		65.1		26.7
Upwelling Zones		0.08		0.1

Percentages based on data presented by H. Lieth at the Second
National Biological Congress, 1971[7].

used very efficiently by preventing its loss to the dry atmosphere
or into the ground.

PLANT ENERGY CONVERSION AND METABOLISM

Efficiency of Plants as Solar Energy Converters. For any utilization of desert plants or plants growing under CEA, it is important
to understand the factors controlling the efficiency of conversion of
sunlight to the chemical energy of biomass. Efficiency is important
but restricted in desert plants, which must capture and store the
sun's energy while at the same time conserving water. To prevent loss
of water to the low humidity sink of the desert in the daytime, but
still be able to take up carbon dioxide is a major accomplishment of
desert plants. There are four aspects of the plant physiology of
green plants that are especially important in this respect: the primary process of photosynthesis and its efficiency,[7] photorespiration,
C-4 metabolism, and Crassulacean Acid Metabolism (CAM).

The Mechanism of Photosynthesis and its Efficiency. Total dry
mass of organic material produced by a land plant, and to a lesser
extent the yield of the harvested organ (seed, root, fruit, etc.)
are related to the efficiency with which the plant uses the energy
of sunlight to drive the conversion of carbon dioxide, water and
minerals to oxygen and organic compounds [8,10]--the process of photosynthesis.

Increased photosynthesis is helpful in most cases in increasing
the yield of harvested organs (seeds, etc.), but an increase in
photosynthesis does not necessarily translate linearly into increased
crops in such cases. When the crop is the whole plant, however, and
that plant is harvested while still growing rapidly (before senescence
sets in) there can be such a relationship. For example, a forage crop
(such as alfalfa) harvested repeatedly so that the plants are always
growing at high rates may well produce total biomass which depends
on rate of photosynthesis. Photosynthesis takes place entirely in
the chloroplasts of green cells. Chloroplasts have an outer double
membrane. Inside the chloroplasts is a complex organization of membranes and soluble enzymes. The membranes are formed into very thin
hollow discs (thylakoids). These inner membranes contain the light-
absorbing pigments, chlorophylls a and b, and carotenes, and various
electron carriers, membrane-bound enzymes, etc. All these components
are required for the conversion of light energy to chemical energy.

As a result of the photochemistry in the membranes, water is
oxidized inside the thylakoids, releasing protons and molecular oxygen, O_2. The electrons pass through the membranes and bring about
the reduction of a soluble, low molecular weight protein called ferredoxin, which contains iron bound to sulfhydryl groups of the protein. The oxidation of two water molecules takes four electrons from

water and these are transferred to four ferredoxin molecules. Each
electron following this course must be transferred through a number
of steps, a photon of light is used with quantum efficiency of 1.0.
The light requirement for the transfer of four electrons is thus
two times four, or eight photons.

$$2H_2O + 4Fd^{+3} \xrightarrow{\text{eight photons}} 4H^+ + O_2 + 4Fd^{+2}$$

Concurrent with the electron transfer, there is a conversion of
adenosine diphosphate (ADP) and inorganic phosphate (P_i) to the biological acid anhydride, adenosine triphosphate (ATP).

$$ADP^{-3} + P_i^{-2} \longrightarrow ATP^{-4}$$

It appears that about three ATP molecules are formed for each
four electrons transferred, so the approximate complete equation becomes:

$$2H_2O + 4Fd^{+3} + 3ADP^{-3} + 3P_i^{-2} \xrightarrow{\text{8 h}\nu} H^+ + 3ATP^{-4} + 4Fd^{+2} + O_2$$

With the utilization of eight einsteins (moles of photons), the
thylakoid photochemical apparatus produces four moles of reduced ferredoxin and about three moles of ATP. These amounts of reduced ferredoxin and ATP are needed to bring about the reduction of one mole
of carbon dioxide to sugar in the dark reactions that follow. This
occurs in the stroma region of the chloroplasts, outside the thylakoids. The early reactions of photosynthesis are complete when car-
bon-dioxide has been converted to the glucose moiety of starch, a
major storage product in chloroplasts. By considering only a sixth
of a mole of such a glucose moiety, one can write a simplified equation for the entire process of photosynthesis:

$$CO_2 + H_2O \xrightarrow{\text{8 photons}} (CH_2O) + O_2$$
$$\text{1/6 glucose}$$
$$\text{moiety}$$

The free energy stored by this reaction is about 114 Kcal per
mole of CO_2 reduced to starch. (There is a bit more energy stored
per carbon in starch than in free glucose). While one-sixth glucose
molecule (CH_2O) has a molecular weight of 30, formation of starch or
cellulose occurs with the elimination of one water molecule (m.w. =
18) per glucose (m.w. = 180) incorporated. The effective weight increase for the reduction of one CO_2 molecule to (CH_2O) in starch or
cellulose is therefore $(180 - 18)/6 = 27$.

Green plants use only light with wavelenghts from 400nm to 700nm.
This photosynthetically active radiation (P.A.R.) constitutes only
about 0.43 of the total solar radiation of the earth's surface at
latitudes common to dry lands. All this light is used no more effi-

ciently by the green plant cells than if it were 700nm light. The
integrated solar energy input between 400 and 700nm at the Earth's
surface is equivalent in energy to monochromatic light at 575nm.
An einstein of light has an energy content given by Avogadro's number
times $h\nu$, where h is Planck's constant and ν is the frequency of the
light. With the appropriate units, E (Kcal/einstein) = $28.6/\lambda$, where
λ = wavelength – c/ν, in nm. An einstein of 575 nm light contains
28.6/.575 = 50 Kcal. At least eight photons of light are required
per molecule of CO_2 reduced; eight einsteins of light are required
per mole of CO_2. Probably the actual efficiency is somewhat less,
but measurements of quantum requirements under optimal conditions
in the laboratory have given quantum requirements in the range of 8
to 10 einsteins required per O_2 molecule evolved.[11] The maximum
efficiency for the primary light-driven formation of $[CH_2O]$ and O_2
is therefore 114/(8 x 50) = 0.286.

The maximum efficiency of 0.286 is for conversion of P.A.R.
The efficiency based on total solar radiation incident on the plants
with total absorption of P.A.R. is 0.43 x .286 = 0.123. This is
the basis for the statement sometimes made that the maximum efficiency
for solar energy conversion by photosynthesis is about 12%.

The maximum net efficiency, over a 24 hour period, and under
field or aquatic conditions, depends on two other factors: The
amount of incident light actually absorbed in the green tissue, and
the cost of energy used in respiration and biosynthesis. For land
plants it has been estimated that the maximum absorption to be expected from an optimal leaf canopy may be 0.80.[9] This is due to some
light being reflected, and some reaching the ground or falling on non-
photosynthetic parts of the plant (such as the bark of trees). With
aquatic plants such as unicellular algae that are totally immersed
there may be less reflection and with sufficient density of algae,
absorption could be essentially complete in green tissues.

A major loss in stored chemical energy results from respiration
which occurs in all tissue not actively photosynthesizing. This includes green cells at night or in dim light, and roots, trunks and
other organs that are not green or only a little green. The energy
derived from respiration is used for various physiological needs of
the plant, transport and translocation, conversion of photosynthate
to protein, lipids (including hydrocarbons in some plants), cellulose
for structures such as stalks and trunks, and so forth. In the green
cells during photosynthesis, some energy from the photosynthetic process itself may be used for such purposes, as mentioned earlier.
Like the light absorption factor, the factor for respiration/bio-
synthesis is extremely variable, depending on the physiological conditions and needs of the plant, but it is estimated that in a typical
case respiration and biosynthesis use up one-third of the energy
stored by photosynthesis.[9] The factor would thus be 0.67.

It may be argued that both the absorption factor and the respiration factor are not true maximum values, since there may be cases where each is exceeded. The product of these two factors, 0.80 x 0.67 = 0.53 probably is close to the maximum, since there is some trade-off between the two factors. For example, for a land plant to have all brightly illuminated leaves and hence lower respiration compared to photosynthesis would mean that its leaf canopy was probably less perfect than required for 0.8 absorption. At the other extreme, when there is dense foliage, little light may reach the ground, but the respiration in the shaded leaves may nearly equal photosynthesis. Similarly, an algae pond may be nearly totally absorbing, but the average light intensity for the cells would then be so low as to allow a high rate of respiration.

If we combine the photosynthetic efficiency, 0.123, with the product of the absorption and respiration/biosynthesis factors, 0.534, we obtain an overall maximum efficiency for photosynthetic/biosynthetic energy storage by green plants of 0.066. This calculated maximum efficiency can be compared with various reported high yield figures from agriculture. Before doing this it is useful to convert the efficiency to expected yield of dry matter.

From the equation and discussion given earlier, the reduction of a mole of CO_2 to the glucose moiety of starch or cellulose stores about 114 Kcal and results in an organic molecular weight of 27. Each Kcal of stored energy thus results in the formation of 27/114 = 0.237 grams of biomass (dry weight), if the biomass were entirely cellulose and starch. Of course, this is not the actual case, but the assumption provides a reasonable approximation.

<u>Calculated Maximum Biomass Production and Reported High Yields</u>. From the foregoing discussion, the upper limit for biomass production can be calculated by multiplying the efficiency, 0.066 times the daily total energy times 0.237. For high solar energy areas with 200 $Kcal/cm^2$ · year, or $2 \cdot 10^6$ $Kcal/m^2$ · year, the maximum energy stored would be 0.066 x $2 \cdot 10^6$ = $1.32 \cdot 10^5$ $Kcal/m^2 \cdot$year. The biomass, if all starch and cellulose, would be $1.32 \cdot 10^5$ x 0.237 = $3.1284 \cdot 10^4$ grams/ year·m^2, 85.7 grams/day·m^2 or 313 metric tons/hectare year. This is the total dry biomass that could be produced with continuous optimal growth.

Since optimal conditions of temperature, light absorption, etc. are never found during all times for crops grown under conventional agriculture, it is obvious that reported crop yields will not approach closely to this maximum on an annual yield basis. Also, crops in the temperate zone are usually grown under lower annual energy inputs.

What are the actual rates measured? The figures in parentheses

TABLE II

MAXIMUM PHOTOSYNTHETIC PRODUCTIVITY AND MEASURED MAXIMUM YIELDS

IN SELECTED PLANTS

	Assumed Radiation Kcal/cm^2.yr	gm/m^2.day	metric tons/hectare yr.	eff. %
Theoretical max. (Table II)				
High Solar Desert ann.	200	86	313	6.6
U.S. Average Annual	144	61	224	6.6
U.S. Southwest Ave. Ann.	168	72	263	6.6
U.S. Southwest, summer	247	106	387	6.6
Maximum Measured				
C-4 Plants				
Sugar cane	247	38	(138)*	2.4
Napier grass	247	39	(139)	2.4
Sudan grass (sorghum)	247	51	(186)	3.2
Corn (Zea mays)	247	52	(190)	3.2
C-3 Plants				
Sugar beet	247	31	(113)	1.9
Alfalfa	247	23	(84)	1.4
Chlorella	247	28	(102)	1.7
Annual Yield				
C-4 Plants				
Sugar cane	168	31	112	2.8
Sudan grass (Sorghum)	168	10	36	0.9
C-3 Plants				
Alfalfa	168	8	29	0.7
Eucalyptus	168	15	54	1.3
Sugar beet	168	9	33	0.8
Algae	168	24	87	2.2

*Parentheses indicate maximum rates. Since these are not sustained over a whole year, they are much higher than annual yields.

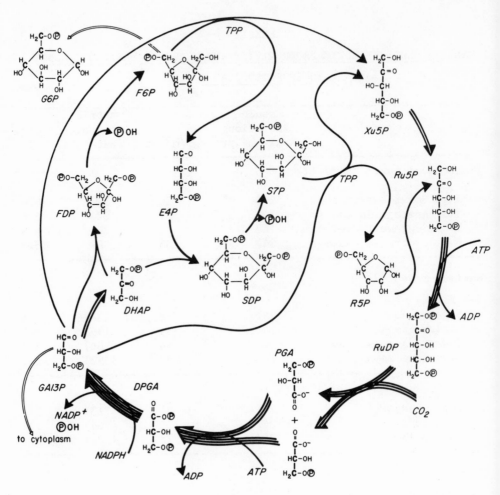

FIGURE 2

The reductive pentose phosphate cycle. The heavy lines indicate
reactions of the RPP cycle; the faint lines indicate removal of intermediate compounds of the cycle for biosynthesis. The number of heavy
lines in each arrow equals the number of times that step in the cycle
occurs for one complete turn of the cycle, in which three molecules
of CO_2 are converted to one molecule of GA13P. Abbreviations: RuDP,
Ribulose 1,5-diphosphate; PGA, 3-phosphoglycerate; DPGA, 1,3-diphos-
phoglycerate; NADPH and $NADP^+$, reduced and oxidized nicotinamide-ade-
nine dinucleotide phosphate, respectively; GA13P, 3-phosphoglycer-
aldehyde; DHAP, dihydroxyacetone phosphate; E4P, erythrose 4-phosph-
ate; SDP, sedoheptulose 1,7-diphosphate; S7P, sedoheptulose 7-phos-
phate; Xu5P, xylulose 5-phosphate, R5P, ribose 5-phosphate; Ru5P,
ribulose 5-phosphate; and TPP, thiamine pyrophosphate.

(Table II)[12] are rates during the active growing season, not annual
rates. For C-4 plants, these maximum rates range from 138 up to 190
metric tons per hectare per year. The highest (190) is about half the
calculated maximum. Similarly, the highest reported annual yield,
with sugar cane in Texas, is 112 metric tons per hectare--again about
1/2 the calculated maximum (263) for the U.S. Southwest. The energy
storage efficiency for these reported yields suggests that 3.3% to
perhaps 5% with CEA as the best we can hope for with land plants in
the future.

The Photosynthetic Carbon Reduction Pathway (Reductive Pentose
Phosphate Pathway). The terms "C-4" plants and "C-3" plants in Table
II refer to important characteristics of photosynthetic carbon metabolism. All known green plants and algae capable of oxidation of water
to O_2 employ the reductive pentose phosphate cycle (RPP cycle).[13,14]
This RPP cycle begins with the carboxylation of a five-carbon sugar
diphosphate (RuDP, Figure 2). The six-carbon proposed intermediate
is not seen, but is hydrolytically split with internal oxidation-re-
duction, giving two moledules of the three-carbon product, 3-phospho-
glycerate (PGA). With ATP from the light reactions, PGA is converted
to phosphoryl PGA, which in turn is reduced by NADPH to the three-
carbon sugar phosphate, 3-phosphoglyceraldehyde (Gal3P). The reduced
two-electron carrier, NADPH, is regenerated by the reaction of the
oxidized form, $NADP^+$, with two molecules of reduced ferredoxin, also
produced by the light reactions in the thylakoid membranes. Five
molecules of triose phosphate are converted to three molecules of the
pentose monophosphate, ribulose 5-phosphate (Ru5P) by a series of con-
densations, isomerizations, and chain length dismutations. Finally,
the Ru5P molecules are converted with ATP to the carbon dioxide accep-
tor, RuDP, completing the cycle.

When the three RuDP molecules are carboxylated to give six PGA
molecules, and these are in turn reduced to six Gal3P molecules, there
is a net gain of one triose phosphate molecule, equivalent to the
three CO_2 molecules taken up. This net Gal3P molecule can either be
converted to glucose 6-phosphate (G6P) and then to starch, or it can
be exported from the chloroplasts to the cytoplasm. Once there, it
may be reoxidized to PGA, yielding in addition ATP and NADH, which
thus become available to the non-photosynthetic part of the cell for
biosynthesis. In an expanding leaf, the material exported from the
chloroplasts may stay in the cell and be used in the synthesis of new
cellular material, leading to cell division. Alternatively some of
this exported carbon and ATP may be converted to sucrose, a sugar
which can then be translocated from the photosynthetic cell into the
vascular system of high plants through which it can move to other
parts of the plant such as the growing tip, seeds, roots, or other
sinks.

The C-4 Pathway. Plants which have only the RPP cycle for CO_2

fixation and reduction are termed "C-3" plants, since the primary
carboxylation product is a three-carbon acid. Certain plants of
supposed tropical origin including but not restricted to a number of
"tropical grasses" such as sugar cane, corn, crabgrass, sorghum, etc.
have, in addition to the RPP cycle, another CO_2 fixation cycle.[15-17]
In this cycle, CO_2 is first fixed by carboxylation of phosphoenoly-
pyruvate (PEPA), to give a four-carbon acid, oxalacetate (OAA), which
is then reduced with NADPH to give malate (or in some cases the amino
acid aspartate) (Figure 3).

The malic or aspartic acids are believed to be translocated into
the chloroplasts in cells near the vascular system of the leaf which
contain the enzymes and compounds of the RPP cycle. There these acids
are oxidatively decarboxylated, yielding CO_2, NADPH, and pyruvate,
which is translocated back out of the chloroplasts containing the RPP
cycle. In another variant, not shown in Figure 3, the malic acid is
converted once again to oxalacetic acid in the vascular bundle chloroplasts, and this acid is decarboyxlated to give PEPA which is then
converted to pyruvate. Finally, the pyruvate is converted in the
outer, mesophyll cells to PEPA by reactions which use up two ATP molecules. Since the first compounds into which CO_2 is incorporated in
this cycle are four-carbon acids, plants with this cycle are called
C-4 plants. The site of the conversion of pyruvate back to PEPA appears to be in specialized mesophyll cells whose chloroplasts do not
contain a complete RPP cycle (RuDP carboxylase is missing).

The net result of the C-4 cycle appears to be the fixation of CO_2
at sites removed from the RPP cycle chloroplasts, the translocation
of the C-4 acid products into these chloroplasts, and the release of
CO_2 close to RuDP carboyxlase. The cost is two ATP's per CO_2 molecule
transported. While at first glance this complex mechanism may appear
to be hardly worth the trouble (after all, C-3 plants do without it),
it turns out that the C-4 cycle performs an extremely valuable function. One reflection of its value is the higher productivity of C-4
plants seen in Table II. C-4 plants are in general capable of higher
rates of net photosynthesis in air under bright sunlight than the
most active C-3 plants. The C-4 plants are believed to have evolved
in the very regions we are interested in: the semi-arid lands with
high solar energy incidence.

Photorespiration.[18] The reason for the difference between C-4
and C-3 plants in maximum rates of photosynthesis lies in the virtual
abolition of the effects of photorespiration in C-4 plants. In C-3
plants, in air under bright sunlight, and especially on a warm day
where growing conditions should be very favorable, a certain part
of the sugar phosphates formed in the chloroplasts by photosynthetic
fixation are reoxidized, and are in part converted back to CO_2. Apparently, the energy and reducing power liberated by this oxidation
are not conserved and the process is energetically wasteful. As
light intensity and temperature increase, any increase in photosyn-

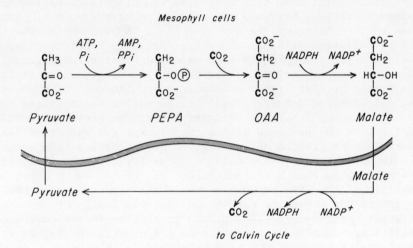

FIGURE 3

The C-4 cycle of photosynthesis. This is one version of the
preliminary CO_2 fixing cycle which occurs in certain tropical grasses
as well as in a scattering of other plant species. This cycle by
itself does not result in any net fixation of CO_2 into organic compounds, but rather serves as a vehicle to move CO_2 from cell cytoplasm and perhaps outer leaf cells into the chloroplasts of the vascular bundle cells in these plants. This CO_2 transport is thought
to be responsible for the minimization of photorespiration in these
cells (see text). In some plants another version (not shown) of the
C-4 cycle is found in which OAA is converted to aspartate rather than
malate for transport. Abbreviations: PEPA, phosphoenolpyruvate; OAA,
oxaloacetate.

thetic CO_2 uptake is negated by increased photorespiration. Net photo
synthesis, the difference between the two processes, cannot increase
beyond a certain point. The limiting effect on C-3 plants can be removed by reduction of the level of O_2 in the atmosphere to 2% or by
elevating the CO_2 pressure, but in the field plants, must live with the
natural atmosphere which contains 0.033% CO_2 and 20% O_2.[18]

There is still some controversy surrounding the detailed mechanis
of photorespiration, but much evidence supports the role of glycolic
acid as the key intermediate compound.[18] It is produced in the chloroplasts by oxidation of sugar phosphate (probably RuDP) to phosphogly-
colate and glycolate which is then oxidized outside the chloroplasts
to give photorespiratory CO_2. The production of glycolate is favored
in C-3 plants by high light, atmospheric or higher O_2, low CO_2 pressures, and elevated temperatures. Its formation is inhibited by elevated CO_2, although there is reported to be some glycolate formation
insensitive to CO_2 pressure inside the cloroplasts where the C-3 cycle
is operating. It is thought that glycolate formation from sugar phosphates is minimized in C-4 plants.[18] Some glycolate is produced even
in C-4 plants, so that a further effect of the C-4 cycle may be due to
the ability of the PEPA carboxylation in the other parts of the leaf
to recapture CO_2 before it can escape from the leaf. C-4 metabolism
is of great importance to many plants growing in desert environments.
C-4 plants are able to continue net photosynthetic CO_2 uptake at much
lower effective internal CO_2 pressures than C-3 plants, due to the
virtual absence of photorespiratory loss of CO_2 from leaves. This is
an advantage when water stress dictates partial or complete closing
of stomata, and at other times permits higher rates of photosynthesis
in bright light so that the C-4 plants can grow faster when favorable
conditions exist. The importance of Zea mays (corn) to Amerindians
of the U.S. Southwest and Mexico stemmed from the ability of this C-4
plant to grow in semi-arid environments.

Crassulacean Acid Metabolism. It is of particular interest to
consider plants native to semi-arid areas and deserts which do not
require irrigation. Not surprisingly, many such plants have evolved
very long root systems for collecting water from considerable depth
and other large areas. They have also developed physiological mechanisms for avoiding water loss. Such mechanisms can conserve water
but sometimes at the cost of limited photosynthetic productivity.
For example, plants with thick waxy cuticles and with stomata that
can be closed during the heat of the day are not able to take in
carbon dioxide rapidly; thus photosynthesis is limited. Many desert
plants exhibit Crassulacean Acid Metabolism (CAM) in which CO_2 is
taken in through the stomata open at night and incorporated by a carboxylation of PEPA to give dicarboxylic acids with four carbon-atoms
(for reviews see Osmond,[19] Ting[20]). During the night this PEPA is
made from sugars stored in the plant (Figure 4). In the daytime,
the stomata are closed, limiting water loss but also CO_2 entry. The

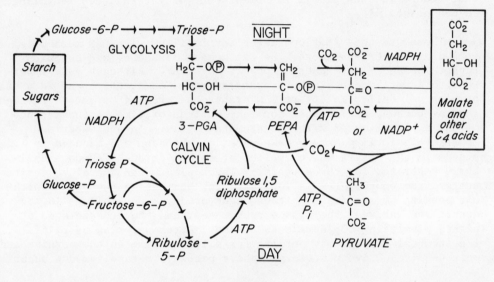

FIGURE 4

Crassulacean Acid Metabolism (CAM). In CAM plants during the night, starch and sugars are converted via glycolosis to PEPA which is then carboxylated to make C-4 acids. The C-4 acids accumulated during the night are converted during the day by transcarboxylation back to PEPA, and the CO_2 is utilized via the Calvin cycle to make sugars. In some plants a C-4 cycle type of metabolism in which the C-4 acids are transcarboxylated to give pyruvate which then is converted with ATP from the light to PEPA.

four-carbon acids are decarboxylated, the CO_2 released is reduced to
sugars by photosynthesis, and the PEPA is also reduced back to sugars.
In the morning and again in the late afternoon there can be intermediate stages when the stomata are open and CO_2 fixation by carboxylation of both ribulose diphosphate (RPP cycle) and PEPA occurs at
the same time.

Plants with CAM also exhibit photorespiration in the heat of the
day when the stomata are closed. The recycling of CO_2 within the leaf
that occurs in such plants is reminiscent of internal CO_2 recycling
in C-4 plants.[19]

There are some 18 flowering plant families with CAM metabolism
including the Crassulaceae, the Cactaceae, Alzaceae, and Succulent
Euphorbiaceae.[20] All cacti probably have CAM. Although very important to desert ecology, there are also many CAM plants found in areas
of high rainfall. In desert CAM plants the cycling of carbon through
the CAM pathway can persist for long periods of time in the absence
of any external water with the stomata closed. In one experiment
Opuntia bigelovii plants were severed at the base and mounted in
stands in the desert where cycling of carbon through CAM on a daily
basis persisted for three years.[20] These plants can therefore derive
energy from photosynthesis for very long periods in the desert without opening of stomata in either night or day. When plants in the
desert are watered, the tissue rehydrates, and the stomata open at
night, permitting CO_2 uptake to resume. After watering by rainfall,
the stomatal opening may persist for a longer time in the morning and
more C-3 (reductive pentose phosphate pathway) metabolism can occur.

Other Physiological Adaptations. The ability to conserve water
is obviously important to desert plants, but there are other requirements as well. In very hot areas, tolerance of high temperature is
required. Desert plants employ a great variety of physical shapes,
reflectances, insulation, etc. to protect themselves from heat.
Since water is a limiting factor, few species can afford the luxury
of extensive cooling by transpiration, as employed by plants accustomed to plentiful water. One studied species, which does grow in
very hot locations with abundant fresh water demonstrates the adaptation of enzyme systems to high temperatures. Tidestroma oblongi-
fola, a C-4 species grows in Death Valley, U.S., at fresh water
springs. Its maximum growth is reached at 45°C, a temperature at
which some temperate zone species greatly decline in growth rate,
even if well watered.[21] Although accustomed to growth in atmospheres at very low humidity, this plant does very well in chambers
maintained at high temperature and high humidity. Plants with such
characteristics could prove to be very useful in desert greenhouses
when maximum growth rates and minimal cooling are desirable.

UTILIZING THE DESERT FOR PLANT GROWTH

Desert agriculture for Food and Chemicals. Many types of
utilization of plants growing in the desert might be imagined, from
the already widespread (and often excessive) grazing of desert or
dryland grasses by livestock to proposals to harvest hydrocarbon-
containing plants growing in dry land as a source of liquid fuels and
chemical feedstocks, proposed by Calvin.[22] A principal problem with
using the desert fringes for grazing livestock is the tendency to
over graze, resulting in the conversion of desert fringe to desert
(desertification). Educational programs for the people living in
these environments might help, but only if alternative sources of
food and wealth from the desert can be developed. Even so, population
control would seem to be necessary, once a stable base for agriculture
and industry adequate to the planned population was developed.

One dry-land plant which has been suggested as a useful source
of food and materials is the common Mesquite (Prosopis species), found
growing wild in many parts of the U.S. Southwest.[23] The pods of this
plant have a high food value as protein and carbohydrate and were
used by American Indians as an important dietary supplement. Possibly
this plant could be used to supply both fuel and food. The plants are
legumes and do not require nitrogen fertilization. An annual yield
of 43 Kg dry weight of pods is harvested from one large tree in
Southern California. The protein has high nutritional value.[26]

Several types of plants well adapted to semi-arid environments
and capable of producing useful chemicals appear to have considerable
potential. Guayule has been raised in Mexico, and at times in the U.S.
for many years as a source of natural rubber.[25] From 1910 to 1946,
the U.S. imported more than 150 million pounds of guayule rubber from
Mexico. Much of this came from wild stands, which eventually could
not support such sustained harvesting. During the 1920's, large plantations were planted in the U.S., and rubber from these plantations
competed with Hevea rubber from Indonesia. Such developments stopped
during the 1930's, and later when the supply of Indonesian rubber was
interrupted, it was necessary to launch a massive new project in
guayule production in the U.S. and in Mexico. Over three million
pounds of resinous rubber were produced. By 1943, synthetic rubbers
were being produced from fossil fuels, and at the end of World War II
these synthetics plus large supplies of Hevea rubber which became
available from Indonesia again removed the necessity for producing
quayule rubber.

In Mexico, however, guayule development has continued, and
agencies of the Mexican government are embarking on rubber production
from guayule plants growing wild over about 4 million hectares. There
is considerable technology available for the production of guayule,
harvesting, extraction and deresination. From work done in Manzinar,

California in 1942-44, it is clear that good yields of guayule can
be grown in semi-arid regions without irrigation. Thus, guayule production may serve as a model for the production of other dry land
plants capable of supplying useful chemicals. Yokayama[26] has been
able to increase the rubber content of harvested guayule by a factor
of 2 to 3 by treating the 4-week old seedlings with 500 ppm each of
2-(3,4-dichlorophynoxy)-triethylamine and 2-diethylamino-ethanol plus
a wetting agent, and harvesting three weeks later.

The direct production of hydrocarbons as liquid fuels and chemical feedstocks by the extraction of latex bearing plants of the
Euphorbia family has been proposed and is being studied by Calvin.
(22,27). Test plots of several species are now being grown in southern California. Preliminary yield figures suggest that the hydrocarbon content of the biomass produced could supply as much as 5 to 10
barrels of oil per acre per year. These species can grow on semi-
arid lands in the U.S. Southwest. Such direct production of liquid
fuels is very attractive since it bypasses the conversion of biomass
to heat and then to electricity, with the resulting losses in efficiency. Moreover, the time will come when supplies of petroleum will
be exhausted, and it may well be necessary to obtain chemical feedstocks from plants as a replacement. Even if that time is 50 years
or more in the future, it is hardly too soon to being to develop the
technology to guarantee continued supplies so vital to modern civilization. Preliminary analyses of the hydrocarbon and lipid materials
in the latex of plants suggest a wealth of useful chemicals may become available.[27]

Another example of a specialty dry land plant is Jojoba. This
plant is now being grown on Indian reservations and in other areas in
Arizona as a source of a valuable lubricant with properties which
allow it to replace oil obtained from whales.

Controlled Environment Agriculture (CEA). With respect to Controlled Environment Agriculture (CEA), a substantial start has already
been made. The Environmental Research Laboratory (ERL) of the University of Arizona has been a pioneer in the development of CEA for
desert environments. Because such environments often contain populations that can otherwise only obtain fresh fruit and vegetables by
having them brought in by air at considerable expense, crops grown
in CEA can have a high local value, contributing to the cost effectiveness of the installation.

Desert greenhouses may be built of air-inflated plastic or combinations of plastic and glass. Sea water can be used for evaporative
cooling, and sea water can be distilled to provide for irrigation.
Problems of salt disposal in the greenhouses can be controlled because the soils are sandy and can be flushed with fresh water.

Given this promising growth in CEA technology, what is needed for the future? Can the application of CEA, now limited to relatively high value crops, and to construction requiring rather high investment capital, be applied to staple crops such as grains and fodder, and can this technology ever be constructed by developing nations not favored by the possession of large deposits of fossil fuel wealth? Finally, can those inputs of energy from fossil fuels, such as the fuel to drive sea water pumps, distillation units, and the hydrocarbons required for the synthesis of plastic be replaced by solar energy? I believe that there is an affirmative answer to these questions.

The status of controlled environment agriculture around the world has been reviewed in 1973 by Dalrymple,[28] and further discussion of CEA with examples of advanced CEA systems has been provided in 1977 by the extensive report by deBivort.[29] In the latter report it was concluded that CEA could substantially alleviate the agro-food problems of environmental degradation regional shortages of arable land, water, and fertilizers, and unreliability of production. The costs of present types of CEA systems were found to be prohibitive for agronomic crops, but acceptable for some high value fresh vegetables, but new types of CEA systems can be conceived for growing crops at considerably lower costs and much less total energy consumption than present CEA. Finally, CEA would appear most attractive if integrated with solar energy and water management systems for community units of several thousand people. It was recognized that CEA benefits are of interest to all concerned with food, energy and water resources and new opportunities for local self-sufficiency.[29]

At the present time, by far the greatest application of covered agriculture is in countries other than semi-arid and desert lands. For example, Japan is by far the largest user of covered agriculture, with over 10,000 hectares under cultivation in 1973.[28] Other leading countries, in terms of area under cover include the Netherlands, Italy, Belgium, France, the United Kingdom, USSR, Romania, Greece and South Korea.

Some of the most advanced CEA systems are to be found in arid lands. Although relatively small in area, these facilities are often very productive. Such facilities are located in Abu Dhabi, Kuwait, Iran, in Arizona in the U.S., and Puerto Pinasco in Sonora, Mexico. The four hectare Environmental Farms, Inc. near Tucson, Arizona, in the U.S. produces more than one million kilograms of tomatoes annually, with the produce being sold at off season times in the U.S. Another example of yield obtainable from such facilities is 538 metric tons (fresh weight) per hectare of cucumbers from the one in Abu Dhabi.[29] This facility includes both an air inflated polyethylene structure covering 2.5 hectares, and structured greenhouses covering one hectare. Cooling is by evaporation of sea water, with fans forcing air through the cooler and the greenhouses. Freshwater is ob-

tained by desalting sea water, and considerable care is taken to use
this costly water as efficiently as possible. The water vapor from
the evaporative cooling by seawater is thus extremely important in
preventing excessive water loss to the air from transpiration. Many
other important details of engineering and horticulture have been
worked out in such installations,[29] and this experience will be a
most valuable resource for the development of larger or more advanced
systems. Further details of CEA, present and proposed, are discussed
by deBivort in this workshop.

Can such systems be applied on a large scale to agriculture in
arid or semi-arid lands? The author [7,30-32] has proposed covering
large areas in dry lands with high greenhouses made from tough, sun-
resistant plastic. The structures might be 1 Km^2 in area and 300
meters high (at maximum extension), perhaps with a capacity to go up
and down daily. A requirement would be to maintain growing temperatures year round. Under this canopy would be grown high-protein
forage legumes such as alfalfa. They would be harvested periodically
during the year, leaving after each harvest enough of the plant to
produce quickly a good leaf canopy. Growth would be year round. The
atmosphere would be enriched in CO_2 and neither water vapor nor CO_2
would be allowed to escape, although some CO_2 would diffuse through
the plastic canopy (Figure 4).

While there are serious problems to be overcome with this system
(economic, engineering, and physiological), there are a number of
important advantages.

1. With year round growth and CO_2 enrichment (photorespiration
eliminated), maximum photosynthetic efficiency should be possible.
At a 5% conversion efficiency the yield would be about 200 metric tons
(dry weight)/hectare-year. The whole plant except for roots would be
harvested and used.

2. Most or perhaps all of the nitrogen requirements in legumes
would be met by N_2 fixation, due to stimulation of these high photosynthetic rates. Enrichment with CO_2 can result in a five-fold increase or more in N_2 fixation in the root nodules of legumes.

3. Alfalfa grown under optimal conditions has as high as 24%
protein content based on dry weight. It is feasible and economic to
remove a part of this protein as a high value product using the
methods developed at the Western Regional Research Laboratory of the
U.S. Department of Agriculture at Albany, California.[34] The residue
is a feed for ruminants, replacing expensive cereal grains which
could be sold for human nutrition in the U.S.A. and abroad where
there is a rapidly growing market. The protein extract of the
alfalfa has a high value as animal (poulty, for example) feed. An

interesting alternative is to convert part of it to a protein product for human consumption.[35] Nutritionally it is as good as milk protein (36) and far superior to soy protein. From the 15 metric tons of dry matter removed as juice from the leaves, it might be possible to recover 5 tons of protein, worth $5,000 at $1 per Kg.

4. Land with relatively low value at present because of lack of water could be used because of water recycling. With water vapor containment, only a few percent of the present irrigation requirements for desert land would have to be met.

5. The modular nature of the system would help in the prevention, containment, and elimination of plant diseases.

Since this scheme is envisaged as applicable to areas far removed from the sea, the use of evaporative cooling with sea water was not assumed. Instead, it was proposed to include a high enough canopy to enclose a sufficient volume of air so that the daytime temperature excursion would not be excessive. This might work in the higher cooler desert areas, especially where nighttime temperatures are very low, and sufficient loss of heat through the plastic at night occurs to bring the internal temperature down by morning. Even so, additional cooling powered by solar energy collectors outside the enclosure would probably be required in hot desert areas.

The advantages of a completely closed system over the air flow-through system would be complete retention of water vapor and more effective enrichment with added CO_2.

Of course, there are many problems; some very serious. The greenhouse effect would have to be controlled, perhaps by allowing daily expansion of the canopy. The plastic would have to be tough, sun-resistant, and not too permeable to CO_2. It would have to be synthesized from materials grown under the canopy, and inexpensive. Use of fossil fuels to synthesize the plastic could be avoided by making the plastic from some of the solid biomass residue, after protein extraction. Cellulose could be converted to glucose by treatment with enzymes from the fungi[37] and the resulting glucose could be fermented to give ethanol. Ethanol in turn could be converted to ethylene and thence to polyethylene or other suitable plastic. The insoluble material of plants also contains polymers of xylose. After acid hydrolysis, the xylose can be converted to furfural,[38] a possible starting material for other plastics.

There are other problems, but they may all be solvable. These very serious engineering and economic problems are not be lightly dismissed, but a discussion of possible solutions will require considerable engineering study to be meaningful.

FUTURE RESEARCH NEEDS

Many areas requiring research and engineering will be evident
from the foregoing discussion. Some are extensions of already abundant knowledge, such as the identification and characterization of
desert plants. Other areas require the development of relatively new
research and engineering areas.

Botany and Taxonomy. Identification and listing of abundance of
various species of plants in dry lands seem to be very complete in
some areas. For example, there is a very detailed study of the plants
of the Central-Southern Nevada area.[39] Probably there are large
areas of dry lands where such detailed studies are not yet available,
but would be useful both for evaluation of the potential of native
plants and for assessing environmental costs of proposed developments.

Plant Physiology. As indicated earlier, the biochemistry of
plant photosynthesis including that of desert plants is widely studied
and appears to be well understood. Other areas of plant physiology
relating to water and mineral use and conservation may be much less
well studied or understood.

Potential of Native Plants to provide Useful Food, Feed and
Materials. While there may be much historical and cultural knowledge
about plants from dry lands as sources of food and feed, there appears
to be a need for extensive analysis of plants to determine the amounts
and identity of useful constituents. A few examples were given earlier of some plants that are only recently being recognized as sources
of valuable substances. Also, there is little information about the
potential of crops from native plants if they were systematically
cultivated.

Plant Improvement and Adaptation. Experience with temperate zone
plants requiring moderate to heavy water application shows that very
great increases in productivity, quality of product, resistance to
disease, and other desirable properties can be achieved through breeding and other agronomic techniques. Although some plants such as Zea
mays, capable of growing under semi-arid conditions have been extensively bred and improved, there may be many other plants from arid
regions that could also benefit from such programs, particularly if
the search is extended to plants useful for chemicals and materials.

Desert Ecology and Management. Although much has been learned,
much more remains to be done, particularly with respect to educational
programs to help inhabitants of dry lands make better use of these
resources without degradation of the land. The problem of desertification of desert fringe land deserves much attention.

Controlled Environment Agriculture. Although use of greenhouses

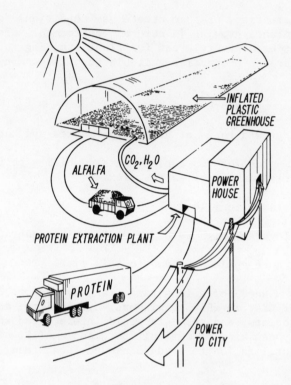

FIGURE 5

Scheme for energy and protein production by covered agriculture. Alfalfa, grown under transparent cover year-round with CO_2 enrichment, would be harvested and processed to remove some protein as a valuable product. The residue would be used as animal fodder or, in the version shown here, as fuel for power plants. Combustion CO_2 and H_2O from this and fossil fuels would be returned to the greenhouses.

goes far back in man's history, advanced CEA is in its infancy. The promising starts made by several countries around the Persian Gulf and in the U.S. and Mexico should serve as a beginning for more extensive and sophisticated projects. In particular, systems should eventually be powered entirely by solar energy. There are complex problems of mechanical, chemical and civil engineering involved.

At the same time, CEA can create new conditions for plant growth for which no plants growing in natural environments are fully adapted. The possibilities are very great for plant breeding to produce plants capable of improved properties suitable for CEA. Among these may be mentioned:

1. High temperature tolerance

2. High growth rates at high temperatures and humidity

3. Maximum use of CO_2 enrichment and ability to tolerate substantial levels of sulfur dioxide

4. Resistance to mildew

5. For legumes, high rates of N_2 fixation under CEA conditions.

No doubt many more could be added to this list.

There is a need for a long range, stably supported research and development of CEA, in which engineers, agronomists, economists, plant physiologists, and chemists would interact and work together towards a really new kind of agriculture capable of highly efficient solar energy utilization to produce needed food, feed, and materials.

REFERENCES

1. "21st Century Bio-Saline Research," draft paper submitted by NSF scientists. J.C. Aller, L.B. Mayfield, and O.R. Zaborsky, Oct. 29, 1976, to Middle East Section (INT/NSF), G.B. Devey, Section Head.

2. Copp Colins, 1977, "Biosaline Research Project and Arid Land Sciences: Application of New Scientific Knowledge and Advanced Technologies to the Arid Lands of the World." (INT/NSF), G.B. Devey, Section Head.

3. Indiana University International Workshop on Bio-Saline Research, Kiawah, S.C., USA, Sept. 15-18, 1977. Anthony San Pietro, Dir.

4. Meigs, P. (1968) in McGinnies, Goldman and Paylore, eds. "Deserts

of the World: An Appraisal of Research into Their Physical and Biological Environments." University of Arizona Press, Tucson, AZ.

5. Based on map in article by Calvin, M., 1977. "The Sunny Side of the Future." Chem. Tech. $\underline{7}$, pp. 252-263.

6. Based on data presented by H. Lieth at the Second National Biological Congress, Miami, FL, 1971.

7. Bassham, J.A., 1977. "Increasing Crop Production through More Controlled Photosynthesis" Science $\underline{197}$, pp. 630-643.

8. R.S. Loomis and W.A. Williams, Crop. Sci. $\underline{3}$, 67 (1963).

9. R.S. Loomis, W.A. Williams, and A.E. Hall, Ann. Rev. Plant Physiol. 22, 431 (1971).

10. R.S. Loomis and P.A. Gerakis in "Photosynthesis and Productivity in Different Environments (Cambridge Univ. Press, London, 1975) pp. 145-172.

11. K.S. Ng and J.A. Bassham (1968) Biochim. Biophys. Acta $\underline{162}$, 254-264.

12. J.A. Alich and R.E. Inman (1974) "Effective Utilization of Solar Energy to Produce Clean Fuel" Stanford Research Institute, Report No. NSF/RANN/SE/GI 38723/FR/2.

13. J.A. Bassham, A.A. Benson, L.D. Kay, A.Z. Harris, A.T. Wilson, M. Calvin, J. Am. Chem. Soc. $\underline{76}$, 1760 (1954).

14. J.A. Bassham and M. Calvin, "The Path of Carbon in Photosynthesis" (Prentice-Hall, Englewood Cliffs, NJ 1957) pp. 1-107.

15. H.P. Kortschak, C.E. Hartt, G.O. Burr, Plant Physiol. $\underline{40}$, 409 (1965).

16. M.D. Hatch and C.R. Slack, Biochem. J. $\underline{101}$, 103 (1966).

17. M.D. Hatch and C.R. Slack, Ann. Rev. Plant Physiol. $\underline{21}$, 141 (1970).

18. For a review of photorespiration and of the possibilities for improving the efficiency of photosynthesis by regulating photorespiration, see I. Zelitch, Science $\underline{188}$, 626 (1975).

19. C.B. Osmond, 1975, "CO_2 Assimilation and Dissimilation in the Light and Dark in CAM Plants." in R.H. Burris and C.C. Black, eds. "CO_2 Metabolism and Plant Productivity." University Park Press, Baltimore, MD, USA pp. 217.

20. I.P Ting, "Crassulacean Acid Metabolism in Natural Ecosystems in Relation to Annual CO_2 Uptake Patterns and Water Utilization," 1975, in R.H. Burns and C.C. Black, eds. "CO_2 Metabolism and Plant Productivity." University Park Press, Baltimore, MD USA.

21. O. Bjorkman, 1975. "Adaptive and Genetic Aspects of C_4 Photosynthesis" in R.H. Burns and C.C. Black, eds. "CO_2 Metabolism and and Plant Productivity," University Park Press, Baltimore, MD, USA.

22. M. Calvin, 1976. "Hydrocarbons via Photosynthesis" Presented to the 110th Rubber Division Meeting of the American Chemical Society, San Francisco, 5-8 October. Available from American Chemical Society, Washington, DC.

23. P. Felker, and G. Waines, 1977. "Potential Use of Mesquite as a Low Energy Water and Machinery Requiring Food Source." Presented at the Conference on "Energy Farms on Cropland," July 14, 1977 Sponsored by R.H. Hodam, Jr., Alternative Implementation Division of the California Energy Resources Conservation and Development Commission, Sacramento, California.

24. P. Felkner and R.A. Bandurski, 1977. Protein and Amino Acid Composition of Free Legume Seeds. J. Sci. Fed. Agric. in Press.

25. National Academy of Sciences, 1977. "Guayule: An Alternative Source of Natural Rubber." Report of an Ad Hoc Panel of the Board on Agriculture and Renewable Resources.

26. H. Yokayama, (1977) "Chemical Bioinduction of Rubber in Guayule Plant" Science (in press).

27. M. Calvin, 1977. "Hydrocarbons via Photosynthesis," International Journal of Energy Research (in press).

28. D.G. Dalrymple, 1973. "Controlled Environment Agriculture: A Global Review of Greenhouse Food Production." Economic Research Service, U.S. Department of Agriculture, Washington, DC 20250.

29. L. deBivort, 1977. "An Assessment of Controlled Environment Agriculture Technology" submitted to the National Science Foundation (RANN) under contract No. C-1026 with International Research and Technology Coporation.

30. J.A. Bassham, 1975 in "Cellulose as a Chemical and Energy Resource" Charles Wilke, ed. Biotechnology and Bioengineering Symposium No. 5. pp. 9-20. John Wiley and Sons, New York.

31. J.A. Bassham, 1976 in Proceedings of "A Conference on Capturing the Sun through Bioconversion," pp. 627-646. The Washington Center for Metropolitan Studies, Washington, DC.

32. J.A. Bassham, 1977 in "Biological Solar Energy Conversion," A. Mitsui, S. Miyachi, A. San Pietro, S. Tamura, eds. pp. 151-166. Academic Press, New York.

33. R.W.F. Hardy and U.D. Havelka, 1975. Science 188, 633.

34. R.R. Spencer, A.C. Mottola, E.M. Bickoff, J.P. Clark and G.O. Kohler, 1971. J. Agric. Food Chem. 19, 3.

35. R.H. Edwards, R.E. Miller, D. DeFremery, B.E. Knuckles, E.M. Bickoff, and G.O. Kohler, 1975. J. Agric. Food Chem. 23, 260.

36. M.A. Stahmann, 1968. Econ. Bot. 22, 73.

37. C.R. Wilke (ed.) 1975. "Cellulose as a Chemical and Energy Resource." Cellulose Conference Proceedings, National Science Foundation, U.C. Berkeley, June 25-27, 1974. John Wiley & Sons, New York.

38. W.J. Sheppard, 1977. "Ethanol and Furfural from Corn." Proceedings of a Conference on "Biomass--A Cash Crop for the Future?" Prepared by Midwest Research and Development Administration Division of Solar Energy. March 2-3, Kansas City, MO.

39. J.C. Beatley, 1976. "Vascular Plants of the Nevada Test Site and Central Southern Nevada: Ecologic and Geographical Distributions." Office of Technical Information, U.S. Energy Research and Development Administration, Washington, DC.

SALT TOLERANCE OF PLANTS: STRATEGIES OF BIOLOGICAL SYSTEMS

D. William Rains

Department of Agronomy & Range Science
and Plant Growth Laboratory
University of California
Davis, California 95616

INTRODUCTION

Large terrestrial areas of the world are affected by levels of salt inimical to plant growth. It has been estimated that as many as 4×10^8 hectares (ha) of the 1.5×10^9 ha land currently under cultivation have enough salt to reduce the agronomic potential of these areas (28). This is exclusive of the regions classified as arid lands (14). With the enormous amount of arid land, it is very possible that the hectares of land affected by elevated levels of salt will exceed 1.5×10^9 ha or approximately 40% of the 4×10^9 ha of potentially arable land.

Salinity has been an important factor in man's history and in the life spans of agricultural systems. Salt encroaching on agricultural soils has created instability and has frequently destroyed ancient and recent agrarian societies. As a result of salt accumulation in soils, the Fertile Crescent area bounded by the Tigris and Euphrates Rivers became less productive and faded as a power in the ancient world. Large areas of the Indian subcontinent have been rendered unproductive through salt accumulation and poor water management (3, 48). In California, the Imperial Valley was threatened by salt encroachment until tile drains were placed under 100,000 ha of land, a project completed in the early 1950's (44).

Other areas of immediate concern might include the Nile Delta in Egypt. The construction of the Aswan Dam has greatly reduced the annual flooding of the very productive river plain. Soil fertility is expected to decrease since new deposits of nutrient rich sediments

carried by flood waters will be markedly diminished. The presence of
flood waters on soil surfaces also promotes leaching of soluble salts.
A reduction in the annual flooding could result in an accumulation of
salt in soils irrigated during the dry season because of the lack of
wet season leaching. Evidence is now available that groundwater is
rising due to maintenance of high water flows from the dam to generate
electricity. This has resulted in encroachment of salts from below
into rooting zones.

In the arid regions of the world, a large percentage of the soils
are inherently high in salt (3). Elevated concentrations of salt
characterize climates where evaporation exceeds precipitation, with
attendant reduced cycling of mineral-elements. These arid regions require irrigation water to be productive. This compounds the salinity
problem since the waters contain dissolved solids of various kinds and
amounts. The agriculturist is keenly aware of the problems associated
with growing plants under irrigation for extended periods. An application of water is also an application of salt, varying in amount with
the quality and quantity of water used. Pure water is evaporated via
transpirational losses from plants and evaporation from soil surfaces,
leaving large quantities of salt in the soil and drainage waters.
Poor water management (excessive applications, improper drainage, inadequate leaching of salt) will lead to salt accumulation and potentially serious agronomic problems.

The potentially damaging effect of salt accumulation in agricultural soils can be reduced by various means. A common and highly successful procedure is the reclamation of salt-affected soils by the
management of water and land. Large volumes of water are leached
through soil profiles to reduce the concentration of salt. Chemicals
are used to accelerate the movement of salt and water and proper
drainage facilitates this leaching (8). These procedures have been
very important in irrigation agriculture and have been responsible
for maintaining large areas and land in high production agriculture
for hundreds of years.

A major aspect of the reduced productivity is correlated directly
with the inability of crop plants to tolerate high levels of salt in
their environments. The deleterious effects of salinity could be significantly reduced if plants capable of tolerating excessive levels of
salt were grown in irrigated regions of the world. The selection and
development of salt-tolerant crops species would enhance agriculture
productivity and make brackish and saline waters usable for irrigation
(see chapters 5, 8 and 9). There are large areas of the world with
climate and soils which are capable of high plant productivity but are
limited by water. The ocean is a vast resource of water and nutrients
and it would be possible to exploit the enormous reservoir of irrigation water if high production, agronomic useful crops can be grown in
the presence of high concentration of potentially inimical salts.

Salinity is not incompatible with many forms of plant life. It
has been estimated that up to 50% of the plant biomass of the earth
is found in the ocean (52). Large areas of the terrestrial environment have various degrees of salinity and yet support considerable
plant production. In desert regions of the world, large plant populations are supported in an environment high in salt and chronically
short of water (51). Mangrove swamps of highly productive large trees
growing in sea water are excellent examples of higher plants adapted
to perform extremely well in saline environments. The Halobacterium
species contain some of the most tolerant forms of organisms identified in the biosphere and, along with the green algae, Dunaliella,
grow in 4 M NaCl solutions (22), a solution 8 times as salty as the
ocean.

The obvious conclusion is that there are large numbers of microorganisms and lower and higher forms of plant life which are very com-
petent in growing in saline environments which would be inimical to
most crop species (35).

The focus of this presentation is on the biological characteristics associated with tolerance of salinity by higher forms of plant
life with a very limited discussion of the physiological response of
green algae to excess salinity.

The questions addressed in this presentation are: a) What are the
physiological characteristics related to survival and productivity of
plants in the presence of high concentrations of salt? b) What are the
differences between salt-tolerant and salt-intolerant organisms? c)
What genetic variability exists in these organisms in their responses
to saline environments?

SALT TOLERANCE OF GREEN ALGAE[1]

Algae have been proposed as a source of food and feed for man and
animals in arid regions of the world. Algae photosynthetically convert
solar energy into high energy foodstuffs such as protein, lipids and
carbohydrates. The primary productivity of these organisms depends on
a readily available supply of water, nutrients and aquaculture techniques. In desert areas near the ocean, large shallow lagoons of seawater are visualized as sites for algal culture. The ocean would

---

[1]Characteristics related to salt tolerance of blue-green algae have
not been presented in this paper as a separate section. It did not
seem appropriate since many of the mechanisms and strategies associated with salt tolerance of blue-green algae are similar to those described for green algae. Other position papers will discuss the performance of blue-green algae in saline environments with detailed discussion on energetics and ion transport processes of algae and blue-
green algae developed in other chapters.

supply water and some of the essential mineral nutrients, but as with
bacteria, the algae must be able to tolerate relatively high levels
of salt and grow rapidly in these saline environments (see Chapt. 16).

Green algae are found over a wide range of environments including
fresh water, brackish water, sea water, and extreme halophilic environments such as the Dead Sea (14).

In one genus, Dunaliella, species exist in marine environments
as well as in the Dead Sea. Dunaliella tertiolecta (marine) performs
well in 0.5–1.0 M NaCl, while Dunaliella viridis (halophilic) grows
and reproduces in 4 M NaCl, a concentration deadly to D. tertiolecta
(1).

The occurrence of one genus in two very different saline environments provides a unique opportunity to compare the possible physiological and biochemical bases for salt tolerance in these organisms.

Osmotic Adjustment

Detailed analysis of the intracellular composition of these organisms exposed to saline conditions indicate that the cell does not
use inorganic ions to maintain osmotic balance. This is observed for
Dunaliella spp., Chlamydomonas spp., and others (1, 14, 32). The outstanding change occurring in the intracellular composition is in the
levels of organic constituents. Dunaliella spp. show increases in
glycerol concentrations almost directly proportional to the increase
in NaCl in the external medium (4). For example, when D. tertiolecta
was grown in 1.36 M NaCl, the glycerol content in the cell was 1.4
molal. When D. viridis was grown in 4.25 M NaCl, glycerol levels
approached 4.4 molal in the algal cells. As seen in Figure 1, the
intracellular glycerol concentration paralleled the increase in concentrations of NaCl in the external medium (1). This indicates that
these organisms achieve a limited but significant adaptive adjustment
to salinity by increasing glycerol levels in response to NaCl. Less
dramatic responses of green algae to increasing salinity include increases in RNA content and protein levels in the cell (14).

Biochemical Properties of Salt Tolerant Algae

The most significant results were obtained from studies on the
biochemical characteristics of the enzymes extracted from the salt-
tolerant algae, Dunaliella. Unlike those of halophilic bacteria, the
enzymes extracted from D. viridis, the most halophilic algae, are inhibited by very low levels of salt (22). The levels of salt which inhibited the enzymes were considerably lower than the levels of NaCl
required in the medium for optimum growth. This suggests that inorganic ions such as NaCl are excluded from the cell and that the internal contents do not come to equilibrium with the external environment.

In this situation, the activity of cell water must be regulated by organic cellular constituents such as glycerol or other osmotically active substances. This would suggest that osmoregulation is more important than the osmostability of enzymes in green algae exposed to saline conditions (45).

Enzymes extracted from algae from a variety of ionic habitats showed little response in activity to a wide range of concentrations of organic constituents (glycerol and mannitol). The presence of these organic osmoticums in reaction media at concentrations greater than 4 molal did not suppress or enhance the in vivo activity of a number of enzymes (1). This indicates that these substances can accumulate to high levels, maintain water relations favorable for the cell, and promote the physiological and biochemical activity of algae exposed to saline conditions without affecting enzymic processes.

In contrast, proteins of halophilic bacteria require high levels of inorganic ions for osmostability and activity of enzymes in both intact cells and in enzymes extracted for the bacteria (See Chapt.13).

Inorganic Ion Regulation

As discussed previously, certain algae maintain intracellular salt concentrations lower than those of the external medium. The membranes of these green algae appear to exclude certain ions and mediate the adsorption of others. A comparison of the external medium with the internal content of the cells showed differences in the levels and compositions of the solutions. The chapter on microorganisms presented information which suggests that selective transport mechanisms are involved in regulating the intracellular ionic environment. These selective transport systems are energy-requiring (25) and can pump ions into a cell, as with bacteria, or out of a cell, as proposed for algae (39,43). The situation for algae must involve the transport of Na from a dilute solution within the cell out into a concentrated external solution if the internal salt level is to be maintained at levels below the external solution. Such transport requires the expenditure of energy.

Both processes -- accumulation of organic osmoticums (glycerol, mannitol, arabitol) and ion transport -- require the expenditure of considerable metabolic energy. The use of carbon compounds in osmotic regulation means that a large amount of potential energy is tied up for as long as the algae are exposed to salinity. These carbon components could be used for growth and development of the organism, and hence increased productivity, if they were not required for osmoregulation.

The metabolic energy expended in transport and regulation of inorganic ions would also be expected to reduce the energy available for productivity of the algae.

It is common observation that algae growing in saline environments have slower growth than fresh-water algae. It is very possible
that this reduced vigor results from the expenditure of metabolic
energy on osmotic adjustment and ionic regulation. This would suggest
that the productivity of these salt-tolerant algae could be incresed
by screening for algae that use primary productivity more efficiently
in the processes.

Research Priorities for Salt Tolerant Algae

Some of the physiological characteristics which might be altered
to improve the performance of algae in saline environments include:
1) increased photosynthetic capacity of algal cells, supplying the
energy necessary for the competing processes of growth and osmotic
and salt regulation; and 2) more efficient channeling of carbon and
energy from photosynthesis to processes involved in increasing salt
tolerance. Evaluation is needed for genetic variability in the processes identified as critical for bestowing salt tolerance on algae.
A breeding program coupled with investigations on the physiological
bases of salt tolerance should be established, and a concerted effort
directed toward improving the productivity of algae in saline environments.

BIOLOGICAL CHARACTERISTICS OF SALT TOLERANCE OF HIGHER PLANTS

Plants exposed to saline environments are faced with two major
problems. Water stress, a constant threat to the plant, varies with
the environment and salt concentration of the soil. Accompanying reduced water availability in saline soils are high levels of ions, potentially inimical to plant growth. These two problems, water availability and high levels of toxic ions, are basic limitations to higher plant productivity in salt-affected soils.

Desert plants have been observed growing in saline environments
which contain salts at concentrations exceeding those found in sea
water. Plants growing in these environments have evolved a number of
physical and physiological mechanisms which enhance survival and productivity. Resistance to salt may involve either tolerating or modifying internal salinity (46).

Leaves of _Salicornia_ and _Nitraria_ growing in salt-affected soils
contain a 10% salt solution and total salts make up 50% or more of the
leaf dry matter. These plants would be expected to have mechanisms
which bestow salt tolerance of the cellular level.

Salt content of tissues can be modified by a number of mechanisms. Certain desert shrubs shed lower leaves when salt accumulates
to a critical level in the leaf. This effectively removes excessive

accumulation of salt from the rest of the plant. The agronomic value
of these plants, however, might be questionable.

Internal salinity can be reduced by other means. Leaves of Atriplex species and other halophytes have specialized structures called
salt glands or salt bladders. The glands collect salt from the surrounding tissue and then secrete high concentrations of salt to the
leaf surface. The salt can then be washed off by rain or dew. Salt
secreting structures are common to a number of plant species found in
saline habitats (i.e., Mangrove, Tamarix, Limonium) (28).

Some plants when exposed to salinity respond by increasing uptake of water into various tissues. This increases the succulence and
effectively dilutes out salt accumulated by the tissue.

A few plants synthesize organic acids, particularly oxalic in
response to high concentrations of NaCl. The negatively charged organic acid balances the excessive accumulation of positively charged
Na (17).

Mechanisms in plants responsible for enhanced resistance to salt
might be incorporated into breeding programs as a means to improve
the performance of important crop plants when grown in saline environments. Some of the mechanisms, for example, salt glands, leaf shed-
ding, succulence, may not be easy to transfer, genetically, from one
plant to another or they may not be a desirable characteristic. One
characteristic that potentially could be genetically exploited is
salt tolerance at the cellular level. The osmotic regulation of water
is easily accomplished by accumulating salt in the cell. If the cell
can maintain physiological and biochemical processes in the presence
of high concentrations of salt, the organism should be successful and
productive in saline environments. It seems important to better understand the mechanisms responsible for salt tolerance at the cellular level and identify physiological markers that can be used in
breeding programs focused on increasing plant tolerance to salinity.

A number of physiological strategies invoked by higher plants to
enhance tolerance to salinity have been identified and are presented
and discussed in the following sections.

Ionic Environment and Regulation

A common observation in describing a saline system is the imbalance of the ionic environment, in particular the predominance of
physiologically unessential nutrients over essential nutrients.
Plants growing in such situations must acquire the necessary nutrients
(i.e., K, Ca, NO3) from a solution containing low levels of required
ions and an excessive level of ions that are required in smaller
amounts or not needed at all.

The plant must regulate not only the kinds of ions absorbed but also the amounts of ions taken into the cell. As discussed previously, the plants are exposed to moisture stress in saline environments, and if cellular water is to be retained the osmotic potential within the cell must be increased. The plant can increase internal osmotic potential by producing intracellular osmotica (sugar, organic acids, etc.) or by absorbing inorganic osmotica (salt) from the surrounding medium (4, 10, 35). The structural and functional characteristics of the ion acquisition processes are of prime importance to plants in saline environments (33). The process is responsible for both osmotic regulation and ionic regulation of the intracellular environment.

1. <u>Ion Regulations</u>. Plants which tolerate and grow in the saline environments quite often have high levels of salt in their cells. A major part of the osmotic adjustment in these halophytes (salt-loving plants) is accomplished by ion uptake (15). Plants which are unable to tolerate high salt (glycophytes), however, show ion exclusion, particularly exclusion of Na from their leaves (20). This seems to be mediated by high rates of accumulation of Na in roots and stems, a situation not evident in plants which do not exclude Na from their leaves (20, 34). Agriculturally, crop plants fall into both of these categories, excluders and nonexcluders.

2. <u>Ion Transport and Osmotic Regulation</u>. The transport of ions into cells is one mechanism whereby osmotic regulation can be accomplished. The ion transport process must not only have the capacity to absorb large amounts of ions but must be characterized by the transport of specific ions required for physiological reactions. A high-capacity selective transport mechanism, described by a number of investigators, has been proposed as an integral component of a plant's response to salinity (10,35,38). The transport system has been designated as the dual mechanism of ion transport (13).

The dual mechanism has a number of specific properties that are necessary for a plant to survive in a saline environment. First, the absorption of an essential ion such as K is selective. The uptake of a specific ion operates in the presence of high levels of chemically similar ions (i.e., Na). K is accumulated in the required amounts from a nutritionally unbalanced medium (e.g., saline soils) (Fig. 2). In this way, the internal ionic content of cells is regulated and the physiological processes proceed in the face of a high salt in the external environment.

Second, at increasing levels of salt, another mechanism becomes operative (see Fig. 3). This mechanism shows little ion selectivity but is capable of accumulating at fairly high rates whatever ions are in the environment (37). The accumulation of ions within the cell enhances the osmotic flux of water into that cell, reducing the possibility of physiological drought (41).

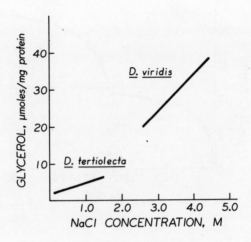

FIGURE 1

The increase of glycerol content in Dunaliella tertiolecta (a
marine species) and D. viridis (a Dead Sea species) as a function of
increasing NaCl. D. tertiolecta responds to a range of NaCl that is
commonly found in brackish and sea water. D. viridis responds to concentrations of NaCl similar to those measured for the Dead Sea.

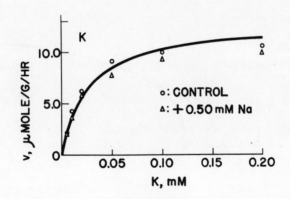

FIGURE 2

The absorption of K as a function of K concentration in the presence or absence of NaCl. The presence of 50 times more Na and K had
no effect on K uptake.

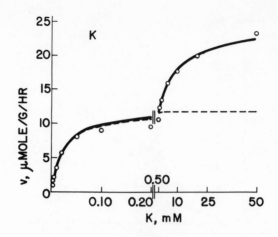

FIGURE 3A

The absorption of K as a function of K concentrations.   The uptake of K over the low range of concentration is the same transport
mechanism shown in Figure 2.   At higher concentrations >0.5mM a second
mechanism is observed and transport large amounts of whatever ions
are present.

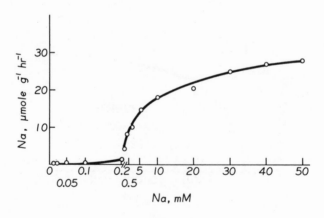

FIGURE 3B

The absorption of Na as a function of Na concentrations.   K is
present at 0.5mM and completely preempts Na transport by the low concentration mechanism (See Fig. 2).   At higher concentrations, both K
and Na are taken up.

The use of inorganic ions for osmotic adjustment suggests that plants must be able to tolerate high levels of salts within their cells. If that is the situation, it is very probable that the biochemical properties of plant enzymes could be similar to those found in the halophilic bacteria since those bacteria can tolerate high levels of inorganic ions within their cells.

Enzymes extracted from salt-tolerant plants have been compared with salt-sensitive plants, and little difference was found in the response of those enzymes to increase concentration of salt (50). In many instances, the enzymes extracted from halophytes were more sensitive to salt than were the enzymes from glycophytes (16). From the literature, it is apparent that enzymes from halophytes are neither salt-resistant nor salt-requiring and in fact, are similar to those from nonhalophytic plants. This is similar to the evidence presented concerning halophytic algae, although in marked contrast to the situation described for halophilic bacteria.

The sensitivity to salt of the enzymes from halophytic plants would suggest that compartmentation of ions must be necessary if the enzymes in the cell are to be protected from the inimcal effects of salt. Plant cells are structurally well suited to the compartmentation of ionic substances. Large membrane-bound vacuoles are very likely the site for a considerable amount of sequestration of ions and other osmotically active substances. Transport mechanisms could actively and selectively move ions into the vacuole, removing these potentially inimical ions from the cytoplasm. In this way, the cytoplasm would be scrubbed clean of unnecessary and possibly toxic ions (21). These ions, in turn, could act as osmotic materials within the vacuole, and the vacuole would then be responsible for maintaining water flow into the cell. The compartmentation of mineral ions and of other organic ions into the vacuole is common in higher plants (5,27).

Osmotic pumping by the vacuole, however, may result in dehydration of the cytoplasmic volume. For vacuolar pumping to be important in maintenance of a favorable water balance in the cell it is necessary to suggest alterations in the cytoplasmic compartment. The cytoplasmic volume could increase in water content by formation of osmotically active organic solutes or by changes in the matrix potential of the cytoplasmic constituents which favor water retention.

Organic Solutes and Osmotic Adjustment

A number of organic materials have been identified as important in osmotic regulation of water in plant cells.

1. Organic acids. Many halophytic plants respond to excessive cation uptake by producing organic acids. As the level of salinity increases, cation uptake increases, and if cations are in excess over

anions, organic acids are produced. Oxalic and malic acids are frequently found to accumulate in halophytes and other plants exposed
to high concentration of salt or water stress (2,17,31).

    2. <u>Nitrogen metabolism</u>. Nitrogen compounds increase to fairly
high levels in resonse to increased salinity. Amino acids accumulate in plants that are increasingly salinized. It has been suggested that malic dehydrogenase is inhibited by salt and that transamination is stimulated. This would allow the channeling of organic
acids into these nitrogenous materials. Proline is one amino acid
that seems to accumulate in response to salt or water stress in a
large number of organisms. It has been determined to represent 10-
20% of the shoot dry weight (30). Tertiary nitrogenous compounds
such as betaine and choline are also involved in osmotic adjustment
(47).

    The evidence available suggests that nitrogenous compounds are
of prime importance in the ability of plants to tolerate salinity.
The dramatic changes observed in amino acid levels (e.g., proline)
and the increased movement of reduced nitrogen via transamination of
organic acids and synthesis of amines would suggest that the assimilation of nitrate to reduced forms may greatly influence the response
of plants to salinity. There are a number of sites where salinity
might influence the assimilation of nitrate to reduced forms. The
diagram below shows the processes involved in nitrate assimilation
(19).

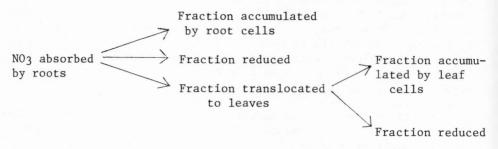

    These processes are tightly integrated and the rates are in-
fluenced by the overall energy level of the plant. Light, carbohydrate level, and photosynthesis all influence the rate of nitrate
assimilation. The effect of saline conditions on these processes and
their overall integrative responses is not yet worked out. This
information is required as background for understanding the biochemistry of tolerance or lack of tolerance to saline conditions.
Nitrate reductase is known to be greatly influenced by environmental
stresses including water stress, heat stress, and low light (30).
A very important area for future research on salt tolerance may be
the reduction of nitrate and the flux of reduced nitrogen through
the plant and the effect of salinity on these processes.

Carbon Metabolism

Carbon metabolism has a central role in the response of a plant to salinity. The formation of carbon materials necessary for enhanced synthesis of organic acids and nitrogenous compounds is critical to osmotic adjustment. The increased energy necessary to drive the ion transport systems must be a high-priority metabolic function. All of these processes are required to regulate cell water and intracellular organic and inorganic ions.

The photosynthetic reactions supplying the carbon materials for the above processes appear to be considerably different in saline environments, and some rather unusual biochemical reactions are associated with photosynthesis in this situation.

In general, plants exposed to saline environments seem to assimilate carbon more efficiently (23). This is accomplished by alteration of the carbon assimilation pathway (18). The following diagram represents CO_2 fixation under saline or nonsaline conditions.

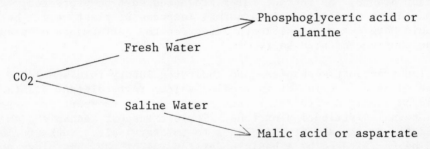

The presence of salt has been shown to increase the activity of PEP carboxylase while reducing the activity of RuDP carboxylase. Plants growing in low salt environments show C_3 metabolism, but if the environment is salinized with NaCl the carbon assimilation of Aeluropus litoralis shifts to C_4-dicarboxylic acid metabolism (42) while Mesembryanthemin crystallinium (26) and Portulacaria afra (49) shift to crassulacean acid metabolism (CAM). In either situation, there is an increase in organic acids and, as discussed previously, these compounds influence osmotic adjustment, an important aspect of salt tolerance in plants. The metabolism of carbon via the C_4 pathway or the CAM pathway is stimulated in the presence of Na, and it has been proposed that Na is an essential micronutrient for these processes (36). As a result of those observations, it has been suggested that β-carboxylation is an adaptive response to increasing salinity. This enhances the synthesis of organic anions to balance the excess accumulation of cations in plants growing in saline environments (24).

Research Priorities on Salt Tolerance in Higher Plants

Increased levels of osmotically active substances contribute to
the osmotic adjustment of plants exposed to saline environments.
These responses vary both with environmental changes and differences
in genetic potential. Genetic variability within species is highly
probable, and major priorities of future research should include a
basic understanding of the physiological responses to plants to
salinity and the use of natural genetic variability and genetic techniques to improve the biological performance of plants in saline environments (11).

   1. Inorganic-ion regulation. Plants differ in the amounts and
species of inorganic ions accumulated in response to increasing salinity. This is an important component of osmotic regulation by plants.
The accumulation of both inorganic and organic osmotica exposes the
enzymes and biochemical processes of the cell to potentially toxic
levels of these substances. Since the proteins of salt-tolerant
plants are frequently as sensitive to salt as are the proteins of intolerant plants, the intracellular processes must be protected. Compartmentation of osmotica is a normal response of plant cells, but
it could have a very significatn effect on the performance of plants
in the presence of high salt.

   Ion transport mechanisms and their regulatory role should be
evaluated as a function of increasing salt. Salt-tolerant plants
should be compared with intolerant plants as to transport processes
and internal distribution of ions. These transport mechanisms should
be related to the ability of plants to tolerate salt. Such studies
would be facilitated by a basic understanding of the physiology and
biochemistry of transport systems.

   Compartmental analyses should be undertaken of the uptake and
distribution of ions within the cell. It is necessary to know
whether there are alterations in the types of ions transported and
whether these ions are localized in various compartments. The effect
of these processes on the ability of plants to tolerate salt should
be evaluated.

   It is extremely difficult to measure the ion content of cellular
compartments. Flux analyses based on ion exchange data have proven
to be unreliable as estimates of the ion content of cellular compartments. New techniques are required with the application of sophisticated equipment and procedures. Electron probe analyzers facilitate
the determination of the amounts and kinds of ions within the cell.
with these techniques, compartmental analyses can be done directly
on the tissue. Additional new techniques include ion probe analyzers.
With this instrument, individual isotopes on ions inside a cell compartment can be measured. The acquisition of these new analytical
instruments is important in the study of ion compartmentation. These

studies would be facilitated by the development of high resolution autoradiography capable of detecting spatial distribution of high energy radioisotopes of mineral nutrients common in saline environments.

An integral part of future research will be determination of the genetic variability in transport mechanisms (11) and their relation to salinity tolerance (40).

2. <u>Carbon metabolism.</u> The effects of carbon metabolism on the response of plants to salinity are manifold. Carbon supplies the energy for ion regulation, the carbon for organic acids, and the carbon skeletons for nitrogenous compounds. Through its effects on these processes, it has an overall effect on osmotic adjustment. An increase in the efficiency of carbon fixation and assimilation would influence biological performance in saline environments.

Alteration in carbon metabolism in response to salinity must be studied systematically. The switch from C_3 to C_4 or to CAM carbon metabolism with increasing salt is a dramatic response having a major effect on organic acid levels and nitrogenous compounds. The C_4 system is also much more efficient in the fixation of CO_2 than is the C_3 pathway. The relationship between alterations in carbon metabolism and salt tolerance needs to be defined.

Physiological markers related to regulation of carbon metabolism must be identified so that genotypes with superior salt tolerance can be selected.

3. Novel approaches toward obtaining salt-tolerant plants must be considered. Much of the research proposed in this paper depends on the comparison of various genotypes and their response to salinity. If variability in salt tolerance exists, then it may be possible to relate this variability to physiological and biochemical processes. Selection procedures which mediate the identification and/or development of genotypes with superior tolerance are very important to many of the studies proposed on salt tolerance and essential to the success of some of the research programs proposed.

Selection and breeding for salt tolerance has had some limited success (7). Epstein and Norlyn (12) applied extreme selection pressure to a mixture of barley genotypes by growing a large number of plants in culture solutions salinized with NaCl. Plants capable of tolerating high concentrations of salt were grown to maturity and seed was collected. These plants were then grown in sandy soils irrigated with seawater. A few genotypes were identified which performed quite satisfactorily in the highly salinized environment. Such investigations need expansion using as screening material the world collection of crop plant genotypes. From the results of Epstein and Norlyn, it is apparent that large scale screening of available plant genotypes is a valid procedure and will be essential

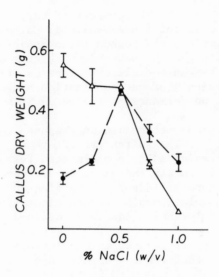

FIGURE 4

The growth of 2 lines of alfalfa cells as a function on increasing concentrations of NaCl in the culture media. Alfalfa cells from
non-selected lines (Δ-Δ) are inhibited by increasing NaCl and at 0.5%
the growth is severely derepressed and the cells appear dead at 1%
NaCl. The salt selected line of alfalfa cells (0---0) show reduced
growth at 0% NaCl and grow the best at 0.5% NaCl. At the 1% NaCl, the
cells are still growing and show no evidence of damage except for reduction in the rate of growth. These cells appear to be more tolerant
of NaCl and require a substantial amount of NaCl for optimal performance.

in supplying plant material necessary for physiological and biochemical research.

New techniques have become available which will enhance the identification of genotypes better suited to salinity. Cell and tissue culture techniques have been used to select lines of cells which grow considerably better at elevated levels of NaCl than cells from the original population. That has been shown for tobacco (9, 29) and pepper (9). More recently, we have been able to select a line of alfalfa cells which not only grow well at 1% NaCl, but actually required at least 0.5% NaCl for optimum growth (Fig. 4) (6). The use of cell culture to select for plant material with enhanced tolerance to salt is essential to research on the physiological and biochemical aspects of salt tolerance. Additionally, the regeneration of these salt tolerance cells into whole plants may in itself provide a new source of salt tolerant crop plants which can be used in breeding programs.

REFERENCES

1. Borowitzka, L.J. and A.D. Brown, 1974. The salt regulations of marine and halophilic species of the unicellular green algae, Dunaliella. The role of glycerol as a compatible solute. Arch. Microbiol. 96:37-52.

2. Caldwell, M.M. 1974. Physiology of desert halophytes. In: R.J. Reimold and W.H. Queen, Eds., Ecology of Halophytes, Academic Press, NY, pp. 355-378.

3. Chapman, V.J. 1974. Salt marshes and salt deserts of the world. J. Cramer, 392 pp.

4. Craigie, J.S. 1974. Storage Products. In: W.D.P. Stewart, Ed., Algal Physiology and Biochemistry, Univ. of Calif. Press, pp.206-235.

5. Cram, W.J. 1976. Negative feedback regulation of transport in cells. The maintenance of turgor, volume and nutrient supply. In: Transport in Plants, II, Part A. Cells. U. Luttge and M.G. Pitman, eds., Encyclopaedia of Plant Physiol., Vol. 2, pp.284-316.

6. Croughan, T.P., D.W. Rains and Suzan J. Stavarek. 1978. Salt tolerant lines of cultured alfalfa cells. Crop Science 18:959-63.

7. Dewey, D.R. 1962. Breeding crested wheatgrass for salt tolerance. Crop Sci. 2:403-407.

8. Diagnosis and Improvement of Saline and Alkali Soils. L.A. Richards, ed., U.S. Salinity Laboratory, Handbook 60, 1969.

9. Dix, P.J. and H.E. Street. 1975. Sodium chloride-resistant cul-
    tured cell lines from Nicotiana sylvestris and Capsicum annum.
    Plant Sci. Letters 5:231-237.

10. Epstein, E. 1972. Mineral nutrition of plants: Principles and
    perspectives. Wiley and Sons., Inc., NY.

11. Epstein, E. 1977. Genetic potentials for solving problems of soil
    mineral stress: Adaptation of crops to salinity. Proc. Workshop
    on Plant adaptation to Mineral Stress in Problem Soils. M.J.
    Wright, ed. A Special Publ. of Cornell University Agric. Exper.
    Sta., Ithaca, p. 73-82.

12. Epstein, E. and J.D. Norlyn. 1977. Seawater based crop production:
    A feasibility study. Science 197:249-251.

13. Epstein, E., D.W. Rains and O.E. Elzam. 1963. Resolution of dual
    mechanisms of potassium adsorption by barley roots. Proc. Natl.
    Acad. Sci. 49:684-692.

14. Flowers, T.J., P.F. Troke and A.R. Yeo. 1977. The mechanism of
    salt tolerance in halophytes. Ann. Rev. Plant Physiol. 28:89-121.

15. Greenway, H. 1973. Salinity, plant growth and metabolism. J. Aust.
    Inst. Agric. Sci. 39:24-34.

16. Greenway, H. and C.B. Osmond. 1972. Salt responses of enzymes
    from species differing in salt tolerance. Plant Physiol. 49:256-
    259.

17. Hellebust, J.A. 1976. Osmoregulation. Ann. Rev. Plant Physiol.
    27:485-505.

18. Hochachka, P.W. and G.N. Somero. 1973. Strategies of biochemical
    adaptation. W.B. Saunders, Co., Philadelphia, PA.

19. Huffaker, R.C. and D.W. Rains. 1978. Factors influencing nitrate
    acquisition by plants; assimilation and fate of reduced nitrogen.
    In: Nitrogen in the Environment, Soil-Plant-Nitrogen Relation-
    ships. D.R. Nielsen and J.G. MacDonald eds., Acad. Press, NY,
    pp. 1-43.

20. Jacoby, B. 1965. Sodium retention in excised bean stem. Physiol.
    Plant. 18:730-739.

21. Jennings, D.H. 1968. Halphytes, succulence and sodium in plants -
    a unified theory. New Phytol. 67:899-911.

22. Johnson, M.K., E.L. Johnson, R.D. MacElroy, H.L. Speer and B.S.
    Bruff. 1968. Effects of salt on the halophilic alga Dunaliella
    viridis. J. Bacteriol. 95: 1461-1468.

23. Joshi, G.V. 1976. Studies in photosynthesis under saline conditions. Report of P.L. 480 Project, Shivraji Univ. Kalhapur, India.

24. Laetsch, M.W. 1974. The C4 syndrome: A structural analysis. Ann. Rev. Plant Physiol. 25:27-52.

25. Luttge, U. and M.G. Pitman. 1976. Transport and energy (see ref. 5), pp. 252-259.

26. Luttge, U., E. Ball and H.W. Trombolla. 1975. Potassium independence of osmoregulated oscillations of malate^{2-} levels in the cells of CAM-leaves. Biochem. Physiol. Pflanz. 167:67-83.

27. Matile, P. 1978. Biochemistry and function of vacuoles. Ann. Rev. of Plant Physiol. 29:193-213.

28. Mudie, P.J. 1974. The potential economic uses of halphytes. In: R.J. Reimold and W.H. Queen, eds. Acadmic Press, NY, pp. 565-597.

29. Nabors, M.W., A. Daniels, L. Nabolny and C. Brown. 1975. Sodium chlordie tolerant lines of tobacco cells. Plant Sci. Letters 4:155-159.

30. Naylor, A.W. 1972. Water deficits and nitrogen metabolism. In: Water Deficits and Plant Growth. III. Plant Responses and Control of Water Balance. T.T. Kozlowski, ed., Academic Press, NY. pp. 241-254.

31. Osmond, C.B. 1978. Crassulacean acid metabolism: A curiosity in context. Ann. Rev. Plant Physiol. 29: 379-414.

32. Okamoto, H. and Y. Suzuki. 1964. Intracellular concentrations of ions in a halophilic strain of Chlamydomonas. I. Concentration of Na, K, and Cl in the cell. A. Algem. Mikrobiol. 4:350-357.

33. Poljakoff-Mayber, A. 1975. Morphological and anatomical changes in plants as response to salinity stress. In: Plants in Saline Environments, A. Poljakoff-Mayber and J. Gale, eds., Ecol. Series #15, Chapt. 6, Springer-Verlag, pp. 97-117.

34. Rains, D.W. 1969. Cation absorption by slices of stem tissue of bean and cotton. Experimentia 25:215-216.

35. Rains, D.W. 1972. Salt transport by plants in relation to salinity. Ann. Rev. Plant Physiol. 23:357-388.

36. Rains, D.W. 1976. Mineral metabolism. In: Plant Biochemistry, J. Bonner and J.E. Varner, eds., 3rd Edition, Academic Press, NY.

37. Rains, D.W. and E. Epstein. 1967. Sodium absorption by barley roots: role of the dual mechanisms of alkali cation transport. Plant Physiol. 42:314-318.

38. Rains, D.W. and E. Epstein. 1967. Preferential absorption of potassium by leaf tissue of the mangrove, Avicennia marina: An aspect of halophytic competence in coping with salt. Aust. J. Biol. Sci. 20:847-857.

39. Raven, J.A. 1976. Transport in algal cells (see Ref. 5), pp. 129-188.

40. Rush, D.W. and E. Epstein. 1976. Genotypic responses to salinity differences between salt-sensitive and salt-tolerant genotypes of tomato. Plant Physiol. 57:162-166.

41. Schrimper, A.F.W. 1935. Pflanzengeographie auf physiologischer grundlag, 3rd ed. Verlag von Gustav, Fischer, Java.

42. Shomer-Ilan, A. and Y. Waisel. 1976. Further comments on the effects of NaCl on photosynthesis in Aeluropus litoralis. Z. Pflanzenphysiol. 77:272-273.

43. Smith, F.A. and J.A. Raven. 1976. H^+ transport and regulation of cell pH (see ref. 5), pp. 317-346.

44. Smith, J.F. 1965. Imperial Valley Salt Balance, Public Information Office, Imperial Irrigation District, Brawley, CA.

45. Soeder, C. and E. Stengel. 1974. Physico-chemical factors affecting metabolism and growth rate. In: Algal Physiology and Biochemistry, W.D.P. Stewart, ed., Univ. of Calif. Press, pp. 714-740.

46. Sommers, F. (ed.). 1975. Seed-bearing halophytes as food plants. Proceedings of a Conference, University of Delaware.

47. Storey, R. and R.G. Wyn Jones. 1975. Betaine and choline levels in plants and their relationship to sodium chloride stress. Plant Sci. Letters 4:161-168.

48. Thorne, D.W. and H.B. Peterson. 1965. Irrigated Soils. New York, McGraw-Hill, Blakiston Division.

49. Ting, I.P. and Z. Hanscom, III. 1977. Induction of acid metabolism in Portulacaria afra. Plant Physiol. 59: 511-514.

50. Ting, I.P. and C.B. Osmond. 1973. Photosynthetic phosphoenolpyruvate carboxylases. Characteristics of alloenzymes for leaves of C_3 and C_4 plants. Plant Physiol. 51:439-447.

51. Waisel, Y. 1972. Biology of halophytes. Physiol. Ecol. Monogr., Kozlowski, ed., Academic Press, NY.

52. Weyle, P.K. 1970. Oceanography. An introduction to the marine environment. Wiley & Sons, NY.

THE POTENTIAL OF NATIVE PLANTS FOR FOOD, FIBER AND FUEL IN ARID

REGIONS

William G. McGinnies

Office of Arid Lands Studies

University of Arizona

There is a wealth of native plant material either not being efficiently used or not used at all, that might make important contributions to the food, fiber and fuel needs in arid areas. These are
non-agricultural plants, that may be grown with limited water supplies
either derived from rainfall or from rainfall augmentation by means
of dams, diversions and other means of collecting and distributing
moisture from precipitation.

First consideration should be given to the environment which in
arid areas is not favorable for plant growth but even so there are
plants that can survive and in many cases provide a source for food,
fiber and fuel. Both the environment and plants are subject to modification, and manipulation including moisture enhancement, cultural
practices and genetic changes.

Not all parts of the landscape in arid areas are equally treated
as far as moisture is concerned, runoff removes moisture from some
areas and may increase it in other locations such as basins and
stream channels. It is possible to take advantage of these inequali-
ties to concentrate moisture in such a manner that areas of low precipitation may support vegetation in the concentrated areas. (1)

Rock dams in wadis and arroyas may result in terraces that will
support vegetation not possible without restraining runoff. Small
drainages may be treated in several ways to increase runoff from the
watershed that may be concentrated for use at the lower part of the
drainage. In larger drainages, diversion dams may be used to spread
water over wide areas to provide moisture for trees, grasses and
other food producing plants. In some instances, particularly with

trees, small basins can be constructed to catch and hold precipitation. All these devices and others including trickle irrigation can
be used to produce food, wood, and fiber throughout wide desert areas
as a supplement to the more concentrated irrigation agriculture where
more water is available. It is in these natural or enhanced moisture
situations that the use of native plants may be of greatest value.(1)

The use of native plants predates agricultural practices and in
many societies still supplies an important part of subsistence either
directly or through the means of grazing animals.(2) Grazing by
livestock has been considered essential as a means of harvesting
scattered plants many of which are unsuitable for direct use by man.
But grazing is inefficient in terms of water use and often destructive to the environment. (3) If a forage plant has a water requirement of 500 kg's and 10 kg's of forage are necessary to produce a
kilogram of meat, it all adds up to 5000 liters of water and perhaps
much more to obtain a kilogram of milk or flesh. For this reason it
is believed that the direct use of plants and plant products for food
should receive more attention than it has in the past.

For convenience of discussion native plants can be divided into
four groups:

(1)   herbs - seeds, foliage, and tubers for food

(2)   woody plants - food and wood

(3)   fiber plants - food and fiber

(4)   succulents - food and other constituents

In the use of herbs several things should be pointed out. First
the water requirement of native plants may not be lower than that for
crop plants. (4) (5) Secondly, in general native plants can not obtain more moisture from the soil than crop plants. If a tomato plant
and a native ephemeral are planted in a desert environment in pots
at field moisture capacity and allowed to use up the moisture supply
they will both wilt at approximately the same moisture content. (6)
The difference is the tomato plant will die and the ephemeral will
survive if more moisture becomes available.

The greatest advantage that the native herb has is the ability
to survive dry spells and to complete its life cycle even though
this may mean a stunted growth of an inch or two and perhaps developing only one flower, which even under such severe conditions may
produce viable seed.

Of the native herbs perhaps the grasses offer the greatest
opportunities for improvement.(7) Several produce large quantities

of seed but unfortunately the seeds drop as they ripen so by harvest time only a small portion of the total seed produced is obtained. Plant breeding may be very productive especially if crosses with domestic relatives are included.

There are many leafy herbs, consumed as greens and they may have a fairly high nutritional value. Many of these can be dried and stored for later use. Some seeds as the "Quinoa" a chenopodium are presently harvested by natives of South America. (8) Other members of Chenopodiaceae, Amaranthaceae and the large number of "salt bushes" have possibilities for both man and domestic animals.

Root and tuber plants are abundant in native environments. One of these the buffalo gourd has been recently exploited and selected plantings have produced 300 kgs of seed per hectare containing 31 percent protein and 27 percent oil, the latter being very similar to that obtained from safflower. (9)

There are many trees and shrubs with a potential for supplying food and wood. Too many of these have been considered only as they relate to grazing-sometimes as a source of animal food, but too often as something to be removed because they compete with grasses. So the trees and shrubs are removed by burning, spraying, cutting or other means, the released grass is often over-grazed and the processes of desertification accelerated. It may be desirable especially under certain conditions to leave the shrubs and trees and make the maximum use of them for human needs first and livestock second.

One of the most interesting shrubs of the American deserts is the creosote bush (Larrea tridentata). (10) It is well adapted to desert conditions. It can complete its life cycle under as little as 50 mm of precipitation and thrives as an ornamental with a total moisture of 300 to 600mm. Its photosynthetic rate and transpiration are high when moisture is readily available, but it can also lose leaves and remain alive for long periods when little or no precipitation is available. It stands pruning and responds with rapid growth when moisture is available. It competes successfully with other desert plants and is usually the dominant species where it occurs.

Add to these growth characteristics its possible value for various chemical products such as terpenes and the antioxidant Nordihydroguiaretic acid (NDGA). (10) (16) The leaves have a protein content roughly equivalent to alfalfa, but the drawback of creosote bush is to find economic uses, so the approach has been to remove it with the hope some other plant will survive and furnish forage. It is possible to remove the resins with solvents, and perhaps we should take a new look to see if some high priced food such as a breakfast food could be obtained from the leaves, and this with other derived substances might make it an economic plant for desert areas.

The mesquite (<u>Prosopis</u> spp) differs from creosote bush in
that it produces food products that can be and are used by man and
livestock, the mesquite is also very much at home in the desert. (11)
It endures drought by losing its leaves. It has a very serious com-
petitor with grass. Because of this mesquite is often exterminated.

The mesquite produces an abundance of edible flowers, which are
also an important source of honey, and up to 1000 kgs of pods per
hectare. The pods including seeds have 17 percent cane sugar and the
pods alone 21 percent. The hulled seeds have more than 50 percent
protein, and the leaves of the mesquite have approximately the same
composition as alfalfa hay.

The native populations of western North and South America have
esteemed the mesquite as a source of food. (11) There are some 40
species, all of which have potential value. Perhaps it is a mistake
of modern man to think of mesquite only in terms of livestock use.
As the pods are relished, it provides some nutritional benefits to
livestock, however, the seeds are not digested and this source of
protein is lost. In addition the undigested seeds germinate readily
resulting in mesquite thickets where little other vegetation can
survive. One proposal would be to keep the mesquite, but fence out
livestock during the fruiting season so that the pods and seeds both
of which are readily consumed and digested by man can be harvested.

There are many other trees and shrubs having potential for food
and fuel, many belong to the same family as the mesquite. Some of
these like the carob are already well known.

The oaks (<u>Quercus</u> <u>spp</u>) were a very important part of the diet
of California Indians and although not as drought tolerant as some
other species have considerable potential.(7)

Pinyon nuts, produced by the pinyon pine, and relished by both
Indians and non-Indians, are said to produce 2,500,000 pounds of nuts
annually in the United States. The wood is also an important source
of fuel. But like the mesquite the pinyon pine is often eradicated
to promote growth of grass.(7)

The food and fiber plants are varied, the agaves and the yuccas
being well known examples. (12)(13)(14)(15)

The agaves are found throughout American deserts and even extend
into the more humid tropics. They vary in size from plants that can
be held in the hand to plants weighing up to 100 kilograms with
flower stalks reaching heights of more than 10 meters. All parts of
the plant are edible; the leaves of the smaller members are eaten,
the flower stalk bases and the flower stalks themselves are relished,
and the flowers and fruits are also a part of the native diet. The

most nutritious and the most highly prized portion is the base of
the flowering stalk as it starts to grow. These base portions are
removed and roasted for food or used to supply the sugars for the
fermentation of alcoholic beverages. Agaves are cultivated in Latin
America and elsewhere for their attractiveness, and utility for food
and fiber.

There are many species of Yucca in the Americas. All have
nutritious flower stalks and the flowers and fruits of many are eaten.
Some provide soaplike material and nearly all have been more or less
used for fiber. In Mexico the fruits of the yucca have been found
to be a promising source of protein and oil and research is underway
to improve the exploitation of these native plants especially to
provide a supplementary income for people in isolated areas. (14)

The cactacese, a large family indigenous to the Americas, because of their ability to store carbon dioxide taken in at night
for utilization in photosynthesis during daylight hours are well
adapted to desert life. They are also an important source of food
among native peoples.

The large columnar cacti of which the saguaro (Carnegiea gigan-
tea) is an example of favorite foods. The flowers may be eaten but
the fruits are preferred. Even today in the highly developed deserts
of southwestern United States the fruit is harvested by native
families using the same devices for removing the out of reach fruits
as in ancient times but they may drive from place to place in a modern pickup. The fruit has a very high sugar content and 13 percent
protein.

Other columnar cacti are also food producers.(2) Sometimes only
the flowers are eaten and sometime the fruits, but usually some product is harvested.

The arborescent cacti locally called "chollas" are perhaps the
least valuable as far as food is concerned, but they have been used
for livestock feed in times of drought. The fruits or joints may
be eaten by animals after the spines have been more or less rubbed
off on the ground. When the spines are burned the joints and fruits
are readily eaten. In some species the fruit may accumulate for
years and are available in times of drought.

The most widely used group of the cacti are the prickly pears.
These flat jointed plants have all the drought enduring family
characteristics, produce edible flowers and fruits and a great ton-
nage of potential livestock feed per hectare. While some are very
spiny, others are almost spineless.

Prickly pears have been introduced in many parts of the world.
Oputia ficus indica is one of the most productive and well known

species. As a word of caution, the introduction of these plants in
new areas should proceed with caution as they may find the new environment too favorable and become a pest. Witness the proliferation
of prickly pear in Australia.

Recently there has been a revival in the production of cochineal
carmine produced by the cochineal insect (Dactylopius coccus) which
feeds on the prickly pear (Opuntia cochinellifera). The red Cochineal dye is used in many food products replacing suspected carcinogen
synthetic dyes.(15)

In conclusion it is believed that the native plants, examples
of which ahve been discussed above can play a much more important
role in providing food supplies than at present, but to achieve this
goal a more systematic approach is needed. This means that experiences and knowledge must be consolidated and steps taken to survey
the most likely plants and the most likely avenues for improvement.
Then a comprehensive research and development program should be
undertaken. This entire program should be considered in the light
of anthropological, sociological and political consideration to in-
sure its acceptance.

REFERENCES

1.   Evenari, L. Shaman, and N.H. Tadmore, 1971. Runoff Agriculture
     in the Negev Desert of Israel in W.G. McGinnies, B.J. Goldman,
     and P. Paylore, Food, Fiber and Arid Lands p. 311-322.

2.   Felger, R.S. and G.P. Nabhan 1976, Depective Barreness: The
     desert conceals sources that prehistoric people knew how to ex-
     ploit. Will modern man do as well CERES Vol. 9 9102:34-39.

3.   Eckholm, Erik 1976 Losing ground: Environmental Stress and World
     Food Prospects. W.W. Norton and Company.

3a. Brown, L.R. 1978, The Twenty Ninth Day, W.W. Norton Co., New
     York, 363 p.

4.   Shantz, H.L. and L.N. Piemeisel 1927 The Water Requirments of
     Plants at Akron, Colorado. Jour. Agric. Res 34:10 (12): 1093-1190.

5.   McGinnies, W.G. and J.F. Arnold 1939 The relative water require-
     ments of Arizona Range Plants, Univ. of Az. Agr. Exp. Station
     Technical Bull No. 80.

6.   Caldwell, J.S. The relation of environmental conditions to the
     pehnomenon of permanent wilting in plants. 1919 Physiological
     Researches 1: 1-56.

7.   Various sources including McGinnies, W.G., B.J. Goldman and P.
     Paylore eds. 1968 Deserts of the World: An Appraisal of Research
     into their Physical and Biological Environments. Univ. of Az
     Press 788 p.

8.   National Academy of Sciences: 1975 Underexploited Tropical
     Plants with Promising Economic Value 189 p.

9.   Bemis, W.P., J.W. Berry and C.W. Weber 1978. Buffalo Gourd a
     Potential Crop for Arid Lands. Arid Lands Newsletter, Univ.
     of Az. Office of Arid Lands Studies no 8:1-7.

10. Mabry, T.J., J.H. Humziker and D.R. Difeo 1977 Creosote Bush:
     Biology and Chemistry of Larrea in New World Deserts. Academic
     Press 284 p.

11. Simpson, B.B. ed., 1977 Mesquite: Its Biology in Two Desert
     Shrub Ecosystems, Halsted Press 250 p.

12. Johnson, J.D. Plants as Potential Economic Resources in Arid
     Lands 1977 Arid Lands Newsletter, Univ. of Az. Office of Arid
     Lands Studies No 6:1-9.

13. Duisberg, P.C. and J.L. Hay Economic Botany of Arid Regions in
     McGinnies, W.G., B.J. Goldman and P. Paylore eds. 1971 Food,
     Fiber and the Arid Lands: 248-270.

14. Comision Nacional de las Zonas Aridas 1976 Informe de Activi-
     dades 1972-1976 (Guayule, Candelilla Yuccas, Canagria, Jojoba,
     Governadora) Mexico D.F. 53 p. and additional reports.

15. Baranyovits, F.L.C. 1978 Cochineal carmine: an ancient dye with
     a modern role. Endeavor 2(2) 85-92.

16. Duisberg, P.C. 1952, Desert Plant Utilization, Texas Jour. Sci.
     4(3): 269-283.

PRODUCTION OF FOOD CROPS AND OTHER BIOMASS BY SEAWATER CULTURE

Emanuel Epstein, R.W. Kingsbury, J.D. Norlyn, and
D. W. Rush

Department of Land, Air and Water Resources
Soils and Plant Nutrition Program
University of California
Davis, California 95616

"The greatest service which can be rendered any country is
to add a useful plant to its culture..."

-Thomas Jefferson

INTRODUCTION

This chapter presents evidence on the feasibility of adapting
crops to seawater culture. The scheme is enticing. The resources of
land, water and nutrients upon which this novel system of crop production would draw - sea and sand - are at present useless for this purpose and indeed inimical to it in many situations. And while traditionally the sea has served as a source of animal protein (fish and
shellfish),raising crop plants by seawater culture would represent
primary production, without the very large energy losses that are
incurred at each successive trophic level.

The life in the open ocean depends on green plants - the marine
phytoplankton. Along the shores, in salt marshes, deltas, and other
land areas intruded on or dominated by seawater, various higher
plants tolerant of seawater contribute to the biological productivity
of their saline habitats. For all these plants, ranging from microscopic algae to sizable trees, seawater serves as a source of both
water and mineral nutrients (Fig. 1).

On land, the main source of water and nutrients are rain and
soil. Table 1 gives concentrations of major mineral nutrients as
they might be encountered in a soil solution, a typical nutrient solu-

FIGURE 1

Mangroves of the species <u>Avicennia</u> <u>marina</u> growing in soil saturated with seawater off the coast of North Queensland, Australia.
After Rains and Epstein (1967).

Table 1

Concentrations of Major Mineral Nutrients
in a Soil Solution, Experimental Nutrient
Solution, and Seawater[1]

Concentration (ppm)

Element	Soil Solution	Nutrient Solution	Seawater
K	30	235	380
Ca	75	160	400
Mg	75	24	1,272
N	100	224	0.001–0.7
P	0.015	62	>0.001–0.10
S	38	32	884

[1] Soil solution after Reisenauer (1966); experimental nutrient solution after Epstein (1972a); seawater after Weast (1975–1976).

tion of the plant nutritional laboratory or greenhouse, and seawater.

The soil solution in most soils is dilute. Concentrations of mineral ions in this solution represent a quasi-steady state governed by withdrawal of ions by plant roots from the soil solution and their release into this solution from the solid phase of the soil. Rain, irrigations, fertilizer applications, and the activities of roots and soil organisms cause much spatial and temporal variation in the properties of the system (Epstein, 1977a).

Concentrations of most ions in conventional nutrient solutions are fairly high in comparison with those of soil solutions, to minimize rapid depletion and the need for frequent replenishment of nutrients.

Seawater has very low concentrations of nitrogen and phosphorus, ample concentrations of potassium and calcium, and very high concentrations of magnesium and sulfur. The most outstanding characteristic, however, of seawater is its high concentration of salt - 10,561 ppm sodium and 18,980 ppm chloride, or approximately 0.5 M NaCl. Of these, sodium is not known to be generally essential for higher plants, and chlorine is a micronutrient.

Collectively, then, plant life is capable of adaptation to nutrient media ranging from dilute soil solutions to water as salty as the sea. But almost without exception those plants that we presently use for food and fiber are sensitive to saline media. Most of them cannot tolerate salinities of 10-20% that of seawater without detrimental effects, and many are sensitive to still lower concentrations of salt.

The conventional statement that crops are generally intolerant of salt carries a dual implication. It suggests (a) that within a given crop, there is little variation in salt tolerance or sensitivity, and (b) that we know the (low) degree of salt tolerance that the crop possesses. Both these implications, though often taken for granted, are demonstrably false. In the present chapter evidence is presented (a) that within a given crop species there may be very large genetically governed variability with respect to salt tolerance and (b) that salinities representing large fractions of the salinity of seawater or even equaling it may be tolerated by some genotypes. This evidence gives encouragement to programs of selection and breeding for salt tolerance. In addition it is shown (c) that where intraspecific differences in salt tolerance are too small to provide an adequate gene pool to select from, salt tolerant wild relatives of the crop may serve as a source of germplasm to transfer salt tolerance into the economic species. And finally (d) wild salt tolerant plants (halophytes) or selections from them might themselves be put to economic uses.

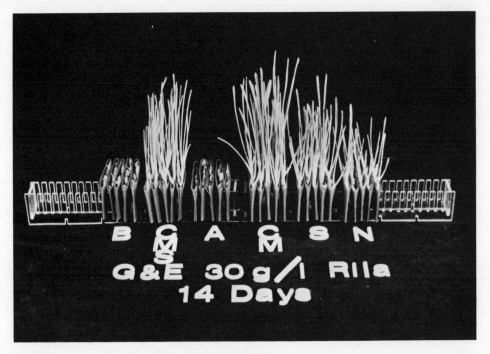

FIGURE 2

Five cultivars of barley and a selection during a test for germination and establishment of seedlings. The salinity was approximately 75% that of seawater. The cultivars B ("Briggs") and A ("Arivat")
were completely suppressed. CM ("California Mariout") and CMS (a
selection from CM), S ("U.C. Signal"), and N ("Numar") established
seedlings but showed differences in vigor under this salt stress.

INTRASPECIFIC DIVERSITY

Figure 2 shows the results of a test on germination and establishment of seedlings of several cultivars of barley, <u>Hordeum vulgare</u>,
under salt stress. The seeds are held between layers of germination
paper (Myhill and Konzak, 1967) soaked in a solution of a synthetic
sea salt mix at 75% seawater salinity. In an unsalinized control set-
up, seeds of all these cultivars germinated and established seedlings
equally well. At this early stage of the life cycle, then, a comparison of just a few genotypes reveals large differences within this species in the response of the plants to salinity, ranging from complete
suppression of germination to performance nearly equaling that of the
controls.

A few cultivars, however, represent only a minute sample of the
genetic diversity in barley. For this and other important crops there
exist collections of seeds representing sources of germplasm from all
over the world (Frankel and Bennett, 1970; Frankel and Hawkes, 1975;
Harlan, 1976; Hawkes <u>et al</u>., 1976; Muhammed <u>et al</u>., 1977; University
of California, 1977; Williams, 1976). These collections serve as
sources of genotypic variability that breeders draw upon to incorporate desirable traits into crops, such as disease resistance and frost
hardiness. In regard to salt tolerance, however, "no species has been
adequately surveyed for this characteristic"(Dewey, 1962). There is
evidence of genotypic variation in respect to several aspects of mineral nutrition and metabolism, including the salt relations of plants
(Table 2). This, together with the considerations discussed in the
Introduction, has led to systematic, large-scale screenings of available gene pools for one such feature, salt tolerance. The specific
aim was selection of genotypes suitable for seawater culture (Epstein,
1977b; Epstein and Norlyn, 1977).

INTRASPECIFIC SELECTION FOR SEAWATER CULTURE

<u>Technique</u>. All initial screening was done by means of solution
cultures which were salinized, usually with a synthetic sea salt mix
(Rila Products, Teaneck, NJ). The main reason for the use of solution culture was that it is the only technique allowing the roots of
plants to be exposed to a mineral medium that can be made up and maintained at a known composition, subject to precise monitoring of all
relevant parameters such as the concentration of individual elements,
salinity, and pH (Asher, 1978; Hewitt, 1966). In addition, there is
in properly maintained nutrient solutions no pronounced spatial variation in these features throughout the volume explored by the roots,
which is inevitable in the presence of a solid matrix such as soil or
sand. Hence, solution culture is the only technique that allows un-
equivocal answers to be provided to questions regarding the ionic
environment to which the roots are exposed. In the present instance,
with salinity the crux of the matter, the concentrations of sodium,

Table 2

Reviews, Chapters, and Summaries Dealing with the Genetic Control
of Mineral Uptake, Transport, and Metabolism of Plants

Author(s) or Editor	Year	Title
Kruckeberg, A.R.	1959	Ecological and genetic aspects of metallic ion uptake by plants and their possible relation to wood preservation.
Myers, W.M.	1960	Genetic control of physiological processes: consideration of differential ion uptake by plants.
Millikan, C.R.	1961	Plant varieties and species in relation to the occurrence of deficiences and excesses of certain nutrient elements.
Epstein, E.	1963	Selective ion transport in plants and its genetic control.
Gerloff, G.C.	1963	Comparative mineral nutrition of plants.
Vose, P.B.	1963	Varietal differences in plant nutrition.
Epstein, E. and R.L. Jefferies	1964	The genetic basis of selective ion transport in plants.
Epstein, E.	1972b	Physiological genetics of plant nutrition.
Klimashevsky, E.L., ed.	1974	Variety and Nutrition.
Lauchli, A.	1976	Genotypic variation in transport.
Wright, M.J., ed.	1977	Plant Adaptation to Mineral Stress in Problem Soils.

chloride, and other ions can be ascertained with precision, and so can other measures of salinity such as the widely used specific electrical conductance, EC (Ayers, 1977; Ayers and Westcot, 1976; Richards, 1954). These and still other data can be obtained by direct sampling of the nutrient solutions, without resort to extractions and other manipulations which introduce a measure of uncertainty whenever soils are sampled for the purpose of gaining information on the chemical composition of the medium in which the roots exist.

Solution culture has the further advantage that the salt concentration can be stepped up or down at will, as quickly or gradually as desired. And finally, the growth of the roots as well as of the shoots can be observed and measured, and the roots themselves can be recovered for chemical analysis or experimentation.

Barley. The prime reason for beginning this program of selection for seawater culture with barley was that it has been historically the most salt tolerant of annual grains (Jacobsen and Adams, 1958). There were additional reasons prompting this choice. Barley is an important grain crop, its mineral nutrition and genetics have been intensively studied, and there are available collections of seeds, expecially the world collection maintained by the U.S. Department of Agriculture, representing reservoirs of genetic diversity. That this diversity applies to the degree of salt tolerance or sensivity was apparent from the work of Ayers (1953), Donovan and Day (1969), and Greenway (1962, 1965); see also Iyengar et al. (1977) and Maddur (1976).

A composite cross of barley was the first source of genetic variation used. It was synthesized by Suneson and Wiebe (1962) from 6,200 barley genotypes which were intercrossed at random yielding a high degree of genetic variability. Of the 7,200 initial entries, 22, or 0.31%, survived the screening in salinized nutrient solutions (Epstein and Norlyn, 1977). Survival here refers to the ability to germinate, establish seedlings, grow, flower, and set seed, that is, to tolerate the salinity imposed (75-90% seawater) throughout the life cycle of the plant. It was found, in line with previous experience (Bernstein, 1964; Dewey, 1962), that success at only one stage of the life cycle - germination, for example - is no assurance of success at other stages, such as flowering or grain production.

To determine the effectiveness of this selection procedure, and the feasibility of irrigating barley with seawater on sandy dune soil, a field trial site was established at the University of California Bodega Marine Laboratory on the Pacific coast 80km north of San Francisco (Figure 3). The most important feature of this site was the soil - dune sand that is deep and highly permeable. At this writing, some of the plots at this site have been irrigated more than sixty times with undiluted seawater over a period of 2.5 years, without progressive salt build-up since the initial three or four irrigations.

FIGURE 3

Overview from a dune of the test site at Bodega Marine Laboratory. Barley in the foreground, plastic greenhouse shelters for tomatoes in back. Bodega Harbor and the small coastal town of Bodega Bay in the background.

The plots were furrow irrigated from a polyethylene lined distribution ditch. The first and second seasons were very dry so that
there was little dilution with rainwater. After each rain the seawater plots were irrigated immediately to maintain the salt concentration close to the test levels. The plots were normally irrigated
once a week to minimize fluctuations in the salt concentration of the
soil solution. The seawater used in the experiments was pumped
directly from the Pacific Ocean at Bodega Marine Laboratory.

The rows were planted to selections from Composite Cross XXI
and several cultivars. Band applications of fertilizer containing
nitrogen and phosphorus were provided (see Table 1 and comment in
the text there).

The results of this experiment have been described by Epstein
and Norlyn (1977). Since then, one selection in one plot irrigated
with undiluted seawater has given a (roughly estimated) yield of
1,580 kg/ha (Figure 4). This compares with 2,070 kg/ha for the 1976
world average for barley (U.S. Department of Agriculture, 1977). The
seed from the undiluted seawater and the fresh water treatments was
subjected to a proximate analysis (measure of feed quality). The
results showed that the seawater irrigated barley was acceptable.

Several conclusions emerge from the results of these preliminary
tests. The basic concept of irrigating barley with seawater is, at
the very least, a biological success, in that an early selection has
been able to go from seed to seed in three successive seasons. Barley
has demonstrated substantial phenotypic plasticity facilitating adjustment to seawater culture, and barley grown in this manner seems
acceptable for feed.

A program is now under way to screen the world collection of
barley (22,000 entries) for the best selections in terms of the ability to germinate and establish seedlings as well as for those that
perform best during vegetative and reproductive growth under high
levels of salt stress.

Wheat. There were two main reasons for extending this project
to wheat, Triticum aestivum. First, wheat is the world's premier
grain crop, and second, unlike barley, wheat has no reputation for
a relatively high degree of salt tolerance, making it a doubly challenging candidate. In addition, the world collection of wheat maintained by the U.S. Department of Agriculture has more than 30,000
entries, providing a broad base of genotypic diversity.

Screening was by means of salinized solution cultures, as with
barley. Over 5,000 entries from the world collection have so far
been screened. Of these, 40, or about 0.8%, are promising, 12 particularly so. It appears on the basis of these tests that wheat

FIGURE 4.

Rows of barley at the Bodega site irrigated with undiluted seawater. The center row is a selection which produced a yield estimated at 1,580 kg/ha in 1977.

lines tolerating at least 50% seawater salinity throughout their life
cycle can be obtained, and further improvement is likely. Figure 5
shows a tank containing a nutrient solution salinized to 50% seawater
salinity. Most of the wheat seedlings have died but one plant has
grown and is producing grain. Seed of the successful plants is sown
in soil to produce more seed of the same genotype,for further testing
and use in a breeding program.

USES OF SALT TOLERANT WILD SPECIES

Wild salt tolerant species were mentioned in the introduction
of this chapter. Such species may be used for the production of
food and other biomass based on seawater in two main ways. First,
some of these species, or selections made from them, might be used
directly. Grain may serve as food or fodder, and vegetative matter
as fodder, fiber, or as a source of energy (fuel). Second, wild salt
tolerant but economically worthless relatives of crop species may be
used as sources of germplasm to transfer salt tolerance into the
economic species.

The direct use of wild salt tolerant plants. On a limited
scale wild plants adapted to saline substrates including seawater
have been and are being used. For example, Felger and Moser (1973)
reported that the Seri Indians of Sonora, Mexico, used seeds of eelgrass, Zostera marina, for food. This species belongs to one of only
a few groups of flowering plants growing fully submerged in seawater.
The grain was gathered; there was no agricultural management or culture of the plant.

Another direct use of plants native to habitats dominated by
seawater is reported by Chapman (1976). Mangroves (Fig. 1), trees
which grow extensively along tropical coasts with their roots in soil
saturated with seawater, are cut and their wood is burned for fuel
or used to make charcoal. Chapter 11 should be consulted on silviculture with saline water.

By appropriate schemes of selection and management wild plants
native to coasts or salt marshes might be rendered economically useful to a greater extent than they now are. Specifically, halophytic
seed-bearing wild plants may furnish grain for food or fodder, or
their vegetative matter may be used for fresh feed or for hay. Reports on explorations along this line have been published (Boyko,
1966, 1968; Mudie, 1974; Mudie et al., 1972; Somers, 1975, 1978).

Wild salt tolerant plants: their use in breeding. Intraspecific
diversity has been discussed above, under that heading. It can be
exploited to best advantage where there are breeding stocks or collections of germplasm with much genetic variability from a wide

variety of geographic origins. Such stocks are not available for
all crops, or if available, might not be variable enough genetically
in salt tolerance to give a sufficiently broad base from which to
obtain salt tolerant lines.

An alternative strategy may be rewarding in such situations.
Rush and Epstein (1976) compared two species of tomato: the domesticated Lycopersicon esculentum and L. cheesmanii, a wild species seed
of which was collected by Rick (1972) on Isla Isabella (Galapagos
Islands) from a plant growing a few meters above high tide and presumably much exposed to salt water. It was assumed that this might
be a salt tolerant genotype and so it turned out to be. Plants grown
from the seed of this plant survived in solution cultures salinized
to full seawater salinity, although their growth was impaired under
this stress. A comparative physiological study revealed marked differences between the two species in aspects of ion transport, amino
acid profiles and still other features (Rush and Epstein, 1976).

The authors surmised that these two tomato species might be useful not only for such comparative studies but also for a program of
breeding for salt tolerance. The two species were crossed and progeny of the cross is presently growing in plastic greenhouse shelters
(see Fig. 3) on dune sand, being irrigated with various dilutions of
seawater. The highest concentration used is 70% seawater. At this
salinity (33.8 mmhos/cm) the fruit produced is the size of cherry
tomatoes; see Fig. 6. Whether tomatoes can be grown at still higher
salinities remains to be seen, but the work demonstrates that salt
tolerance can be genetically transferred from a wild species to a
domesticated one. Inasmuch as this approach has barely begun, prospects for success in further efforts along this line seem bright.
For a discussion of genetic resources in wild relatives of crops see
Harlan (1976).

COMMENT

No attempt is made here to provide a survey of the literature on
seawater irrigation. Work done in the 1960's and before is available
in two volumes edited by Boyko (1966, 1968). More recently, Mudie
(1974) has reviewed the potential economic uses of halophytes and
Mudie et al. (1972) have reported preliminary studies on seawater
irrigation. Somers (1978) has reviewed a good deal of recent research in a paper devoted primarily to the direct utilization of
halophytes.

As for attempts to generate salt tolerant crops by genetic manipulation, which is emphasized in the present chapter, the work reported here differs from earlier efforts in a number of ways. 1.
Instead of comparing a few existing cultivars under saline conditions,

large collections of germplasm were screened. 2. Where that was
not feasible (in the case of the tomato) exotic germplasm was used
for transfer of salt tolerance into the crop species. 3. All
initial screening was done in homogeneous solution cultures of known
composition. Spatial variation and such ill-defined phenomena as
"partial root contact," "subterranean dew" and still others (Boyko,
1966, 1968) could play no role. 4. Promising selections of progeny
expressly screened for seawater culture were eventually tested in a
maritime environment using seawater or dilutions of it. Both recourse to the existing collections of germplasm and rigorous screening under defined conditions of stress are routine procedures in
plant breeding for all kinds of desirable traits; it is not apparent
why the specific desideratum of salt tolerance should be achievable
by lesser means.

PRODUCTION OF FOOD AND OTHER BIOMASS
BASED ON SEAWATER: RESEARCH PRIORITIES

Adaptation of barley, wheat, and tomato to seawater culture.
Only a beginning has been made in adapting barley, wheat, and tomato
to seawater culture. The work of selecting and breeding has not en-
compassed more than a fraction of the germplasm available for testing.
Even adequate facilities for screening of this type do not exist:
greenhouses with good climate controls to minimize environmental
"noise" and equipped with large solution culture installations for
screening thousands of entries under defined conditions of salt
stress. Such facilities, whose cost is small when considered in
terms of the potential returns, are needed for rapid progress in
this new departure in research and development.

Field tests with seawater or dilutions of it have been made at
only a single location on the Northern California coast, an area probably by no means typical of the sandy coastal lands, mainly in the
developing countries, where the scheme may have its greatest potential (see Geographic-economic studies, below). There is an urgent
need for conducting field tests along a coast like that of Baja
California, Mexico, under hotter and drier conditions than those
prevailing on the coast north of San Francisco.

Once genetic lines capable of being grown by seawater culture
have been generated, plant breeders and agronomists will have to do
additional work to make sure that conventional criteria of yield,
quality, disease resistance, etc. are met before seeds are released.

Technology transfer and international cooperation. Adaptation
of newly developed cultivars to locations and technologies differing
from those for which they were initially generated almost inevitably
requires that local adaptability research be done. Many developing

FIGURE 5

A lone wheat plant survives and produces grain in the stressful
environment of a nutrient solution kept at 50% seawater salinity in
a 700-liter tank. Dead plants litter the top of the tank beyond.

countries have cadres of highly qualified plant breeders and agronomists. Organizations devoted to international agricultural development and cooperation, both public and private, may play a useful
catalytic role in both the development of seawater-based crop culture
and the subsequent transfer of materials (mainly seed stocks) and
know-how to countries where this scheme might be applicable.

Other crops. The early results of the work on barley, wheat,
and tomato, while not conclusive, are sufficiently encouraging to
prompt investigations of other crops with a view to growing them
in seawater culture. Fiber crops such as cotton, Gossypium spp.,
should not be overlooked in this connection. The papers by Ayers
and Westcot (1976) and Mudie (1972, 1974) can serve as useful points
of departure in the search for candidate crops.

Other techniques. The selection and breeding strategies discussed in this chapter are essentially conventional. The novelty
lies mainly in the very purpose that is attempted - the creation
of crop strains suited to seawater culture, and in the techniques -
large-scale screening under closely defined conditions of salinity
and the transfer of salt tolerance by genetic means from one species
to another. Large-scale screening, and breeding, have been eminently
successful tools of the breeder in bestowing a multitude of desirable
traits on crops; there is reason to believe that they can be successful in endeavors to create salt tolerant crops.

There are, however, other techniques to bring into play, ranging from graftage (Cook, private communication) to cell culture (Chapter 6), somatic hybridization, mutation breeding and genetic engineering (Chapter 21). These possibilities should be pursued.

Geographic-economic studies. Coastal deserts extend for nearly
30,000 km (Meigs, 1966), and sand dunes cover about 1.3×10^9 ha
(Mudie, 1974). What fraction of this immense area (about 9% of the
land surface of the earth) might be irrigated with seawater is unknown, but it is evident that seawater culture of crops on only a
fraction of that area might make an appreciable contribution to world
food supplies. Now that the biological feasibility of such schemes
can no longer be dismissed out of hand, studies are needed to identify the areas best suited for this purpose in terms of climate, soil
characteristics, topography, drainage, and social and economic factors. A detailed study like that of Meigs (1966), but devoted specifically to the prospects for seawater culture, is called for; see
also Chapter 4.

Soil-water-salt relations. Studies of the soil-water system
have revealed a formidable complexity (Nielsen et al., 1972). It has
solid, liquid and gaseous phases, plant and animal components, and
chemical, physical, and biological characteristics. Soil being the

FIGURE 6

Fruit of progeny of a cross between the commercial tomato,
Lycopersicon esculentum, and the wild L. cheesmanii. The two fruit
on the left are still attached to a plant irrigated with 70% seawater.
The fruit on the right was taken from a plant of the same genotype
irrigated with fresh water.

basis of life on land, this system has been much studied, but numerous questions remain. If crop production is to be extended to sandy coasts under a regime of seawater culture, new problems will be. encountered concerning the interactions among water, salt, and sand, the effects of temperature and osmotic gradients, and many more. At least some of these problems will have a bearing on crop culture; appropriate investigations will have to be undertaken.

Basic research. The biological strategies that have evolved in plants in response to the selection pressure of salinity are discussed in Chapter 6. Some plants are adapted to higly saline conditions, others are intolerant of even slightly saline substrates, and every degree of intermediate response may be found.

The studies discussed in the present chapter have shown extreme variations in salt tolerance even within species. In order to provide sharply contrasting genotypes for studies on the mechanisms of salt toleration,selections are being made not only for salt tolerance but also for salt sensitivity (D.L. Fredrickson, private communication). In barley, salt tolerant isolines have been obtained whose seeds germinate and establish seedlings at a salt concentration six times higher than that causing death in selected salt sensitive ones. Isolines differing to this extent have been secured by selection within a given cultivar.

The availability of intraspecific genetic lines which have been deliberately "selected apart" with respect to this trait represents an unprecedented opportunity for research on the mechanisms of salt tolerance and sensitivity. Such contrasting lines may differ in aspects of physiology, metabolism, enzymes, membranes, cytoplasmic organelles, ion transport, water relations, and morphology and ultrastructure. Any such differences between closely related genotypes that may come to light are likely to have a bearing on their differential responses to salinity, a statement that cannot be made for comparisons between widely different kinds of plants.

Basic comparative investigations as envisaged here hopefully will in turn facilitate the applied research-and-development discussed in this chapter. For example, they may lead to the discovery of "salinity markers" - readily identified characters consistently associated with salt tolerance. Discovery of such markers could speed up the work of selection and breeding.

Like the aspects of physiology discussed above, the genetic control of salt tolerance can be studied in genotypes sharply contrasting in this trait. Knowledge gained is apt to be helpful to salt tolerance breeders. Comparisons are also needed between genetic lines developed for seawater culture and salt tolerant strains meant for the very different situations inland where soils and irrigation waters are saline (Chapter 10).

SUMMARY

This chapter discusses the possibility of adapting crops to sea-water culture. The scheme would use for crop production resources of sea and sand that are now useless for this purpose and even inimical to it in many situations. Evidence concerning the feasibility of the scheme is presented. It consists mainly of demonstrations that when resources of germplasm in barley and wheat are screened on an adequate scale, much intraspecific diversity with respect to salt tolerance is uncovered - so much as to make it possible to select and breed strains adapted to seawater culture. Where intra-specific diversity is inadequate, as was found to be the case in the tomato, exotic salt tolerant species may be used to transfer salt tolerance into the commercial species. Wild salt tolerant plants may also be used directly. Adaptation of crops to seawater culture is a new departure; needs for research and development are outlined.

ACKNOWLEDGMENTS

Research by the authors referred to here was supported by the National Sea Grant Program, U.S. Department of Commerce, and the National Science Foundation. C.H. Hand, Jr., Director, Bodega Marine Laboratory, has been very helpful.

REFERENCES

Asher, C.J. 1978. Natural and synthetic media for spermatophytes. In: CRC Handbook Series in Nutrition and Food, Section 6, Vol. III. Culture Media for Microorganisms and Plants. M. Rechcigl, Jr., ed. CRC Press, Cleveland. pp. 575-609.

Ayers, A.D. 1953. Germination and emergence of several varieties of barley in salinized soil cultures. Agron. J. 45:68-71.

Ayers, R.S. 1977. Quality of water for irrigation. J. Irrig. Drainage Div., Amer. Soc. Civ. Engineers 103:135-154.

Ayers, R.S. and D.W. Westcot. 1976. Water Quality for Agriculture. Irrigation and Drainage Paper 29. Food and Agriculture Organization of the United Nations, Rome.

Bernstein, L. 1964. Salt tolerance of plants. Agric. Inf. Bull. 283. U.S. Department of Agriculture, Washington, DC.

Boyko, H., ed. 1966. Salinity and Aridity: New Approaches to Old Problems. Junk, The Hague.

Boyko, H., ed. 1968. Saline Irrigation for Agriculture and Forestry. Junk, The Hague.

Chapman, V.J. 1976. Mangrove Vegetation. J. Cramer, Vaduz.

Dewey, D.R. 1962. Breeding crested wheatgrass for salt tolerance. Crop Sci. 2:403-407.

Donovan, T.J. and A.D. Day, 1969. Some effects of high salinity on germination and emergence of barley (Hordeum vulgare L. emend Lam.). Agron. J. 61:236-238.

Epstein, E. 1963. Selective ion transport in plants and its genetic control. In: Desalination Research Conference. National Academy of Sciences - National Research Council Publication 943. Nat. Acad. Sci. - Nat. Res. Council, Washington, DC.pp. 284-298.

Epstein, E. 1972a. Mineral Nutrition of Plants: Principles and Perspectives. John Wiley and Sons. New York.

Epstein, E. 1972b. Physiological genetics of plant nutrition. In: Mineral Nutrition of Plants: Principles and Perspectives. E. Epstein. John Wiley and Sons. New York. pp. 325-344.

Epstein, E. 1977a. The role of roots in the chemical economy of life on Earth. BioSci. 27:783-787.

Epstein, E. 1977b. Genetic potentials for solving problems of soil mineral stress: adaptation of crops to salinity. In: Plant Adaptation to Mineral Stress in Problem Soils. M.J. Wright, ed. A Special Publication of Cornell University Agricultural Experiment Station, Ithaca, NY. pp. 73-82.

Epstein, E. and R.L. Jefferies. 1964. The genetic basis of selective ion transport in plants. Annu. Rev. Plant Physiol. 15:169-184.

Epstein, E. and J.D. Norlyn. 1977. Seawater-based crop production: a feasibility study. Science 197:249-251.

Felger, R. and M.B. Moser. 1973. Eelgrass (Zostera marina L.) in the Gulf of California: discovery of its nutritional value by the Seri Indians. Science 181:355-356.

Frankel, O.H. and E. Bennett,eds. 1970. Genetic Resources in Plants - Their Exploration and Conservation. Blackwell Scientific Publications, Oxford.

Frankel, O.H. and J.G. Hawkes, 1975. Crop Genetic Resources for Today and Tomorrow. Cambridge University Press, Cambridge.

Gerloff, G.C. 1963. Comparative mineral nutrition of plants. Annu. Rev. Plant Physiol. 14:107-124.

Greenway, H. 1962. Plant response to saline substrates. I. Growth and ion uptake of several varieties of Hordeum during and after sodium chloride treatment. Aust. J. Biol. Sci. 15:16-38.

Greenway, H. 1965. Plant response to saline substrates. VII. Growth and ion uptake throughout plant development in two varieties of Hordeum vulgare. Aust. J. Biol. Sci. 18:763-779.

Harlan, J.R. 1976. Genetic resources in wild relatives of crops. Crop Sci. 16:329-333.

Hawkes, J.G., J.T. Williams, and J. Hanson. 1976. A Bibliography of Plant Genetic Resources. International Board for Plant Genetic Resources, Rome.

Hewitt, E.J. 1976. Sand and Water Culture Methods Used in the Study of Plant Nutrition. Revised 2nd ed. Commonwealth Bureau of Horticulture and Plantation Crops, East Malling. Tech. Communication No. 22.

Iyengar, E.R.R., J.S. Patolia, and T. Kurian. 1977. Varietal differences in barley to salinity. Zeitschrift Pflanzenphysiol. 84:355-361.

Jacobsen, T. and R.M. Adams. 1958. Salt and silt in ancient Mesopotamian agriculture. Science 128:1251-1258.

Klimashevsky, E.L., ed. 1974. Variety and Nutrition. Academy of Sciences of the USSR. Siberian Branch. Sib. Inst. Plant Physiol. and Biochem., Irkutsk, Siberia, USSR. (In Russian).

Kruckeberg, A.R. 1959. Ecological and genetic aspects of metallic ion uptake by plants and their possible relation to wood preservation. In: Marine Boring and Fouling Organisms. D.L. Ray, ed. University of Washington Press, Seattle. pp. 526-536.

Lauchli, A. 1976. Genotypic variation in transport. In: Encyclopedia of Plant Physiology, New Series, Vol. 2, Part B. U. Luttge and M.G. Pitman, eds. Springer-Verlag, Berlin. pp. 372-393.

Maddur, A.M. 1976. The inheritance of salt tolerance in barley (Hordeum vulgare L.). Ph.D. Thesis, Michigan State University, East Lansing.

Meigs, P. 1966. Geography of Coastal Deserts. Arid Zone Research XXVIII, UNESCO, Paris.

Millikan, C.R. 1961. Plant varieties and species in relation to the occurrence of deficiencies and excesses of certain nutrient elements. J. Aust. Inst. Ag. Sci. 27:220-233.

Mudie, P.J. 1974. The potential economic uses of halophytes. In: Ecology of Halophytes. R.J. Reimold and W.H. Queen, eds. Academic Press, New York. pp. 565-597.

Mudie, P.J., W.R. Schmitt, E.J. Luard, J.W. Rutherford, and F.H. Wolfson. 1972. Preliminary Studies on Seawater Irrigation. Foundation for Ocean Research Publication No. 1; Scripps Institution of Oceanography Ref. 72-70, La Jolla, California.

Muhammed, A., R. Aksel, and R.C. von Borstel, eds. 1977. Genetic Diversity in Plants. Plenum, New York.

Myers, W.M. 1960. Genetic control of physiological processes: consideration of differential ion uptake by plants. In: A Symposium cn Radioisotopes in the Biosphere. R.S. Caldecott and L.A. Snyder, eds. University of Minnesota, Minneapolis. pp. 201-226.

Myhill, R.R., and C.F. Konzak, 1967. A new technique for culturing and measuring barley seedlings. Crop Sci. 7:275-276.

Nielsen, D.R., R.D. Jackson, J.W. Cary, and D.D. Evans, eds. 1972. Soil Water. American Society of Agronomy - Soil Science Society of America, Madison.

Rains, D.W. and E. Epstein. 1967. Preferential absorption of potassium by leaf tissue of the mangrove, Avicennia marina: an aspect of halophytic competence in coping with salt. Aust. J. Biol. Sci. 20:847-857.

Reisenauer, H.M. 1966. Mineral nutrients in soil solution. In: Environmental Biology. P.L. Altman and D.S. Dittmer, eds. Federation of American Societies for Experimental Biology, Bethesda. pp. 507-508.

Richards, L.A., ed. 1954. Diagnosis and Improvement of Saline and Alkali Soils. Agriculture Handbook No. 60. United States Department of Agriculture, Washington, DC.

Rick, C.M. 1972. Potential genetic resources in tomato species: clues from observations in native habitats. In: A.M. Srb, ed. Genes, Enzymes, and Populations. Plenum, New York. pp. 255-269.

Rush, D.W. and E. Epstein. 1976. Genotypic responses to salinity: differences between salt-sensitive and salt-tolerant genotypes of the tomato. Plant Physiol. 57:162-166.

Somers, G.F., ed. 1975. Seed-bearing halophytes as food plants. University of Delaware, Newark.

Somers, G.F. 1978. Production of food plants in areas supplied with highly saline water. Problems and prospects. In: Stress Physiology in Crop Plants. H. Mussell and R.C. Staples, eds. Wiley Interscience, New York. In press.

Suneson, C.A. and G.A. Wiebe. 1962. A "Paul Bunyan" plant breeding enterprise with barley. Crop Sci. 2:347-348.

University of California. 1977. California Agriculture Special Issue: Germplasm. Calif. Agric. 31, No. 9.

U.S. Department of Agriculture. Agricultural Statistics. 1977. U.S. Government Printing Office, Washington, DC. p. 46.

Vose, P.B. 1963. Varietal differences in plant nutrition. Herb. Abstracts 33:1-13.

Weast, R.C., ed. 1975-1976. Handbook of Chemistry and Physics. 56th ed. Chemical Rubber Publishing Company, Cleveland. p. F-199.

Williams, J.T. 1976. A Bibliography of Plant Genetic Resources. Supplement. International Board for Plant Genetic Resources, Rome.

Wright, M.J., ed. 1977. Plant Adaptation to Mineral Stress in Problem Soils. A Special Publication of Cornell University Agricultural Experiment Station, Ithaca, NY.

NATURAL HALOPHYTES AS A POTENTIAL RESOURCE FOR NEW SALT-TOLERANT

CROPS: SOME PROGRESS AND PROSPECTS

G. Fred Somers

School of Life and Health Sciences
  and College of Marine Studies
University of Delaware
Newark, DE

The desirability of salt-tolerant terrestrial crops has been recognized for some time. The vast areas of the world which are essentially unproductive of food crops because of the high salinity of the water and/or soil which characterize them have long challenged man to find ways to utilize them to meet his needs for food and fiber. Historically the production of these staples has been dependent upon a supply of fresh water. Recently the supply of this resource has become ever more critical as populations and technologies continue to expand. The problem with respect to crop production was stated suc- cintly by Flowers et al. (1977):

> To plant life, salinity is just one inimical factor of the
> environment. To man, salinity creates a problem due to its
> effects on his crops which are predominantly sensitive to
> the presence of high concentrations of salts in the soil.

If at the same time the germ base for crop production could be expanded, other potential problems might be circumvented. A U.S. National Academy of Sciences study panel (Handler, 1970) expressed concern for the limited germ plasm base of man's food supply:

> We still have urgent need . . . to find suitable alternatives
> to the dependence of man, globally, on just a few staple
> crops--rice, wheat and corn. This dependence on three
> cereal types offers the terrifying prospect of a worldwide
> pandemic of a virus to which no strain of one of these
> species might be resistant.

In the 1960's some attempts were made to use very saline water to grow crops (Boyko, 1966, 1968). Epstein, Jeffries and Rains

repeatedly called attention to the potential of using saline water
for crop production (Epstein and Jeffries, 1964; Rains and Epstein,
1967; Rains, 1972; Epstein, 1972; Rush and Epstein, 1976). Similar
views were published by Bronowski (1969), Waisel (1972), Mudie (1974),
Somers (1975), and Nabors (1976). Others have expressed a similar
view, usually less formally. It would appear that the time has come
to address seriously the question: Are there seed-bearing halophytes
which can be grown using water approaching, or equal in, salinity to
that of the oceans and which could be used as food for man and/or
domesticated animals, or which could be modified by selection and
breeding to serve such use?

That abundant growth of angiosperms is possible in the presence
of high salinity is evidenced by the high productivity of coastal
salt marshes which are supplied with water ranging from brackish to
essentially ocean salinity (see for examples: Squiers and Good, 1974;
Kirby and Gosselink, 1976; Reimold and Linthurst, 1977). Annual production of dry matter in some of these areas is comparable with that
of cultivated crops dependent upon fresh water and added fertilizers.
One should not be misled, however, into thinking that all salt marshes
are equally productive, that all of the plants from them have equal
capabilities under comparable conditions (Somers, 1978), or that such
populations are the only potential sources of salt-tolerant crops.
Recent success in selecting barley for a high degree of salt-tolerance (Epstein, 1977; Epstein and Norlyn, 1977) and in incorporating
salt tolerance from a natural population into the tomato (Epstein,
1977) give encouragement that some conventional crops might be developed with a high degree of salt tolerance. However, earlier findings
indicated that such prospects were not bright (Richards, 1954; Bernstein, 1964, 1975; O'Leary, 1975; Maas and Hoffman, 1977).

With very few exceptions, one of which is the beet, our conventional crops have been derived from plants which were not notably
salt-tolerant. They grew naturally in fresh-water habitats and were
selected to succeed in such an environment. Constraints of excessive
salinity were usually avoided, or, when they were encountered, the
approach used to ameliorate the situation was to manage the soil and/
or the water to minimize the stress on the crop. With relatively
abundant supplies of fresh water this was an obvious, economic approach.

Perhaps there is an alternative. Salt tolerance, at least in
some measure, is not limited to a few plants (Mudie, 1974). Flowers
et al (1977) enumerate 94 orders of flowering plants and note that
38 include halophytic species. Obviously salt tolerance per se is
not a unique attribute, though tolerance to salinities equal to, or
approaching that of the oceans is undoubtedly much more limited. Intensive screening of the total available germ plasm pool may provide
additional salt-tolerant lines of conventional crops (see Epstein et

al., this volume). Unfortunately the mechanism of salt tolerance in plants is not well understood (Flowers et al., 1977). Hence, if one is to proceed to select potential crop candidates from among salt-tolerant species the approach will have to be largely pragmatic. The goal should be to obtain crops which can be grown successfully using sea water, or water approaching it in salinity, either totally or as a supplement to whatever fresh water might be available. But it may not be necessary to depend solely upon highly saline water in all cases. Many semi-arid areas have periods of rainfall which might provide sufficient fresh or minimally saline water to start a crop and grow it through the seedling stage when some potential crop plants appear to be most sensitive to excessive salinity (Somers et al., 1978). Once established the crop can then be grown using exclusively very saline water if that is all that is available. In no small measure, approaches similar to those used in the past for crop improvement should be used. One does not search for a crop suitable to all situations. Rather the approach has been to select and breed for more or less restricted soil properties and climate. First, however, the potential of a high degree of salt tolerance must be demonstrated.

Populations of plants which grow naturally in highly saline environments offer the potential of introducing entirely new germ plasm for salt tolerance into agricultural production. Selection for salt tolerance has already occurred. However, the problem now becomes one of a different kind. Salt-tolerance alone is, of course, not sufficient. Other attributes are also essential. Among other things these include (Somers, 1975):

1. Yield of fruit or other edible protein.

2. General characteristics of the edible portion, e.g. dry or fleshy, size, etc.

3. Quality of the edible portion, to include nutritional quality, palatability, similarity to existing foods, etc.

4. Potential for adapting to commercial production.

In at least a preliminary way some of these criteria can, and should be applied in screening potential candidates in natural populations. For example, if one seeks an edible grain or other dry fruit the initial selection criteria should be concerned with seed size. Palatability of the edible portion should be given special attention in selecting potential food plants. Food habits are so highly enculturated that there is likely to be high resistance to an entirely new food. Greater success will be likely if the new material resembles an existing food or if it can be blended with it. The introduction of soybean products into the diet of the United States should be instructive in this regard.

Given a high degree of salt tolerance and a suitable potential with respect to the other selection criteria, examples of the development of present crops can serve to illustrate what might be accomplished. Compare for example wild <u>Daucus carota</u> characterized by a small, very fibrous tap root with the cultivated varieties of carrot. The development of maize is another dramatic example. The many cultivars of <u>Brassica oleracea</u> are illustrative also.

The first step is to screen the many halophytes for their potential as crops. Uses as forage and grains for animals should be included and may be easier to accomplish than food for man. Natural populations must be examined carefully with the above criteria in mind. Care must be taken to examine minor components of the natural ecosystem. Experience has shown that these may have more potential than dominant species. When isolated and grown in monoculture they frequently demonstrate a capacity not appreciated in the wild. This means repeated visits to saline habitats at different seasons to recognize potential candidates and to obtain seed, or other suitable propagules. This selection process can be aided by published evaluations of salt tolerance (e.g. Mudie, 1974; Chapman, 1975; Waisel, 1972; Ranwell, 1972; Pihl et al., 1978), but there is no substitute for careful observations in the field.

That success in growing salt-tolerant crops will depend upon appropriate soil and water management should also be considered. Some halophytes grow in soils which are continuously saturated with water and, hence, are anaerobic. <u>Spartina alterniflora</u>, the dominant grass of tidemarshes of eastern United States, with a C_4 photosynthetic pathway (Smith and Epstein, 1971) is one example. Associated with the low oxygen potential in the soils where this plant grows are reduced compounds of iron, sulfur, etc. This no doubt influences the mineral nutrition of the plants. On the other hand, other halophytes grow in better-drained habitats. <u>Atriplex patula</u> var. <u>hastata</u> (sensu Fernald, 1950 and Gleason, 1952) for example, grows on higher portions or along the edges of the same marshes as <u>Spartina alterniflora</u>. This plant does not tolerate prolonged immersion in saline water, whereas <u>S. alterniflora</u> prefers it (Somers, 1978; Parrondo et al., 1978). While observations of natural populations can be instructive in matters of this kind there is no substitute for careful experimentation in field plots in a variety of climates and soils guided by research under the more controlled situation provided by the laboratory and greenhouse.

But enough of generalities. Can one be more specific? For this purpose it might be useful to consider potentially useful species under headings identified roughly by type of crop with which we have had some experience: 1) grains and other seed crops, 2) forages, 3) leafy vegetables, 4) specialty crops. Recourse to published literature could lead to information on other crops for saline habitats, e.g. timber (Walsh, 1974), fiber [e.g. <u>Phragmites communis</u> (Haslam,

1972) and the nipa palm], etc.

Grains and other seed crops. A food plant of primitive peoples, Chenopodium quinoa, appears from limited evaluations to be moderately salt tolerant (Somers et al., 1978). In view of the high quality of its seed protein (Lopez, 1973) and the fact that it was a staple food of American Indians and is still an important source of food in the high Andes (Heiser, 1973; Vietmeyer, 1975) this plant probably deserves further testing. This view is reinforced by the fact that a strain of Chenopodium album L. has been found which grows successfully in water of 25-30 o/oo once established with fresh water. The seeds are small but they constitute up to 30% of the dry weight of the plant with about 17% protein which has a good spectrum of essential amino acids (Somers et al., 1978). Hybrids of Ch. album and C. quinoa appear likely. Both of these taxa are members of the Chenopodiaceae, a family which includes a number of plants which inhabit highly saline habitats, e.g. species of Salicornia, Suaeda, Kochia, Allenrolfea, Salsola and Atriplex. One of these genera deserves special mention: Atriplex. Species of this taxon are widespread in semi-arid and saline habitats. Atriplex patula, var. hastata (sensu Fernald, 1950 and Gleason, 1952) appears particularly promising. The seeds are small, but dimorphic with regard to size, which suggests the possibility of selecting for a line with large seeds. The plants can be grown using water of from 25-32 o/oo salinity, but grows better if first established with brackish water. The yield of seed is equivalent to about 1.2 T/hectare; the protein content is about 16% and the spectrum of essential amino acids is good (Somers et al., 1978). There is a wide variation among natural populations which persists when selections are grown at a single site. Selection of superior lines from among this germ plasm pool would appear to be promising.

Another promising prospect from amongst the native populations of tide marshes of eastern United States is Kosteletzkya virginica (L.) Presl. (Malvaceae), a perennial which produces relatively large seeds (about 20 mg each), which are about 33% protein with a good spectrum of essential amino acids. This plant has the added attribute that its seeds do not shatter badly when ripe, as contrasted to most wild plants. Moreover, the seed contains an abundance of a gum or mucilage which might have technological usefulness. [The thick mucilaginous roots of Althaea officinalis L., another taxon in this family were used in confectionary (Gleason, 1952).] This plant, although a perennial, grows from seed to seed in a single season and, after being established with fresh water, grows well in water of at least 20 o/oo salinity (Somers et al., 1978).

Various members of the poaceae offer possibilities as grain crops. An attractive inhabitant of brackish tide marshes is Zizania aquatica L., the domestication of which appears to be proceeding successfully in fresh water habitats (Oelke, 1975). Collections of this species have proved to be only moderately salt-tolerant, but the testing has been superficial at best (Somers et al., 1978).

In some measure Spartina alterniflora offers possibilities as a grain crop. The seeds are relatively large compared to most wild grasses and vary considerably in size (Garbisch, 1976), but are rather high in moisture content when mature. Dried seeds from Delaware weight about 3 mg each. However, to maintain its viability the seed must be stored wet (Mooring et al., 1971; Broome et al., 1974). The seeds are about 15% protein with an essential amino acid spectrum similar to that of wheat gluten (Somers et al., 1978).

A more promising halophytic cereal might be Distichlis palmeri (Vasey) Fassett [= Uniola palmeri Vasey (Shreve and Wiggins, 1964)], the seed of which was used as a food by the Cocopa Indians (Chase, 1950). This plant grows in the highly saline estuaries of the Gulf of California. The seed weigh about 10 mg each. The plant is dioecious which might facilitate breeding, but could complicate the production of high yields of seed. Some selections appear to be growing successfully at Puerto Penasco in plots established on dune sand and irrigated with water of 36 o/oo salinity.

Reliable yield data for seed of most of these species are not yet available.

Forages. It has been a long-standing practice to pasture coastal tide marshes or to cut portions for hay (Daiber, 1974; Ranwell, 1972). Some have been harvested for silage (Ranwell, 1972), but these uses for the most part have not depended upon deliberate agronomic management such as is used in producing more conventional forage crops. Of the grasses native to saline habitats, species of Spartina offer substantial potential. S. alterniflora and S. patens can be grown successfully using highly saline water. Seed of the former have been germinated and grown to maturity using full strength sea water during early stages and using water of 30-32 o/oo salinity later. (This was the highest salinity available to our field plots). As mentioned above, this species grows better in soils that are not well drained. On the other hand, S. patens grows better on well-drained soils. There is considerable genetic variability in the growth habits of these species which should facilitate selection of superior strains.

Distichlis spicata and the closely related D. stricta (Mason, 1969) both also grow in highly saline habitats. The latter is grazed by cattle and horses in marshes on the borders of the Gulf of California, Sonora, Mexico. D. palmeri grows in intertidal zones of the Gulf of California, is a more robust plant than D. stricta and, at least in some localities, appears to be less palatable as a forage. Though some selections growing in our plots appear tender when young.

FIGURE 1

    <u>Atriplex patula</u> var. <u>hastata</u> growing in plot flooded three times
weekly with estuarine water of 25,000-30,000 ppm. total salinity.
At this stage of growth this crop appears attractive as a potential
forage crop.

FIGURE 2

Young plants of <u>Atriplex</u> <u>patula</u> var. <u>hastata</u> at an age when they are edible as potherbs or as salad greens.

Chenopodium album has been found to be highly palatable as a
forage (Marten and Andersen, 1975). Atriplex patula var. hastata
might be useful as a forage (Figure 1). In one experiment the yield
of dry matter in mid-August, when much of the plant probably would
be readily digestible by ruminants, the yield was equivalent to 7.8
tons/hectare. Various species of Atriplex grow in saline habitats.
Some of them produce forage of high nutritive value (Vietmeyer, 1975).

Panicum dichotomiflorum is an annual weed of wet areas. At
least one strain of it grows well in water of 20 o/oo salinity. What
use it might have as a forage remains to be determined.

Leafy vegetables. Chenopodium album, C. quinoa, and Atriplex
patula have been used as potherbs and, at least the latter, as a
salad green when the plants are young and tender (Figure 2).

Specialty crops. No systematic survey of halophytes for compounds for pharmaceutical use has come to the author's attention.
However, casual observations indicate that plants of this kind might
have some usefulness as specialty crops. Two species of Brassicaceae,
Cakile edentula (Bigel.) Hook., and Lepidium virginicum L. grow well
in highly saline water (Somers et al., 1978). The fruits are highly
flavored as are other species of this familty. The leaves of the
former have an attractive spicy flavor.

Present status. These few species are an indication that a substantial potential for success exists in a search for terrestrial
halophytes which might serve, or could be developed to be used, as
food for man and/or domesticated animals. In excess of 350 selections from more than 60 species have been examined in our laboratory
(Figure 3) and/or in field plots (Somers et al., 1978). Many more
have been rejected on the basis of casual observations in the field.

But much remains to be done. There is a continued conflict
between the desirability to find even better candidate species and
the need to focus upon a very limited number of the "best" species.
It is not clear that the best taxa for this purpose have been identified. A continued search in more locations should be part of a continuing effort in this program just as it has been in other efforts
aimed at crop improvement. Except in this case, selection needs to
be made not only for better strains of a species, or desirable species
of a genus, but also for additional genera. These need to be evaluated not only in terms of present crops, but also for the possibility
of entirely new crops. Germ plasm pools need to be established. To
some extent existing stocks in plant introduction stations and in
other holdings can serve as a basis for material to be tested, but
such holdings of halphytes are usually very limited. They need to be
augmented.

FIGURE 3

Laboratory facility used for growing halophytes. Artificial seawater, amended as appropriate, is prepared in plastic tanks (A) and
pumped into storage tanks (B); from there it is siphoned into shallow,
plastic-lined tanks (C) in which plants are grown in peat pots filled
with sand. The growing tanks are filled until the pots are completely immersed and then the seawater is siphoned back into the original
tanks (A) where any loss in volume is restored by adding fresh water.

   More extensive testing of selected taxa needs to be conducted
in a variety of locations throughout the world. Some organization
needs to be established to coordinate such efforts and to direct
promising selections into an evaluation system. A single, experienced coordinator could do much to promote such a program.

   But all this is still only prelude to what must be done. The
most promising candidates must be thoroughly studied with respect to
their biology and the technology required for their production and
commercial use. An outline of suggested steps to be taken in developing food plants capable of being grown in saline water has been
provided (Somers, 1975). Suffice it to say that the whole technology
of food production, storage, processing, and distribution will be involved, not de novo, but any new crop must be fitted into the existing structure. What will be needed is specific information about
the characteristics of the crop and any adjustment needs in the present production and marketing structures to accomodate it. This
provides substantial new opportunities for both basic research and
development. But the potential is real and the stakes are high!
Only a substantial, coordinated effort can lead to success.

   The cooperation of Donna Grant, Miguel Fontes, and others is
gratefully acknowledged. This research was supported by the National
Sea Grant College Program, NOAA, U.S.D.O.C. and the University of
Delaware Research Foundation.

REFERENCES

Bernstein, Leon. 1964, Salt tolerance of plants. U.S. Department of
   Agriculture Inf. Bull. 283.

Bernstein, Leon. 1975. Problems in managing saline soils. In Seed-
   bearing halophytes as food plants. Proceedings of a conference,
   G.F. Somers (ed.), DEL-SG-3-75 College of Marine Studies, Univ.
   of Delaware, Newark: 120-133.

Boyko, H. (ed.). 1966. Salinity and Aridity: A new approach to old
   problems. W. Junk, the Hague.

Boyko, H. (ed.). 1968. Saline Irrigation for Agriculture and Forestry. Proceedings of the international symposium on plant growing
   with saline or sea-water without desalination. W. Junk, The Hague.

Bronowski, J. 1969. The impact of new science. In The environment of
   change. A.W. Warner, D. Morse and T.E. Cooney (eds.), Columbia
   University Press, NY: 67-95.

Broome, S.W., W.W. Woodhouse Jr., and E.D. Seneca. 1974. Propagation
   of smooth cordgrass, Spartina alterniflora, from seed in North
   Carolina. Chesapeake Sci. 15: 214-221.

Chapman, V.J. 1975. Terrestrial halophytes as potential food plants. In Seed-bearing halophytes as food plants. Proceedings of a conference, G.F. Somers (ed.), DEL-SG-3-75, College of Marine Studies, Univ. of Delaware, Newark: 75-87.

Chase, A. 1950. Manual of the grasses of the United States, 2nd rev. ed. (A.S. Hitchcock, 1935). Dover publ., New York, Reprint 1971.

Daiber, F.C. 1974. Salt marsh plants and future coastal salt marshes in relation to animals. In R.J. Reimold and W.H. Queen (eds.), Ecology of Halophytes, Academic Press, NY: 475-508.

Epstein, E. 1975. Mineral Nutrition of Plants: Principles and Perspectives. John Wiley and Sons, New York.

Epstein, E. 1977. Genetic potentials for solving problems of soil mineral stress: Adaptation of crops to salinity. In Proceedings of a workshop on plant adaptation to mineral stress, M.J. Wright (ed.), Cornell Univ. Agr. Expt. Station Special Bull: 73-82.

Epstein, E. and R.L. Jeffries. 1964. The genetic basis of selective ion transport in plants. Ann. Review Plant Physiol. 15: 169-184.

Epstein, E. and J.D. Norlyn. 1977. Sea-water based crop production: A feasibility study. Science 197: 249-251.

Fernald, M.L. 1950. Gray's Manual of Botany. 8th ed. Corrected printing 1970. VanNostrand Co., New York.

Flowers,T.J., P.F. Troke, and A.R. Yeo. 1977. The mechanism of salt tolerance in halophytes. Ann. Review Plant Physiol. 28: 89-121.

Garbisch, E.W. 1976. Spartina alterniflora seed collections in the vicinity of Branford Harbor, Conn. and various seed characteristics, tests and comparisons. Environmental Concern Inc., St. Michaels, MD.

Gleason, H.A. 1952. The New Britton and Brown Illustrated Flora of the Northeastern United States and Adjacent Canada. Hafner Publ. Co., New York and London.

Handler, P. (ed.). 1970. Biology and the Future of Man. Oxford University Press, New York.

Haslam, S.M. 1972. Biology flora of the British isles. Phragmites communis Trin. J. Ecology 60: 585-610.

Heiser, C.B., Jr. 1973. Seed to civilization: The story of man's food. Freeman, San Francisco.

Kirby, C.J. and J.G. Gosselink. 1976. Primary production in a
  Louisiana Gulf Coast Spartina alterniflora marsh. Ecology 57:
  1052-1059.

Lopez, J.G. 1973. Evolution of protein quality of Quinoa by protein
  efficiency ratio, biological values and amino acid composition.
  M.S. thesis, Utah State University, Logan, Utah.

Maas, E.V. and G.J. Hoffman. 1977. Crop salt tolerance; Evaluation of
  existing data. In Managing Saline Water for Irrigation, H.E. Dregne
  (ed.), Proceedings of international salinity conference, Texas Tech.
  Univ., Lubbock, TX: 187-198.

Marten, G.C. and R.N. Andersen. 1975. Forage nutritive value and
  palatibility of 12 common annual weeds. Crop Science 15: 821-827.

Mason, H.L. 1969. A Flora of the Marshes of California. Univ. of
  California Press, Berkeley.

Mooring, M.T., A.W. Cooper and E.D. Seneca. 1971. Seed germination
  response and evidence for height ecophenes in Spartina alterniflora
  from North Carolina. Amer. J. Bot. 58: 48-55.

Mudie, P.J. 1974. The potential economic uses of halophytes. In
  Ecology of Halophytes, R.J. Reimold and W.H. Queen (eds.), Academic
  Press, New York: 565-597.

Nabors, M.W. 1976. The use of spontaneously occurring and induced
  mutations to obtain agriculturally useful plants. Bioscience 26:
  761-768.

Oelke, E.A. 1975. Wild rice domestication as a model. In Seed-bearing
  halophytes as food plants. Proceedings of a conference, G.F.Somers
  (ed.), DEL-SG-3-75 College of Marine Studies, Univ. of Delaware,
  Newark: 47-56.

O'Leary, J.W. 1975. Potential for adapting present crops to saline
  habitats. In Seed-bearing halophytes as food plants: Proceedings
  of a conference, G.F. Somers (ed.),DEL-SG-3-75 College of Marine
  Studies, Univ. of Delaware, Newark: 91-114.

Parrondo, R.T., J.G. Gosselink, and C.S. Hopkinson. 1978. Effects
  of salinity and drainage on the growth of three salt marsh grasses.
  Bot. Gaz. 139: 102-107.

Pihl, K., D.M. Grant, and G.F. Somers. 1978. Germination of seeds of
  selected coastal plants. Del-SG-11-78, College of Marine Studies,
  Univ. of Delaware, Newark.

Rains,D.W. 1972. Salt transport by plants in relation to salinity.
    Ann. Review Plant Physiol. 23: 367-388.

Rains, D.W. and E. Epstein. 1967. Preferential absorption of potas-
    sium by leaf tissue of the mangrove, Avicennia marina: An aspect
    of halophytic competence in coping with salt.   Austr. J. Biol.
    Sci. 20:847-857.

Ranwell, D.S. 1972. Ecology of Salt Marshes and Sand Dunes. Chapman
    and Hall, London.

Reimold, R.J. and R.A. Linthurst. 1977. Primary productivity of minor
    marsh plants in Delaware, Georgia and Maine.   Technical Report
    D-77-36, U.S. Army Corps. of Engr., Waterways Expt. Station,
    Vicksburk, MS.

Richards, L.A. (ed.). 1954. Diagnosis and improvement of saline and
    alkali soils. U.S. Dept. Agr. Handbook No. 60.

Rush, D.W. and E. Epstein. 1976. Genotypic responses to salinity:
    Differences between salt-sensitive and salt-tolerant genotypes
    of the tomato. Plant Physiol. 57: 162-166.

Shreve, F. and I.L. Wiggins. 1964. Vegetation and flora of the
    Sonoran Desert, Vols. I and II. Stanford Univ. Press, Stanford, CA.

Smith, B.N. and E. Epstein. 1971. Two categories of C^{13}/C^{12} ratios
    for higher plants.   Plant Physiol. 47:380-384.

Somers, G.F. (ed.). 1975. Seed-bearing halophytes as food plants.
    Proceedings of a conference: DEL-SG-3-75, College of Marine Studies,
    Univ. of Delaware, Newark.

Somers, G.F. 1979. Production of food plants in areas supplied with
    highly saline water. Problems and prospects. In Stress Physiology
    in Crop Plants, H. Mussell and R.C. Staples (eds.), Wiley-Inter-
    science, New York: in press.

Somers, G.F., D.M. Grant, and R.D. Smith. 1978. Domestication of
    halophytes as potential crops for food and/or feed or for marsh
    improvement: Summary of progress 1974-1977. DEL-SG-12-78, College
    of Marine Studies, Univ. of Delaware, Newark.

Squiers, E.R. and R.E. Good. 1974. Seasonal changes in the produc-
    tivity, caloric content and chemical composition of a population
    of salt marsh cord-grass (Spartina alterniflora). Chesapeake Sci.
    15: 63-71

Vietmeyer, N.D. (Study Staff director). 1975. Underexpolited Tropical
   Plants with Promising Economic Value. Nat'l. Acad. Sci., Wash., DC.

Waisel, Y. 1972. Biology of Halophytes. Academic Press, NY.

Walsh, G.E. 1974. Mangroves: A review. In R.J. Reimold and W.H. Queen
   (eds.). Ecology of Halophytes, Academic Press, NY: 51-174.

SILVICULTURE WITH SALINE WATER

Howard J. Teas

Biology Department
University of Miami
Coral Gables, FL 33124

INTRODUCTION

Mangroves are the trees and shrubs that grow in the edge of the sea and thus are cultivated with saline waters. Their principal product is wood, a renewable resource that substitutes for fossil fuels. Management of mangroves requires little fossil fuel expenditure and mangrove forests produce prawns, finfish and shellfish as byproducts. It is proposed that attention to aspects of mangrove management, physiology and ecology, breeding and selection, and biological spin-off products should make it possible to increase mangrove area and productivity for the benefit of people living along many tropical and subtropical shores of the world.

MANGROVES

Geography and General

The major part of the world's coastlines between 20° north and south latitude is dominated by mangroves (McGill, 1959). Actually, one or another species of mangrove is found as far north as 35° on Kyushu Island in Japan (Kandelia candel) Van Steenis, 1962) and 37° south at Auckland, New Zealand (Avicennia marina) (Chapman and Ronaldson, 1958). Mangroves are characteristic of low energy, sloping shorelines, or estuaries where there are muddy banks at low tide, although they are not limited to such habitat. Near the higher north and south latitudes of their occurrence, mangroves typically are smaller and less vigorous (Chapman, 1970; Teas, 1977) and they are often mingled with saltmarsh grasses such as Spartina. As Harshberger (1914) noted, mangroves are found in areas on tropical coasts that would be occupied by saltmarshes on temperate shores. Mangroves appear to be effective competitors of such grasses in the more favorable (tropical) portions of their ranges.

The most extensive mangrove forests are found in the Indo-Pacific area where they are associated with the deltas of the rivers entering the Bay of Bengal, the Straits of Malacca, Southern Borneo, New Guinea, Thailand, and Vietnam (Macnae, 1968).

Mangroves are botanically diverse, including at least 12 plant families and more than 50 species (Chapman, 1970). It is apparent from their nearest relatives that mangroves are land plants that have invaded the saline environment rather than marine plants that have moved into the shore and estuarine area. It should be noted that a number of species of herbaceous plants and lianes that are found along with mangrove are not considered to be mangroves because of the definition. Examples of the latter are the fern, Acrostichum aureum (pantropical); a milkweed, Finlaysonia (Vietman), the nypa palm (Nypa fruticans) (Southeast Asia); and a vine, Rhabdadenia biflora. The mangroves of significance for silvicultures, i.e., the reasonably common ones that reach tree size, are listed in Table 1. The most important mangroves silviculturally are species of Rhizophora and second in importance are some species of Bruguiera. Locally, others than those listed in Table 1 may be utilized. Some inland arboreal halphytes are known, such as the Tamarix, but, although salt tolerant (Ungar, 1974), they apparently have not made the transition to the coastlines to become mangroves.

A group of plants termed "back mangroves" occupy a position adjacent to, but upland of mangroves. They have at least limited salt tolerance. Examples of back mangroves are tropical almond, Terminalia catappa; mahoe, Thespesia populnea; and pandanus, Pandanus pectorius.

Salinity Preference and Tolerance

Mangrove species generally do not appear to require salt. They compete successfully in the saline tidal areas but in non-saline waters or where they become seeded upland they are not successful competitors, probably because they are shaded out by faster growing species (Teas, 1977).

Tidal circulation is not necessary for mangroves. Stoddardt et al. (1973) reported isolated stands of Rhizophora mangle growing in upland saline ponds on the island of Barbuda in the Lesser Antilles. In western Australia Beard (1967) found Avicennia marina growing along a salt creek 40 km inland from the sea. The author has grown three Florida mangroves, Avicennia germinans, Rhizophora mangle, and Laguncularia racemosa, in pot culture to flowering and fruiting with-

TABLE 1. Principal Mangrove Genera of Significance for Silviculture

Family	Genera	Number of Species*
Rhizohporaceae	Rhizophora	7
	Bruguiera	6
	Ceriops	2
	Kandelia	1
Lythraceae	Sonneratia	5
Avicenniaceae	Avicennia	11
Meliaceae	Xylocarpus	10
Combretaceae	Laguncularia	1
	Lumnitzera	2
Pellicieraceae	Pelliciera	1
Sterculiaceae	Heritiera	2
Euphorbiaceae	Excoecaria	1
Bombacaceae	Camptostemon	2

*From Chapman (1970) and other sources. Scientific names used in
 this paper are from Ding Hou (1958) and more recent publications,
 e.g., Chapman (1970).

out tidal change, and at several points in Florida mangroves grow on
dry land (Teas, 1977).

Mangrove species differ in their salinity preferences (Macnae,
1968). For example, Sonneratia caseolaris grows only if the salinity
is low, less than 10 o/oo (parts per thousand); Bruguiera parviflora
reaches its maximum development at about 20 o/oo; and Rhizophora
mucronata, Sonneratia alba, S. griffithii and S. apetala prefer
waters near normal seawater, i.e., ca 35 o/oo. Aegiceras corniculatum
is an indicator of low salinity water.

Some species are notably salt tolerant. Macnae (1968) reported
that Avicennia marina and Lumnitzera racemosa grew at salinities
greater than 90 o/oo. He also reported that Ceriops tagal grew where
salinities exceeded 60 o/oo. In the Florida Keys and in Puerto Rico,
living, but dwarfed and gnarled, Rhizophora, Avicennia and Laguncularia are sometimes found where interstitial salinities are 60-80 o/oo
(Teas, unpublished).

Silviculture

Silviculture of mangroves was practiced in the Andaman Islands
before the beginning of this century (Banerji, 1958) and in Malaya
since 1904 (Watson, 1928). Noakes (1955) estimated that 82% of the ca
145,000 ha (hectares) of mangrove forest in Malaya was under sustained
yield management in Forest Reserves. Mangrove forests also have been
managed in Thailand, the Philippines and parts of Vietnam. As reported by Watson (1928), managed mangrove plats are harvested under
lease or permit from the government. Royalty payments are made by
the licensee and there is a body of regulations governing managing
and harvesting.

A variety of mangrove species have been managed in Malaya
(Watson, 1928). The principal uses of mangroves are for firewood
and charcoal, for which Rhizophora species are preferred. Rhizophora
is so much more valuable than some other species that in some areas
Bruguiera gymorrhiza, B. sexangula, B. parviflora, and Xylocarpus
granatum are removed by hand from mixed Rhizophora forests (Macnae,
1968). Dixon (1959) tested poisoning and ringbarking of Bruguiera
parviflora to improve Rhizophora stands. In Thailand Banerji (1958)
recommended that seedling trees of Ceriops decandra and Bruguiera
cylindrica be removed from Rhizophora forests.

Under ordinary cutting practice, the genera of Rhizophoraceae,
i.e., Rhizophora, Bruguiera, Ceriops, and Kandelia, do not regrow
by coppice shoots (Watson, 1928), so that forest regeneration is a
matter of seed production and dispersal.

The Rhizophoraceae have large seedlings that are ideal for

establishing themselves beneath the parent tree, and, in the case of
Rhizophora mangle, can be carried by ocean currents at least 100 Km
(Davis, 1940). However, the propagules are poorly suited for water
dispersal within the forest, and may be completely prevented from
reaching a felled area by accumulations of slash which, according to
Macnae (1968), requires 3 years to decay. Watson (1928) believed
that even wet areas (subject to frequent tidal inundation) could not
be relied upon to regenerate naturally.

Assistance in regeneration has been provided by regeneration
thinning about 2 years before the area is to be cut over. This procedure, or modifications of it, thin the canopy enough to allow natural development of seedlings to that, after clear cutting, a sufficient crop has already started for stand regeneration. A method that
is often used to enhance regeneration is the clear-cutting of narrow
plots (sub-coupes) so that nearby trees can provide seeds.

An alternative to the shelterwood and other systems for encouraging natural regeneration is hand planting. Watson (1928) stated that
clear-felling followed by hand planting of seedlings is useful, but
usually is ruled out by the expense. Vu Van Cuong (1964) reported
that Rhizophora species and Bruguiera gymnorrhiza were hand planted.
Lang (1974) cited references to earlier reports that in 1934 there
was a major mangrove forest program in South Vietnam that brought
about planting of 38,000 ha with mangrove, mainly R. apiculata.

A novel means of planting Rhizophora propagules was developed by
Teas and Jurgens (1978). Individual propagules attached to paper
bags of sand, with "tails" to orient them in flight, were planted from
a helicopter.

Even with a seed supply and hand planting there are sometimes
problems in obtaining mangrove forest regeneration after cutting.
Animal hazards have been listed as crabs (Sesarma sp.) and mischievous
monkeys (Watson, 1928). Plant pests that sometimes develop in clear-
out areas are dense stands of the fern Acrostichum aureum and the
shrubby mangrove Acanthus, both of which grow up through the slash
and pose a physical barrier to water transport of seed as well as
shading mangrove tree seedlings. Drew (1974) studied the defoliated
zone of the Saigon River delta and concluded that Acrostichum aureum
invasion might be a serious factor where conditions favorable for
spore germination and gametophyte development were present. Locally,
in areas where Acanthus and Acrostichum have become established, hand
clearing and planting may be necessary, and the longer propagules of
Rhizophora mucronata have been used to escape competition (Macnae,
1968).

Recommendations for harvesting cycles are variable. Watson
(1928) suggested cutting at 40-45 years, based on optimal wood pro-

duction, but Noakes (1955) recommended shorter cycles of 22 to 25 or 30 years for Malaya, and Holdridge (1940) suggested a rotation cycle of 25 years for exploitation of mangroves in Puerto Rico. Shorter cycles have been followed in the Philippines where Rhizophora, principally R. mucronata, were hand planted. Blanks were filled in after the first year and harvest of the crop for small posts or stakes was carried out at approximately 12 years (Watson, 1928).

Harvesting of Rhizophora on rotation basis in Malaya at 45 years yielded 6,720 cu ft/a (cubic feet per acre) (134 tons); and at 22 years a wood volume of 3,106 cu ft/a (62 tons) (Watson, 1928). A Laguncularia stand in Puerto Rico, 22 years old, produced 2,680 cu ft/a (57 tons) (Wadsworth, 1959).

Uses of Mangroves

Mangroves are managed because they are an important source of wood and a variety of other products in the tropical regions where they occur. The following listing covers uses of species in addition to those of silvicultural importance. Reported uses of mangroves variously include: timer, railroad ties, boat ribs and planks, furniture, scaffolding, dock pilings, tool handles, poles for fish traps, fence posts, posts for house construction, matchsticks, packing boxes and pencils. The wood of several mangrove species (e.g., Rhizophora) is highly valued as firewood because it burns evenly with little smoke, and as charcoal because of its high caloric value. Mangroves that produce smoke when their wood burns (e.g., Avicennia) are used for smoking sheet rubber, smoking fish, and for burning brick. The wood of one mangrove (Excoecaria) has been used for incense. In recent years, mangroves wood has been used for chipboard manufacture, newspaper, and as a source of cellulose for rayon.

Mangrove bark and other parts have variously been reported to be used for fish poison (Aegiceras); tanbark; dye for cloth; food (condiments from the bark and sweetmeats from the propagules of Bruguiera, a vegetable from propagules of Bruguiera or fruits of Sonneratia); oil for burning or cooking (Xylocarpus seeds); preparation of a fermented drink (extract of Rhizophora fruit); making a "wine" (Rhizophora propagules fermented with raisins); a treatment for preserving fishline and nets; as a source of pectin and plywood glue; as a tea substitute (Rhizophora leaves); and as a cigar substitute (dried Rhizophora propagules).

The leaves of mangroves were found to be suitable as cattle food in Florida and Avicennia leaves are regularly fed to water buffalo and cattle in India and to camels in East Africa. Parts of mangroves have been used for a variety of medicinal preparations such as poultices, treatment for diarrhea, diabetes, and an array of other conditions.

Information on uses was obtained principally from Watson (1928),
Morton (1965), and Bhosale (1978).

Biological Value

Mangroves are important biologically as producers of food materials and as habitat for a variety of organisms. Mangroves are a
source of reduced carbon for estuaries and shoreline communities.
They produce leaves, twigs, fruit, flowers, trunks, and roots that
contribute to the detrital food chain of the swamp. The details of
such a food chain were worked out by Odum and Heald (1972) for a
Florida Rhizophora forest. Utawale et al. (1977) have reported evidence for a detrital food chain in an Indian mangrove estuary.

Mangrove swamps support a diverse fauna that includes crabs,
prawns, mollusks, and other invertebrates, as well as fishes, birds,
amphibians, reptiles, and mammals. Many commercially important fish
use mangrove waterways as nursery grounds. In Florida the mangrove
habitat has been identified as a nursery and feeding ground for such
commercially valuable marine species as the pink shrimp (Penaeus
duorarum), mullet (Mugil cephalus), gray snapper (Lutjianus griseus),
red drum (Sciaenops ocellata), sea trout (Cynoscion nubulosus) and
blue crab (Callinectes sapidus) (Odum and Heald, 1972).

Macnae (1974) credited Hall in 1962 with having been the first
to suggest that there was a correlation of prawn catches with mangrove
forests. Macnae summarized mangrove area and prawn fishing yield for
portions of the Indo-Pacific in which he had catch records, i.e.,
Mozambique, Madagascar, West Thailand, Malaya, and the combined Irian,
Papua and north Australia. He obtained an average of ca 4 tons of
prawns/yr/km^2 of mangroves, which is ca 4 gm/m^2/yr. He noted that
the fin fish data were at least qualitatively parallel to those for
prawns. However, as he reported, the picture is not simple because
some prawn species are not mangrove-dependent. The mangroves of most
of western India have been removed and yet the offshore prawn fishery
is still productive.

The destruction of large areas of mangrove forests in South Vietnam provided an interesting potential test of the role of mangroves in
fishery productivity (de Sylva and Michel, 1974). They compared the
herbicide-defoliated part of the Saigon River delta with a nearby mangrove estuary that had not been defoliated. In the herbicide sprayed
area, the mangrove forest had been almost completely killed and had
decayed or been removed (Lang, 1974). de Sylva and Michel classified
and counted the phytoplankton, zooplankton, and the larval and adult
fishes. They found that mangrove removal resulted in higher siltation
and turbidity in the rivers and creeks and concluded that the defoliation may have resulted in a reduction in spawning habitats for fishes
and reduction in the variety of phytoplankton, zooplankton and larval

fishes, and, to a limited extent, adult fishes. However, because of uncontrolled variables, they could not definitively tie mangroves to the fishery productivity.

Shoreline Protection

Mangroves have been widely recognized as playing an important role in protection of shorelines from erosion and storms. Mangroves were planted among the ballast stones along the Oversea Extension of the Florida East Coast Railway early in this century to protect the road from storm erosion (Bowman, 1917). Rhizophora mangle was introduced into Hawaii in 1902, where it was planted on the lee shore of Molokai to control erosion (MacCaughey, 1917). Today the mangrove forests in this area are extensive. Mangroves have been planted in Sri Lanka (Ceylon) to stabilize banks of fish ponds and canals (Macnae, 1968). Fosberg (1971) suggested that past removal of thousands of hectares of mangroves in Bangladesh may have been partly responsible for the heavy loss of life in the 1970 storm and tidal wave.

Mangrove species differ in their ability to become established at high energy sites. In Biscayne Bay in Florida, Avicennia and Laguncularia, but not Rhizophora, were found on rocky open bay sites (Teas, 1977). In western Costa Rica Laguncularia, but not Rhizophora, or Avicennia, was found along high energy shorelines, and in southern New Caledonia only Avicennia was found growing on the rocky high energy shore (Teas, unpublished). In each of these cases other mangrove species were found nearby at lower energy sites.

Hand planting of mangroves to control erosion is not always practical on high energy shorelines. Ding Hou (1958) mentioned an only partially successful attempt to control erosion in the Deli River by planting mangroves. Teas et al. (1975) found that the most common cause of failure of establishment of experimentally planted Rhizophora propagules and seedlings in Florida was wave or current energy.

PHYSIOLOGY OF MANGROVES

Productivity

The net primary photosynthetic productivity (NPP) of mangroves is represented by the photosynthetically fixed carbon that is accumulated in leaves, branches, trunks, etc. Mangrove NPP has been determined by enclosing portions of plants in plastic bags for measurement of carbon fixation.

NPP measurements by gas exchange have been carried out in Puerto Rico (Golley et al., 1962) and in South Florida (Miller, 1972; Hicks

and Burns, 1965; Lugo and Snedake, 1975). Results of these measurements are shown in Table 2. The Florida mixed mangrove forest was
more efficient in accumulating dry weight than the Puerto Rico Rhizophora mangrove forest or the Avicennia forest. The low productivity
of the Avicennia is to be expected. Watson (1928) reported that even
in pure stands Avicennia naturally adopt wide spacing. The scrub or
dwarf Rhizophora forest listed in Table 2 is made up of sparsely growing, stressed, small trees that showed the anticipated low productivity.

    NPP of mangrove forests can also be estimated indirectly from
litter collection or leaf biomass data if litter component relationships (as shown in Table 2) are known and assumptions such as those
used in Table 3 are made about the ratio of litter or a litter component to NPP. For example, it is possible to make rapid estimates of
NPP for mangrove forests by measuring leaf area index (LAI) or by
measuring light transmission of the forest canopy, if calibration data
on the relationship of LAI to litter, or light transmission to litter,
are available for the forest type. LAI, which is the m^2 of leaves
over a m^2 of substrate, can be estimated by making counts of the
numbers of leaves touching a line passed through the forest canopy at
a series of random points. The LAI is then used to calculate the leaf
biomass, and from that the total litter and the NPP are estimated.
Alternatively, measurements of the light transmission of the canopy
(compared to simultaneous open sky measurements made nearby) can be
made and their averages used to estimate leaf biomass. A curve relating % light transmission to amount of litter is shown in Figure 1.
Calculations based on leaf biomass involve the assumption that the
leaves average 1 year on the tree. This has been reported to be approximately true for Rhizophora mangle by Gill and Tomlinson (1971).
Pool et al. (1975) found that the annual leaf turnover rate for 7
Florida sites averaged 0.97.

    The amount and distribution of litter and calculated NPP for 3
types of Florida mangrove forests are shown in Table 3. The data
reported are based on litter amount and distribution among the components from Teas (1974, 1976) and Pool et al. (1975). The NPP calculated for these mangrove forests varied from 3.8 mt/ha/yr (metric
tons per hectare per year) for Rhizophora scrub to 32.1 mt/ha/yr for
the Coastal Band community. The value for Coastal Band mangroves, a
highly productive community, is about 25% more than the average for
tropical forests, which are generally acknowledged to be more productive than temperate forests. Noakes (1955) reported wood production
for the best Malayan mixed Rhizophora forests that corresponded to a
NPP of ca 59.6 mt/ha/yr. Golley and Lieth (1972), summarizing literature values, listed tropical forests as averaging 12 and 28 mt/ha/yr
for deciduous and coniferous forests, respectively.

    Annual wood increments in mangrove forests are shown in Table 4.

TABLE 2. Mangrove Productivity Estimates from Gas Exchange Measurements (mt/ha/yr)*

Species or Type	Location	GPP*	R*	NPP* Litter	NPP* Wood	Total	References
Mixed Rhizophora, Laguncularia and Avicennia	Florida	37.6	41.6	9.5[a]	36.5[b]	46.0	Hicks and Burns (1975)
Rhizophora	Puerto Rico	59.9	66.4	9.5[a]	3.1	–	Golley et al. (1962)
Rhizophora, scrub	Florida	10.2	14.6	1.0[a]	$\pm$ 0	–	Burns (cited by Lugo and Snedaker, 1974)
Avicennia	Florida	65.7	45.3	9.5[c]	11.0	20.5	Lugo et al. (cited by Lugo and Snedaker, 1974)

*mt/ha/yr = metric tons per hectare per year; GPP= gross photosynthetic productivity; R= respiration; NPP= net photosynthetic productivity

[a] Average litter for mangrove forests, excluding scrub (Pool et al., 1975)

[b] By difference

[c] Data from Teas (1974)

TABLE 3. Litter Measurements and Productivity Estimates for Florida Mangroves (Teas, 1974, 1966; Pool et al., 1975)

| Principal Species and Type | Average Litter g/m²/da | | | | Litter prod. mt/ha/yr | NPP* mt/ha/yr |
	Leaves	Fruit, Flowers and Propagules	Twigs, Bark, misc.	Total		
Rhizophora mangle (mature)	2.20	0.56	0.17	2.93	10.7	32.1
Rhizophora mangle (scrub)**	0.24	0.08	0.03	0.35	1.28	3.8
Avicennia germinans (basin forest, open)	0.69	0.02	0.07	0.78	2.85	8.6

*Based on NPP = litter x 3 (Bray and Gorham, 1964; Golley, 1972)

**Scrub (dwarf, stressed) forest was ca. 1-1.5 m tall. Data estimated from plant harvest

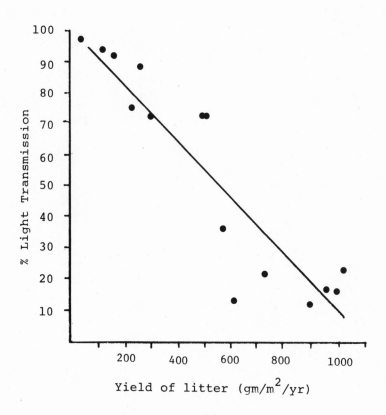

FIGURE 1

Litter/light transmission relationships for predominantly
Rhizophora mangle forests in Florida (Teas, 1974).

TABLE 4.   Annual Wood Increment in Mangrove Forests

Species	Location	Forest Type	Wood mt/ha/yr	Reference
Rhizophora apiculata	Phuket, Thailand	regenerated trees 15 yrs. after cutting	20.0	Christensen (1978)
Rhizophora mangle	Florida	Coastal Band, mature	21.3	Teas (calculated from litter data)
Rhizophora Mixed	Peruk, Malaya	25 yrs. old, "best in Malaya"	39.7	Noakes (1955)

The high value of Rhizophora in Malaya indicates the potential productivity of managed mangrove forests.

Physiology of Roots

Mangroves are frequently found growing in water-logged organic
soils that contain free hydrogen sulfide (Scholander et al., 1955).
Thus, the roots of mangroves, unlike most plants, cannot obtain oxygen
from the soil in which they grow. As an adaptation to anaerobic
soils, mangroves have developed methods of root aeration that involve
"snorkels" or "ventilators" in above-ground structures, which communi-
cate with the subterranean roots through air tubes. The above-ground
openings of these air tubes are the lenticels that are located on the
prop roots, pneumatophores or trunks. In Rhizophora, the air-con-
ducting tissue is found in the prop or stilt roots. In other mangroves (Avicennia, Bruguiera, Heriteria, and Xylocarpus) there are
specialized structures, pneumatophores, that project into the air,
and in others (e.g., Ceriops and Kandelia) that lack pneumatophore-
like structures, lenticels are located on the lower part of the trunk.

Scholander et al. (1955) demonstrated that the lenticels of
Avicennia pneumatophores and Rhizophora prop roots are connected to
the underground roots. Furthermore they found that when the above-
ground breathing organs (pneumatophores or prop roots) were sealed
off, the oxygen concentrations in roots fell and carbon dioxide concentrations increased.

These adaptations for root aeration leave mangroves vulnerable
to flooding that covers lenticels of the roots or trunks, or to other
factors such as thick petroleum residues or siltation that can cover
or block lenticels. Craighead (1971) reported extensive kills of mangroves in Florida following hurricane Donna in 1960, which deposited
a layer of fine silt over the soil shutting off oxygen to mangrove
roots. Diking of mangrove forests followed by pumping and maintaining water levels above pneumatophores and well up on prop roots has
been utilized by real estate developers in Florida to kill mangroves.
The same technique applied in mosquito control impoundments has accomplished a similar result in 2 to 3 weeks, although the plants did not
die for several weeks or months after the flooding (Provost, 1973).
Vu Van Cuong (1964) reported the death of Rhizophora and Sonneratia
in fish ponds in Vietnam when the water level was raised above their
prop roots or pneumatophores. Nearby trees of the same species,
where roots were subjected to normal tidal emergence, prospered.

Salt Requirement

Most mangroves grow in brackish or saline waters, but are probably not obligate halophytes, indeed, several species have been grown
for long periods in freshwater. The late Dr. John Davis told the

author that in 1933 he had taken Rhizophora mangle propagules to the
National Botanic Garden in Washington, D.C. where they were planted
in freshwater. In 1973 those plants, or their progeny, were found
to be alive and growing well in Washington, D.C. tapwater. Records
indicate that several species of mangroves have grown and reproduced
for more than a century along a freshwater lake at Borgor, Indonesia
and in a botanical garden at Hamburg, Germany (Ding Hou, 1958; Macnae,
1968). Vu van Cuong (1964) listed the following species are able to
grow in freshwater over a period of many years: Rhizophora mangle,
R. apiculata, R. mucronata, Sonneratia caseolaris, Aegiceras majus,
Bruguiera sexangula, Lumnitzera sexangula, and L. racemosa.

        Not all mangroves are able to grow in the complete absence of
salt. Vu Van Cuong (1964) stated that Ceriops tagal and Avicennia
officinalis had not been successfully grown in freshwater. These
species, and possibly others that have not yet been tested, may prove
to be obligate halophytes as defined by Barbour (1970).

        Several laboratory experiments have shown that mangroves grow
better in dilute than in full strength seawater. Connor (1969) found
that Avicennia marina grew taller in 1.5% than in 3% NaCl. Pannier
(1959) reported that Rhizophora mangle seedling shoot growth in 25%
seawater was 5 times as great as in full strength seawater. Both of
these experiments were designed to avoid major element deficiencies
that might confound interpretations.

        Field observations as well as some laboratory experiments indi-
cate that mangroves generally grow better with some salt than they
do in freshwater. Bowman (1917) noted that Rhizophora mangle trees
became smaller and the water less saline as he traveled up the Miami
River in a boat. Recent observations on R. mangle trees that had
been carried as propagules into a sawgrass (Cladium jamaicensis) area
in a south Florida by an unusually high tide (F.C. Craighead, personal
communication) showed that they grew as scrubby plants in this freshwater environment and fruited only after more than 10 years. By
comparison, seedlings of Rhizophora mangle planted along the Florida
coast, where they were exposed to bay salinities, fruited in 4 years
(Teas, unpublished). Mineral deficiency may play a role in the slow
rate of maturing and the small size of Rhizophora mangle plants growing in freshwater.

        Both Davis (1940) and Pannier (1959) reported that mangroves grew
most luxuriantly in the middle reaches of an estuary, where the salinities were appreciable, but considerably less than those in seawater.

        An interesting but not understood feature of Rhizophora mangle
and R. mucronata salt relationships is illustrated in Table 5. This
shows the chloride content of leaves on plants growing in a range of
salinities. Even where the salt concentration was very low

TABLE 5.   Chloride Content of Water and Leaves of _Rhizophora_ Species in Different Habitats

Species	Plant Material Source	Chloride (ppm) Leaves (dry wt)	Chloride (ppm) Water	Reference
Rhizophora mangle	Florida (fresh-water, sawgrass marsh)	15,980	15	Teas, this report
"	Florida (N. Fork of St. Lucie River)	18,360	193	"
"	Florida (Biscayne Bay)	18,730	18,700	"
"	Florida	17,400*	-	Walsh (1974)
Rhizophora mucronata	India (Ratnagiri estuary)	50,900	14,870	Joshi, et a. (1975)
"	India (Chiplun, on the Vashisti River)	29,000	40	"
"	India	35,660*	-	Sidhu, cited by Walsh (1974)

*Calculated from sodium values, assuming equimolar quantities of sodium and chloride

in the ambient water these Rhizophora species accumulated large
amounts of chloride. In R. mangle, a difference of ca 1250 times in
surface water chloride concentration resulted in leaf concentration
differences of only 1.2 times, and in R. mucronata, a species with
higher leaf chloride concentration, a difference of 372 times in surface water chloride was reflected in leaf differences of only 1.8
times. Joshi et al. (1975) also listed K and Ca values for the freshwater in the Vashisti River at Chiplun and for the saltwater at
Ratnagiri. Potassium and Ca concentrations of the water at the two
sites differed 200 and 285 times, but the respective K and Ca contents
of the R. mucronata leaves differed only 1.1 and 1.5 times. The
accumulation of chloride is thus parallel to thatof K and Ca, which
suggests that chloride must play a role in the physiology of the
plant.

                    Physiology of Salt Tolerance

     Mangroves grow in saline waters but have sodium chloride con-
centrations in their xylem that are only small fractions of those in
surrounding saline media (Scholander et al., 1962, 1966). Scholander
(1962) reported experiments on mangroves which indicated that there
are two major methods of mangrove ion regulation: by salt exclusion
and by salt excretion. At least two additional means are used by some
mangroves: succulence and the discarding of salt-laden organs or
parts.

     Salt-excluding species of mangroves include: Rhizophora mangle,
R. mucronata, Ceriops tagal, Bruguiera gymnorrhiza, B. parviflora,
and Kandelia candel. Salt secreting species are Avicennia marina,
A. officinalis, A. Germinans, A. Alba, Aegialitis annulatum, Aegiceras
majus, A. corniculatum, Acanthus ilicifolius, and Laguncularia racemosa (Scholander, 1968; Joshi et al., 1975). Species that shed salt-
laden leaves include: Ceriops tagal, Sonneratia acida, Avicennia alba,
Rhizophora mucronata, Excoecaria agallocha, and Lumnitzera racemosa
(Jamale and Joshi, 1976).

     Salt-excluding species - In salt-excluding species, the plants
separate freshwater from saline water by a non-metabolic ultrafiltration powered by the high negative pressure in the xylem (Scholander,
1968). Salt-excluding plants thus obtain freshwater by a "reverse
osmosis" process that is operated by solar energy (transpiration).
Scholander's experiments demonstrated that in the root the separation
process was insensitive to cold, dinitrophenol, or carbon monoxide.
Parallel experiments with mangrove leaves indicated that their ultrafiltration membranes (variously in Avicennia, Osbornia, Lumnitzera,
and Sonneratia) were insensitive to heat (to 50°C), to cold (ice-
chilling), to carbon monoxide or to dinitrophenol, but were destroyed
by ether or chloroform.

The author carried out experiments in which the roots of small seedlings were dipped into 0.1 seawater containing radiochlorine (chlorine-36). In the experiment reported here the plants were exposed to chlorine-36 for 24 hours (in the case of Avicennia) or 72 hours (Rhizophora and Laguncularia). The plants were harvested, the roots washed and radioactivity of the tissue extracts and leaf rinsate (excreted salt) measured. The range of radioisotope taken up per seedling was ca 50,000 to 85,000 cpm. Individual Laguncularia and Avicennia seedlings in the laboratory differ greatly in the amount of salt they secrete on their leaves: The plants used for the experiment had been selected as active salt-secreters. The uptake of radioactivity is shown in Table 6. This experiment demonstrated that Rhizophora is a more effective salt excluder than are Avicennia and Laguncularia, both of which passed large amounts of chloride and secreted it on their leaves. The amount of chloride secreted on the Avicennia leaves suggested, as did Atkinson et al. (1967) for Aegialitis annulata, that salt in the xylem can pass to the salt glands without equilibrating with the main chloride pool of the leaf. The counts in the root fractions are ones that persisted through washing. They are high in all three species, so they represent uptake by root cells that are external to the separating membrane, which is probably located in the Casparian strip of the endodermis (Scholander et al., 1965). Mallery and Teas (1979) obtained other evidence for such a nonspecific uptake of large amounts of chloride by mangrove roots. They loaded seedling roots of the Florida mangrove species with chlorine-36 and measured and analyzed efflux rates by the method of Pitman (1963). Their results were parallel for the three species, indicating that there is a pool of counts taken up by mangrove roots in species that exclude salt as well as those that only partially exclude salt.

In other experiments, seedlings of Rhizophora mangle and Avicennia germinans, the roots of which had been exposed to solutions containing very small amounts of toluene or diesel fuel, showed 240 to 260 times as much chlorine-36 uptake into the stems and leaves as did controls. Thus, the salt-separating membranes of roots were destroyed by lipid solvents, as Scholander (1968) has found for mangrove leaf membranes using chloroform and ether.

Salt secretion--Several lines of evidence point to the salt secretion by the salt glands being very different physiologically from the salt exclusion by roots. Salt secretion by the salt glands was studied in detached leaves of Aegialitis annulata (Atkinson et al., 1967). When carbonyl cyanide 3-chlorophenyl hydrazone, an uncoupler of oxidative phosphorylation, was applied to cut petioles of Aegialitis leaves there was a marked inhibition of chloride secretion by the salt glands. In earlier experiments with salt-secreting non-mangrove halophyte, Limonium latifolium, Arisz et al.(1955) found that cyanide and other respiratory poisons stopped salt secretion. Cells of salt glands of a variety of halophytes have been studied microscopically

TABLE 6. Uptake of Chlorine-36 by the Three Florida Mangrove Species

Species	Plant Part	Radioactivity distribution %
Rhizophora mangle	root	97.8
	stem*	2.0
	leaves	0.2
	leaf rinsate	0.0
Avicennia germinans	root	55.7
	stem	7.7
	leaves	2.4
	leaf rinsate	34.2
Laguncularia racemosa	root	43.9
	stem	40.9
	leaves	6.0
	leaf rinsate	9.2

*Hypocotyl and stem

and found to have cytoplasm with large numbers of mitochondria and
dense endoplasmic reticulum. Secretion of salts by such salt glands
is markedly temperature-sensitive, indicating the involvement of
enzymatic rather than physical processes (Atkinson et al., 1967).
The salt exclusion of roots had the temperature characteristics of
a physical process (Scholander, 1968). Thus, the evidence clearly
indicates that secretion of salt by salt glands involves active transport with a requirement for biochemical energy input.

Succulence--Succulence is the state of having thickened leaves,
which comes about in halophytes that take up water to dilute salt
accumulated by the tissues. Succulence is a solution for the uptake
of excess salt by a plant, but the amount of salt that can be stored
in a given thickened leaf is limited. Joshi et al. (1975) listed
Sonneratia acida, S. apetala, Lumnitzera racemosa, and Excoecaria
agallocha as salt-accumulating, i.e., succulent species.

Discard of organs--Salt is stored in tissue that are to be discarded, such as leaves. Joshi et al. (1975) reported that leaves had
increasing chloride concentrations as they aged.

The author has measured chloride concentrations in Rhizophora
mangle propagules, fruits (just before the propagules were shed) and
leaves. The average chloride concentrations, in ppm on a dry weight
basis, were: 6,000, 32,250, and 17,690, respectively. It can be
argued that both the fruits and leaves are loaded with salt before
their discard by the plant as a means of reducing its salt load.

Mangroves cope with the saline environment by a number of techniques; indeed, individual species probably utilize a variety of
methods as suggested by Albert (1975). All mangroves probably have
root membranes that exclude salt to a degree (Scholander et al., 1966).
Rhizophora mangle excludes salt well, but not perfectly, and so takes
in small amounts of salt with its transpiration water, which it apparently disposes of by storing in the leaves and fruits. Laguncularia
racemosa is a semi-efficient salt excluder, and excretes salt through
glands on its leaves, but also, under hyper-saline conditions, Laguncularia develops thickened succulent leaves and may also lose salt in
the discard of its senescent leaves.

Effect of Salt on Transpiration

The transpiration stream of mangroves is responsible for the negative xylem pressure that provides the pressure differential for the
separation of freshwater from seawater by root membranes (Scholander,
1968). Walter and Steiner (1936) found that the transpiration of
mangrove leaves was about 1/3 that of ordinary plants, which has been
confirmed by others. Scholander et al. (1965) found that at night,
when transpiration stopped, the negative xylem pressure of Avicennia

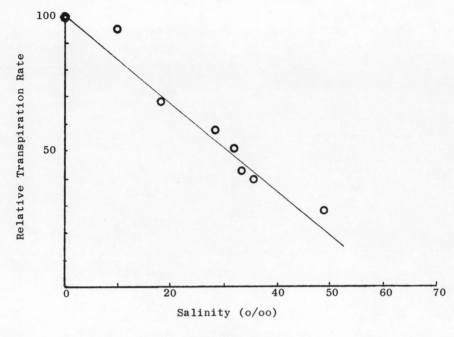

FIGURE 2

Effect of salinity on transpiration in <u>Rhizophora</u> <u>mangle</u> seedlings (calculated from data of Bowman, 1917).

<u>germinans</u> and <u>Laguncularia</u> <u>racemosa</u> dropped from values of 35-50 at-
mospheres to close to that of the osmotic potential of seawater, 25
atmospheres.

Bowman (1917) measured transpiration of <u>Rhizophora</u> <u>mangle</u> seedlings cultured in dilutions of seawater. His data, plotted in Figure
2, show that transpiration is reduced as salinity increases. From
the response curve, zero transpiration would be expected at ca 60-75
o/oo salinity. It is interesting that Cintron et al. (1978) found the
numbers of dead <u>Rhizophora</u> <u>mangle</u> trees were greater than live ones
in forests where the interstitial salinities exceeded 65 o/oo. High
salinities must impose severe restraints on the transpirations capability of mangroves, and thus on their physiological functioning and
survival.

Effects of Salt on Respiration and Photosynthesis

Hicks and Bruns (1975) measured respiration and photosynthesis
on mangroves in a southwest Florida estuary. They used clear plastic
sleeves to enclose leafy branches, trunk segments, etc., and maintained sufficient air flow through the sleeves to assure that the

TABLE 7. GPP of Mangrove Species in a Florida Estuary (Hicks and Burns, 1975)

Species	Estuary Position	Average surface water salinity (o/oo)	Species GPP (g C/m^2/day)
Rhizophora mangle	upper	7.8	8.0
	middle	21.1	3.9
	lower	26.6	1.6
Avicennia germinans	upper	7.8	2.3
	middle	21.1	5.7
	lower	26.6	7.5
Laguncularia racemosa	upper	7.8	–
	middle	21.1	2.2
	lower	26.6	4.8

difference in temperature of air entering and leaving the chamber was less than 4°C. Measurements of carbon dioxide concentration was made hourly on the intake and exhaust ports of the plastic sleeve over a 24 hour period.

They utilized three sites within an estuary system: an upper estuary station where the surface water salinity averaged 7.8 o/oo; a middle estuary station where surface water salinity was 21.1 o/oo; and a lower estuary station where surface water salinity averaged 26.6 o/oo. Their results for individual species are shown in Table 7. At the species level, Rhizophora GPP decreased as salinity increased, but gross photosynthetic productivity (GPP) of Avicennia and

TABLE 8. Community Metabolism in a Florida Estuary (Hicks and Burns, 1975)

Position in Estuary	Surface Water Salinity (o/oo)	Community Metabolism (g C/m²/day)		
		GPP	NPP	R
Upper	7.8	10.3	6.6	3.7
Middle	21.1	11.8	7.5	4.3
Lower	26.6	13.9	4.8	9.1

Laguncularia increased with increasing salinity. The calculated mangrove community metabolism at each of the three sites is shown in Table 8. They reported that although the GPP increased with the increase of salinity for the mangrove communities at the three sites, the NPP dropped because of increased respiration (R). The highest mixed community NPP was the middle portion of the estuary, but the greatest NPP for Rhizophora occurred at the lowest salinity. They suggested that channelizing the water through an estuary might result in a decline of community NPP because of increased salinity through a greater part of the system.

ECOLOGY OF MANGROVES

Establishment

Mangrove seedlings may differ in their physiological requirements from mature trees of the same species. Important factors in

mangrove seedling establishment and early development include:
salinity, light levels, temperature tolerance, anaerobic substrate,
and wave energy.

McMillan (1971) reported that young Avicennia germinans seedlings
were killed by temperatures of 39-40°C for 48 hours, whereas, larger,
established Avicennia seedlings or trees were not damaged by 39-40°C.
The seedlings floated in fresh or saltwater, but did not become established where water depths were greater than 5 cm. He concluded that
these characteristics were appropriate for a species where the seeds
are dispersed by water in the fall and begin to develop when they
reach moist soil or shallow water during the winter or spring.

Propagules of Rhizophoraceae can be dispersed great distances
by water. Davis (1940) reported that Rhizophora mangle propagules
could remain alive for at least 1 year while submerged in seawater
in the laboratory. When this experiment was repeated by the author,
it was found that within a few weeks the water gave off hydrogen
sulfide, indicating that the water had become anaerobic, as would be
expected from respiratory studies of Avicennia marina, Rhizophora
mucronata and Bruguiera sexangula (Brown et al., 1969). No roots
developed during the year's immersion.

It is common in Florida to find large numbers of Rhizophora propagules rooted in shallow water, as though to establish a dense forest
in the future. However, over a period of years only rarely does one
of these seedlings develop into a tree. The author has studied aerial
photographs and measured elevations of substrate at a site in Biscayne
Bay where large numbers of seedlings become rooted. The mean tidal
range in the area is ca 60 cm. Over a period of 10 years the only
Rhizophora trees that developed were ones at substrate elevations
higher than 9 cm below mean sea level, although tens and sometimes
hundreds of propagules were imbedded at greater depths. It is notable
that the roots of mature Rhizophora trees, which have air-filled
tissues that are not present in young seedlings can be found imbedded
in substrate at considerably greater depths than those at which seedlings are successful. Gill and Tomlinson (1977) have found that as
much as 50% of the space in the prop roots of mature Rhizophora plants
is air. It seems likely that the inundation and/or anaerobic substrate
is responsible for the failure of the too-deeply-placed seedlings to
become established and develop into trees. Larger plants with prop
roots would have no such constraints because of the air conducting
tissue in their roots.

Light levels are critical for developing mangrove seedlings.
Avicennia seedlings die under the shade of their parents, indeed,
older Avicennia trees die when overtopped (shaded) by Rhizophora and
Bruguiera trees (Macnae, 1968). Noakes (1955) noted that Bruguiera

parviflora could form a complete under-story beneath a mature Rhizophora forest at light levels where seedlings of Rhizophora itself were
not successful. Rhizophora mangle seedlings at the low light inten-
sities under a mature canopy of parent trees die after forming a few
leaves. In south Florida, when the canopy of mature Rhizohpora is
opened by a lightning strike, seedlings quickly form a dense stand
(Teas, 1974).

     Rhizophora propagules are able to float for long periods of time,
but rarely are such floating propagules found to be rooted, unless as
Davis (1940) noted, they are entangled in a thick floating mass. The
author carried out experiments with recently fallen Rhizophora mangle
propagules in which one series was exposed to fluorescent light at
ca 200 fc for 12 hours per day, and the other was maintained in complete darkness. Both groups were maintained in sealed plastic bags
that were opened and aerated in the dark, daily. At the end of 46
days the propagules in the dark had the same number of roots as those
given light, but their roots were more than 10 times as long as those
of the light treated ones (averages 14 mm vs 1.3 mm/root). The number
of root primordia is probably established before the propagule falls
from the parent tree: the role of light is to inhibit root development. Light inhibition of the growth of dormant root primordia has
long been known (e.g., Shapiro, 1958). In R. mangle, light inhibition
of root development would keep propagules that are being dispersed by
water from having their delicate roots damaged by wave action as they
came into contact with objects. When the bases of floating propagules
become imbedded in mud or moist wrack their roots develop rapidly.
Light levels are apparently the key to root development in R. mangle
propagules. Banus and Kolehmaninen (1975) obtained rooting of floating propagules that were heavily covered with algae and partially
shaded.

     Too high interstitial salinity can inhibit seedling development.
In well-lighted areas of hypersaline soils of some islands in Puerto
Rico, Rhizophora seedlings die after developing only a few leaves.
However, the salinity in these locations is high enough to stress to
mature trees.

Soils

     Mangrove soils are characterized by high water content, low oxy-
gen and often by high salinity and free hydrogen sulfide. Such soils
are sometimes semi-fluid and poorly consolidated. Schuster (1952)
has reported that biological processes are necessary before fresh
alluvium becomes suitable for the growth of higher plants including
mangrove. These processes include the activities of blue-green and
green algae, diatoms, and bacteria. Nitrogen-fixing bacteria and
algae are found in and on mangrove soils (Kimball and Teas, 1975).
Schuster noted the role of invertebrates that work the soil and con-

tribute nitrogen and phosphate in their feces.

Freshly deposited alluvium at the seaward fringe of Indo-Pacific mangroves, in the Avicennia zone, may contain only 5-15% organic matter (measured as loss of weight on ignition), whereas the more mature mangrove soils in the upper levels of a forest may be 65% or more organic matter (Macnae, 1968). In South Florida where the calcareous sand along the shore can be 7% organic, the soil of mature Rhizophora forests averaged 28% organic, an Avicennia forest was 25% organic and an induced Laguncularia forest on the marl soil of former tomato fields was 13% organic matter (Teas, 1974).

Well developed soils are not required for all mangroves. At some sites in south Florida, mangroves of all three species can be found growing in karst limestone formations where there is no evident sediment accumulation. Elsewhere magnroves have been reported as growing in coral rubble (Macnae, 1968).

Interstitial Salinity

Baltzer and Lafond (1971) listed two extremes of coastal environment with respect to interstitial water salinity: 1) areas of high rainfall that are permanently wet, where the soil salinity decreases progressively from the sea towards the upland and the vegetation ranges from mangrove to wet tropical forest without discontinuity, and 2) areas of low rainfall where dry and wet seasons alternate, and where dry, hypersaline fringes appear between the mangrove swamp and the upland.

Soil water or interstitial salinity is regulated by a number of factors that include: tidal circulation; soil type and contours, depth to rock or permeability-limiting substrate and its contours; amount and seasonal distribution of rain; fresh water discharge of rivers and overland runoff; evapo-transpiration rates; local factors, such as salt pans; and formation of sandbars or shore erosion.

Interstitial salinity can be measured in the field by use of a Goldberg refractometer on samples of water that infiltrate into a hole bored with a soil auger or corer. Salinities may also be measured on samples titrated in the laboratory or by use of a conductivity meter. Giglioli and King (1966) developed a method of obtaining interstitial water samples where there is standing water. They used a length of 10-15 cm diameter pipe which was forced into the soil. The surface water within the pipe was aspirated with a stirrup pump, and a soil water sample obtained by boring within the pipe cofferdam.

Soils differ in their rates of salinity equilibrium with surface waters. In Florida, Teas et al. (1976) reported an interstitial

salinity of 20 o/oo in porous peat soil. At the same time, after
rain within the previous 24 hours, the surface salinity in a nearby
Rhizophora scrub forest on a fine marl soil was 16 o/oo, and the
interstition salinity 34 o/oo. Giglioli and King (1966) found that
in a Gambian mangrove forest the salinity of the surface water increased during the dry season but decreased sharply after the rains
began; however, the interstitial salinity of the clay soils at a
depth of 30 cm varied little during the year.

Fine marl, like clay soils, can retain salt. An example of surface and interstitial salinities along a gently sloping 4000 m transect from a Florida bay upland is shown in Table 9. The transect
line crossed marl lands that had become saline from storm tides some
years earlier (Reark, 1975; Teas et al., 1976). The soil at stations
2 and 3 was peat, the soil at stations 4-13 was marl. At the time
of the transect bay salinity was low from rainfall. With increasing
distance upland the surface water salinity in the mangrove zone ranged
from the bay salinity of 31 o/oo to 46 o/oo and then decreased,
whereas the interstitial salinity, which was always higher, varied
from 39 o/oo to 63 o/oo and decreased. The results confirm the finding of Scholl (1965) who determined the equilibration of soil water
salinity with that of surface water was very slow in such fine textured marl soil.

A characteristic of soils that may be involved in mangrove growth
in saline waters was studied by McMillan (1975b). He found the seedlings of Avicennia germinans and Laguncularia racemosa survived ex-
posures to 3X to 5X seawater if they were in river sand or sandy loam
that contained clay, but failed to survive if their roots were in
washed river sand, or beach sand. In each case, 7-10% clay was protective. Depression of the pH of the saline solution that occurred
when it wetted clay soil suggested that cation exchange may be involved in this interaction of soil texture and salinity. He reported
that natural soils where Avicennia germinans was growing contained
14% or more clay.

Effects of Salinity Changes

Increases and decreases in the soil salinity of mangrove forests
are known from both man-induced and natural causes.

Natural increases in salinity--High interstitial salinity is
usually found in areas of low rainfall. An example of naturally increased salinity was studied in Puerto Rico, where the rainfall along
the coast varies from ca 1600 mm near San Juan to less than 700 mm
along portions of the south coast. Evaporation exceeds rainfall by
a small amount at San Juan, but is more than 3 times rainfall on the
arid southwestern coast.

TABLE 9.   Salinities of Soil and Surface Water, Card South Florida
           (Teas et al., 1976)

Station No.	Distance from shore (m)	Plant Community Type	Water Salinity o/oo	
			Surface	Interstitial
1	0	Biscayne Bay	31	–
2	90	Scrub mangrove*	37	39
3	150	Scrub mangrove*	31	43
4	1070	Scrub mangrove*	31	66
5	1220	Scrub mangrove*	29	50
6	1370	Scrub mangrove*	46	63
7	1830	Scrub mangrove*	31	60
8	2130	Scrub mangrove*	27	41
9	2130	Juncus Sp.	8	15
10	2290	Juncus Sp.	5	7
11	2975	Cladium jamaicensis	–	9
12	3810	Cladium jamaicensis	–	8
13	4000	Cladium jamaicensis	–	4

*Principally Rhizophora mangle

Mangroves are found along the shore at low energy sites around the island of Puerto Rico. Along the drier shores there is typically a fringe of mangroves with a vegetation-free hypersaline lagoon or salt flat behind it. Such vegetation-free zones along arid shore- lines are known from many parts of the world (Fosberg, 1961; Guilcher, 1963) and are variously termed salines, salitrals, salinas, salterns, salt flats and salt barrens. The low, seasonal rainfall and high ev- aporation rates along such shores give rise to characteristic com- pressed vegetation transitions between the bay or sea and upland. In- terstitial salinities along the dry Puerto Rico coast increase rapidly with distance into the mangroves, and, as salinity increases, the tree sizes, species, and survivals change.

A transect through the mangrove fringe of a saline pond island on the Puerto Rico south coast is shown in Figure 3. Within 3 m of the shore the interstitial salinity was higher than that of bay water, and it increased along the transect to the vegetation-free saline pond in the center, where it was 107 o/oo. The shoreline trees were all Rhizophora, which decreased in height with distance from open water. At 9 m from the shore there was a band of Laguncularia, and a few Avicennia. Progressively toward the center of the island there were live Avicennia and dead Laguncularia and Rhizophora, then live Avi- cennia and dead Rhizophora, and, near the saline pond, there were only dead Avicennia and Rhizophora.

Soil auger borings of the substrate along a transect across the island showed a layer of Rhizophora peat and below that there was coarse sand and shell. From estimates of the rate of Rhizophora peat deposition in Florida (Egler, 1952), it was calculated that the is- land might be on the order of 800 years old.

The history of such an island can be reconstructed from present- day evidence, as follows: at some time, likely more than 800 years ago, a sand-shell bank formed on which Rhizophora became established. As the Rhizophora increased in number, flotsam, jetsam, silt, shell and sand were trapped by the outer prop roots of the island, devel- oping a rim or berm. Because of this berm there was reduced tidal circulation to the center of the island and seawater left by the high tides concentrated by evaporation.

The increased salinity in the center of the island stressed the mangroves so that they lost leaves and thus came to have reduced LAI's. The reduced LAI's resulted in less shading and thus in a higher heat load at the soil surface. The increased heat load in turn resulted in higher evaporation rates, which further increased the salinity and stressed the Rhizophora mangroves. The annual temp- erature range under a Rhizophora mangrove canopy in this part of Puerto Rico is ca 25.5-30.5°C (Mattox, 1949); however, the author found temperatures of 42-45°C in open saline lagoons in the islands

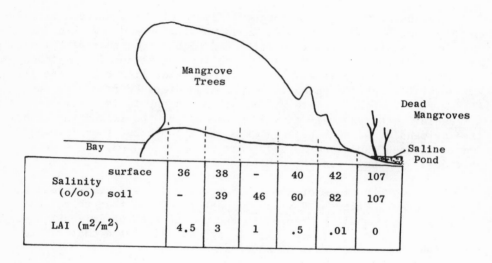

FIGURE 3

Section through half of a saline pond island.

and along the south coast of Puerto Rico. As the interstitial salinity at the island's center increased, the reduced LAI of the Rhizophora canopy permitted Laguncularia and Avicennia, which have greater
salt tolerance than Rhizophora, to become established. The Rhizophora
continued to expand at the periphery of the island, further building
up the berm. With opened canopies and water temperatures of 42-45°C,
the salinity increased further and killed the Rhizophora in the center, and, in time, also killed the Laguncularia and Avicennia. The
peat in the open center of the island was subjected to heat and oxidation so that a permanent pond was formed. Figure 4a is an aerial
view of a Puerto Rican island with a hypersaline pond. Some of the
changes in vegetation that occur in the transition from bay salinity
to the hypersalinity of the central pond can be seen in the photograph.

In Bahia Montalva and Bahia Fosforescente on the south coast of
Puerto Rico there are small islands that show a berm at the periphery,
and increased interstitial salinity and sparse-growing small mangroves
of all three species in the center. Figure 4b is an aerial photograph of one such island. The field evidence supports the hypothesis
that, as these islands increase in size, they will develop hypersaline
central ponds.

Mangrove islands with central hypersaline ponds are not limited
to the arid south coast in Puerto Rico. Two such islands can be
seen near San Juan, a short distance east of the Isla Verde airport,
where there is relatively high rainfall. Investigation of one of
these islands showed that it had a pond with dead mangroves at the
center, but that interstitial and pond salinities were not so high
as those on the south coast, and the band of live mangroves was much
wider than on the south coast. It is suggested that the formation of
hypersaline ponds in high rainfall areas is the consequence of unusual
weather conditions, i.e., a rare prolonged period of low rainfall.

Natural salinity increases can occur in a short time as well as
on a time scale of centuries. For example, storms can build up berms
that bring about hypersalinity. Vu Van Cuong (1964) reported the
case of a violent storm that caused sandbanks to build up which isolated portions of the mangrove forest from tidal circulation. The
rise in salinity of the tidally isolated forest brought about proges-
sive disappearance of Sonneratia alba, Rhizophora mucronata, R. apiculata, Ceriops tagal, Avicennia officinalis, and A. marina.

The reduced growth of mangroves associated with naturally increased interstitial salinity appears to have a parallel in Spartina
alterniflora marshes in Georgia. The form of Spartina on the low
elevation soil along the creeks, the so-called "low marsh," grows
taller and has more leaves and greater biomass per unit area than do
the plants of the "high marsh" farther from the creeks at higher soil

FIGURE 4a

Aerial view of a saline pond mangrove island.

FIGURE 4b

Aerial view of a small mangrove island that is developing into a saline pond island.

elevation. Nestler (1977) has recently provided evidence that interstitial salinity is the basis for the differentiation into high marsh
and low marsh. He found that the growth of Spartina alterniflora was
inversely related to interstitial salinity. Less frequent tidal
inundation and increased evaporation are responsible for the higher
salinity and reduced growth of the high marsh Spartina.

Anthropogenic salinity increases--Man-induced salinity increases
that affect mangroves are numerous. Two examples are associated with
saline intrusion in south Florida where it has been estimated that
the water table in the Miami area has been lowered at least 6 feet
since the year 1900 (see Teas et al., 1976). Soil maps from early in
the century show a narrow band of mangroves along southern Biscayne
Bay with saltmarsh and sawgrass behind the mangroves. Agricultural
drainage ditches were opened to the Bay to drain excess water from
the fields. Subsequent hurricane high tides salinized low land near
the shore and carried mangrove propagules into former agricultural
fields (Reark, 1975). The combination of lowered water table, reduced surface run-off and hurricane seeding have caused large areas
to develop into types of mangroves communities that did not exist
when Davis carried out the field studies for his 1940 paper (Teas,
1974).

Lowering of the water table in southern Florida also resulted
in increased salinities in the lower sawgrass Everglades, which
selected against the freshwater sawgrass (Cladium jamaicensis) and
resulted in mangrove encroachment into sawgrass lands (Craighead,
1971).

Another man-induced change in the Miami area of south Florida
was the opening in 1925 of Baker's Haulover, a passage-way between
the northern part of Biscayne Bay and the Atlantic Ocean. The influx
of saltwater into formerly fresh-brackish saltmarsh and sawgrass
areas was responsible for mangrove invasion of large areas of northern
Biscayne Bay (Teas et al., 1976).

Bacon (1970, 1975) has documented the attempted "reclamation"
and partial recovery of Caroni Swamp, a Trinidadian mangrove swamp.
The reclamation, which was abandoned about the year 1925, included
canalization, diverting rivers that previously entered the swamp and
digging new channels. He reported that the highly saline "dead
lagoons" where the mangroves had been killed had been caused by em-
bankments that cut off free drainage. Killing of mangroves in hypersaline areas caused by earthen diking has also been reported at
Guayanilla on the dry south coast of Puerto Rico (Teas, 1978).

Artificial lowering of salinity--Lowering of salinity in mangrove areas by human activities is much less common than are increases in salinity. Joshi and Shinde (1978) reported on changes in

the small Vashisti river in Western India. Approximately 12 years before their study the tail waters from the large Koyna dam power station were diverted into the Vashisti river, so that it flowed freshwater 12 months of the year rather than only during the monsoon season. Formerly there were well developed mangrove forests at Chiplun, 32 km up the Vashisti River from the Arabian sea. Now, because of the freshwater condition, the mangroves face competition from Cyperus, Scirpus, etc. Although the present mature mangroves in the Chiplun area are in no danger from this competition, the invading freshwater hydrophytes seriously limit mangrove regeneration.

One means of lowering interstitial salinity in a mangrove forest is by channelization. Noakes (1955) reported that channelizing of mangrove forests to improve tidal flushing retarded succession from Rhizophora apiculata to Bruguiera species. Vu Van Cuong (1964) stated that Rhizophora apiculata developed well and maintained its abundance-dominance ratio if the forest was criss-crossed with small irrigation canals to enhance tidal circulation. It is commonly seen in aerial photographs of Florida, taken with False Color Infra-Red (FCIR) film, that well flushed Rhizophora along a tidal creek show up scarlet-red rather than the dark red of nearby trees of Rhizophora mangrove not so well tidally flushed. Criss-crossing of mangrove forests with drainage ditches has been practiced near Miami since the 1930's as a means of controlling mosquitoes. More vigorous and taller Rhizophora mangroves that are scarlet-red on the FCIR aerial photographs can be seen along some of these Florida "artificial tidal creeks." The beneficial effect of increased tidal flushing on Rhizophora, as recommended in Malaya, appear to be principally because it reduces interstitial salinity, thereby relieving salinity stress on the Rhizophora trees.

A possible parallel to the improvement of Rhizophora by criss-crossing it with drainage ditches is the report of Shisler and Jobbins (1977) who ditched Spartina alterniflora high marsh, which Nestler (1977) had found to have high interstitial salinity. Shisler and Jobbins reported that the ditched marsh, with its improved tidal circulation, produced 77% more plant biomass than did control high marsh.

Effects of Clear-Cutting

Clear-cut mangrove forests may fail to regenerate because slash excludes seed supply, but other factors can be involved. Holdridge (1940) noted in Puerto Rico that in some clear-cut mangrove areas there was no natural regeneration and that, furthermore, he had little or no success in establishing handplanted seedlings. He identified such areas as hypersaline salitrals. Macnae (1968) reported that the areas where mangroves had been clear-cut often became deserts, useful only for salt production. He cited examples along the Gulf

of Thailand, east and west of Bangkok. Local vegetation-free areas
called salt pans are known to occur in salt marshes where piled-up
decaying vegetation killed the plants locally and conditions became
unfavorable for their re-establishment (Chapman, 1974; Pethick, 1974).
Under some conditions high temperatures and increased evaporation in
clear-cut mangrove areas could raise the soil salinity so that mangroves are unable to become re-established.

The failure of revegetation of clear-cut mangrove areas is probably not due to the loss of mineral nutrients. Studies in the defoliated mangrove forests of Vietnam showed that nutrient losses were
similar in hand-cut forests, but in neither case was the loss considered to be sufficient to limit mangrove growth (Lang, 1974).

A factor other than salinity that may be involved in the inhibition of mangrove regrowth in clear-cut forest areas is the chemical
changes that can occur in exposed soils. Baltzer and Lafond (1971)
reported that bare hypersaline areas in saltmarshes were frequent
sites of acid sulfate soils, i.e., cat-clays (Moorman and Pons, 1975).

Macnae (1968), in reviewing the history of mangroves provided by
early writers, noted accounts of mangroves along the Red Sea and
Arabian Gulf more than 2,000 years ago and suggested that formerly
mangroves may have occurred much more extensively than they do at
present. It may be that mangroves along some of these arid shores
were over-exploited by clear-cutting and failed to regenerate. Mangroves might be planted and managed along parts of the Arabian Sea,
Arabian Gulf and Red Sea that presently lack mangroves, but where they
have grown in former times. Indeed, by enlightened methods, mangroves
might be made to prosper in areas where they have never grown before.

Zonation and Succession

In many of the mangrove forests there are found well-developed
zones of particular species or associations of species. Zonation of
mangroves has been attributed to such factors as the frequency of
tidal inundation, the salinity of the interstitial water, and waterlogging of the soil (Watson, 1928; Macnae, 1968). The correlation of
habitat and species occurrence is not, in all cases, a necessary one.
Rabinowitz (1975) planted propagules of Rhizophora, Pelliciera, Laguncularia and Avicennia in Panama into stands of each of the other
genera. She found that seedlings of all genera grew well in all habitats, and proposed that tidal sorting based on propagule size, rather
than habitat adaptation, was the important mechanism for control of
zonation. Macnae (1968) reported on the colonization of new sand or
mud banks by whichever of several species happened to be producing
seedlings when the banks formed. It is this author's opinion that
what mangrove species or mixture of species are found at a particular
site involves tidal sorting of propagules, salinity preference and

tolerance, interstitial salinity, competition for light in a developing stand, predation and parasitism, and especially opportunism in
the form of propagule availability.

DISCUSSION

Mangroves provide fuel in parts of the world that are chronically
short of firewood. For the most part mangroves grow on soils that
are too saline for other crops, and in areas that would otherwise be
non-productive wastelands.

Experience with exploitation of the tropical tree crop, rubber,
suggests that it should be possible to introduce mangroves into new
ares, increase yields, develop locally adapted varieties, and improve
harvesting technology. Areas to be considered or tested for feasibility of mangrove cultivation are located along thousands of miles
of the world's shorelines. Some of these may have supported mangroves in earlier times, others are ones where mangroves may never
have colonized or been introduced, and still others are areas where
shoreline modification would be required to make the land suitable
for mangroves.

Increased exploitation of mangroves as a fossil fuel substitute
will require consideration of management, engineering, physiology
and ecology, breeding and selection, associated biological products,
and social and economic impacts.

Management

Mangrove cultivation generally involves planting, thinning,
cropping and regeneration of mangrove forests. Mangroves have been
managed as a crop in southeast Asia for more than 75 years, so that
there is available extensive experience with cutting cycles, harvesting, yield, regeneration systems, and regulation of state-leased
mangrove lands.

Recent increases in the value of wood affect the value of mangrove forests so that the economics of mangement need to be re-exam-
ined. For example, because of the higher value of the crop today,
it may be practical to hand plant in order to obtain early and uni-
form regeneration, to eliminate weedy mangrove species by hand, to
carry out salinity control by improving tidal circulation, and to
investigate an array of practical matters such as the possible value
of fertilizer for obtaining more rapid seedling growth and the value
of partial shade in establishing some species.

Engineering

Wave and current energy are known to hinder mangrove establishment (Teas, 1977). Zahran (1975) pointed out that the well-developed

mangroves of the Red Sea are found along mersas (lagoons). Engineering would be utilized in developing the conditions of the mersas or
other environments favorable to mangroves, which could involve earthworks, jetties, or the placement of wavebreakers such as floating
reefs of discarded rubber tires (Shaw and Ross, 1977).

Technology developed in the polders of Norther Europe for re-
covering saline land from the sea (Davis, 1956; Chapman, 1974) might
be adapted to the renovation of hypersaline lands to make them suitable for mangroves.

A variety of mangrove planting test sites need to be developed
in order to assess the feasibility and economics of the process.

Physiology and Ecology

The physiology and ecology of mangroves need to be studied as
a background for species and variety testing and as an adjunct to
engineering aspects of mangrove establishment. Features that should
be investigated include: basic aspects of mangrove ion exclusion and
salt secretion; physiological differences in species and varieties;
the role of major and minor elements in mangrove growth; overall
aspects of salinity tolerance; the effect of salinity on photosynthesis, transpiration and growth; the role of light levels in seedling
establishment; the role of ion balance in growth; the role of mycor-
rhiza-like organisms; the role of nitrogen fixation in mangrove
swamps; and the role of bacteria, algae, fungi and invertebrates in
the maturing of mangrove soils.

Breeding and Genetics

The history of agricultural and forest science offer many examples of crop improvement by application of plant breeding and
genetic methods, especially where they have been combined with knowledge of the basic physiology of the organism.

Experience has shown that natural populations have genetic diversity for almost any character that is investigated. Mangroves
are likely no exception. McMillan (1975a) found that _Avicennia_
germinans and _A. marina_ seedlings collected from the higher latitudes
of their ranges showed greater chill resistance than did those from
sites nearer the equator, suggesting that inheritance is involved in
cold tolerance.

Salt excluding species depend on root cell membranes for effectiveness in their ion exclusions. In other organisms the structure
of membranes is subject to genetic control, so that breeding and
selection for effective ion excluders in mangroves may be practical.
As noted in this report, individual seedlings of _Rhizophora mangle_

and <u>Laguncularia</u> <u>racemosa</u> differ greatly in the amount of salt that is secreted by their salt glands. It may be possible to obtain genetic lines that have more effective salt secreting systems, which would let them prosper in higher salinity environments. Some background information on the genetic system of <u>Rhizophora</u> <u>mangle</u> is available (Teas and Handler, 1979).

Associated Biological Products

The available evidence points to a correlation of mangroves with prawn and other fisheries; however, details of the relationship of mangroves to exploitable fisheries need to be investigated. It is possible that production of a specific food resource such as crabs or prawns could be enhanced selectively by such mangrove forest features as tidal flow patterns, soil elevation or fertilization. Also, mangrove forests might be utilized for limited aquaculture by planting selected larval forms.

Calculations by the author, based on Macnae's 1974 summary of prawn catches and mangroves, indicate that the dollar value of fisheries associated with mangroves may equal that of the mangrove wood.

Additional spin-off values from establishing new mangrove forests include the possibility of climatic modification, the attenuation of storm damage, and the retarding of shore erosion.

Sociological and Economic Impacts

Planting and management of mangroves will undoubtedly require government capital investment. Consideration needs to be given to alternate uses of the land, to the human resources that would be involved in managing forests at remote locations and analyses need to be made of the energy and capital effectiveness of extending mangrove cultivation as a crop.

ACKNOWLEDGMENTS

The author wishes to express his thanks to the Oak Ridge Associated Universities for support of his investigations in Puerto Rico, to Dr. J. G. Gonzalez of the Marine Ecology Division of PRCEER for providing facilities in Puerto Rico, and to Patricia Jennings for assistance in some of the laboratory studies.

REFERENCES

Albert, R. 1975. Salt regulation in halophytes. Oecologia <u>21</u>: 57-71.

Arisz, W. H., Chapman, I.J., Heikens, H. and van Tooren, A.J. 1975. The secretion of the salt glands of Limonium latifolium Ktze. Acta Botan. Neerl. 4: 322-338.

Atkinson, M.R., Findlay, G.P., Hope, A.B., Pitman, M.G., Saddler, H.D.W. and West, H.R. 1967. Salt regulation in the mangroves Rhizophora mangle Lam. and Aegialitis annulata R. Br. Australian Journal of Biological Sciences 20: 589-599.

Bacon, P.R. 1970. The ecology of Caroni Swamp. Spec. Public. Central Statistical Office, Trinidad. 68 pp.

Bacon, P.R. 1975. Recovery of a Trinidadian mangrove swamp from attempted reclamation. pp. 805-815. In Proc. Internat. Sympos. on Biology and Management of Mangroves. Univ. Florida Press.

Baltzer, F. and Lafond, L.R. 1971. Marais Maritime Tropicau. Revue de Geographie Physique et de Geologie Dynamique 13: 173-196.

Banerji, J. 1958. The mangrove forests of the Andamans. World Forest. Congress 3: 425-430.

Banus, M.D. and Kolehmainen, S.E. 1975. Floating, rooting and growth of red mangrove (Rhizophora mangle L.) seedlings: Effect on expansion of mangroves in Southwest Puerto Rico. pp. 370-384. In Internat. Sympos. on Biology and Management of Mangroves. Univ. Florida Press.

Barbour, M.G. 1970. Is any angiosperm an obligate halophyte? Amer. Midlands Naturalist 84: 105-120.

Beard, J.S. 1967. An inland occurrence of mangrove. Western Australia Naturalist 10: 112-115.

Bhosale, L.J. 1978. Ecophysiological studies of the mangroves from the western coast of India. Dept. of Botany, Shivaji University, Kolhapur, India. 81 pp.

Bowman, H.H.M. 1917. Ecology and physiology of red mangroves. Proc. Amer. Philosophical Soc. 56: 589-672.

Bray, J.R. and Gorham, E. 1964. Litter production in forests of the world. pp. 101-157. In Advances in Ecological Research. Vol. 2

Brown, J.M.A., Outred, H.A. and Hill, C.F. 1969. Respiratory metabolism in mangrove seedlings. Plant Physiology 44: 287-294.

Chapman, V.J. 1970. Mangrove phytosociology. Tropical Ecology 11:1-19.

Chapman, V.J. 1974. Salt Marshes and Salt Deserts of the World. Verlag von J. Cramer, Germany. 2nd Edition. 392 pp.

Chapman, V.J. and Ronaldson, J.W. 1958. The mangrove and salt-grass flats of the Auckland Isthmus. New Zealand Dept. Scient. Indust. Research, Bull. 125, 79 pp.

Christensen, B. 1978. Biomass and primary production of Rhizophora apiculata Bl. in a mangroves in southern Thailand. Aquatic Botany 4: 43-52.

Cintron, G., Lugo, A.E., Pool, D.J. and Morris, G. 1978. Mangroves of arid environments in Puerto Rico and adjacent islands. Biotropica 10: 110-121.

Connor, D.J. 1969. Growth of grey mangroves (Avicennia marina) in nutrient culture. Biotropica 1: 36-40.

Craighead, F.C., Jr. 1971. The Trees of South Florida. University of Miami Press. 212 pp.

Davis, J.H., Jr. 1940. The ecology and geological role of mangroves in Florida. Carnegie Inst. Wash. Publ. 32: 305-412.

Davis, J.H., Jr. 1956. Influences of man on coast lines. pp. 504-521. In Man's Role in Changing the Face of the Earth. Univ. Chic. Press.

de Sylva, D.P. and Michel, H.B. 1974. Effects of mangrove defoliation on the estuarine ecology of South Vietnam. pp. 710-728. In Internat. Sympos. on Biology and Management of Mangroves. University of Florida Press.

Ding Hou. 1958. Rhizophoraceae. Flora Malesiana 5: 429-493.

Dixon, R.G. 1959. A working plan for the Matang Forest Reserve. Forest Dept. Perak. 70 pp.

Drew, W.B. 1974. The ecological role of the fern (Acrostichum aureum) in sprayed and unsprayed mangrove forests. The Effects of Herbicides in South Vietnam. National Academy of Sciences, Wash., DC 13 pp.

Egler, F.E. 1952. Southeast saline Everglades vegetation, Florida and its management. Vegetatio: Acta Geobotanica 3: 213-265.

Fosberg, F.R. 1961. Vegetation-free zones on dry mangrove coasts. U.S. Geological Survey Professional Papers. No. 365. pp. D-216-218.

Fosberg, F.R. 1971. Mangroves versus tidal waves. Biological Conservation 4: 38-39.

Giglioli, M.E.C. and King, D.F. 1966. The mangrove swamps of Keneba, Lower Gambia River basin, III. Jour. of Applied Ecology 3: 1-19.

Gill, A.M. and Tomlinson, P.B. 1977. Studies on the growth of red mangrove (Rhizophora mangle L.) 4. The adult root system. Biotropica 9: 145-155.

Golley, F., Odum, H.T. and Wilson, R.F. 1962. The structure and metabolism of a Puerto Rican red mangrove forest in May. Ecology 43: 1-19.

Golley, F.B. 1972. Tropical Ecology with an Emphasis on Organic Productivity. Univ. Georgia. pp. 407-413.

Golley, F.B. and Leith, H. 1972. The bases of tropical production. pp. 1-26. In Tropical Ecology. F.B. Golley and R. Misra, eds., Univ. of Georgia, Athens, GA 418 pp.

Guilcher, A. 1963. The Sea. Vol. 3., pp. 620-654. Interscience Publishers. 936 pp.

Harshberger, J.W. 1914. The vegetation of South Florida. Trans. Wagner Free Inst. 7: 49-189.

Hicks, D.B. and Burns, L.A. 1975. Mangrove metabolic response to alteration of natural freshwater drainage in southwestern Florida estuaries. pp. 238-255. In Proc. Int. Sympos. on Biol. and Management Mangroves.

Holdridge, L.R. 1940. Some notes on the mangrove swamps of Puerto Rico. Carib. Forester 1: 19-29.

Jamale, B.B. and Joshi, G.V. 1976. Physiological studies in senescent leaves of mangroves. Indian J. Exptl. Biol. 14: 697-699.

Joshi, G.V., Jamale, B.B. and Bhosale, L.J. 1975. Ion regulation in mangroves. pp. 595-607. In Inter. Sympos. on Biol. and Management Mangroves.

Joshi, G.V. and Shinde, S.D. 1978. Ecogeographical studies. Terekhol and Vashisti Rivers. Shivaji Univ. Press, Kolhapur, India. 56 pp.

Kimball, M. and Teas, H.J. 1975. Nitrogen fixation in mangroves areas of South Florida. pp. 654-661. In Proc. First Internat. Symp. on Biology and Management of Mangroves. Univ. Florida Press.

Lang, A. 1974. The effects of herbicides in South Vietnam. Part A. National Academy of Sciences, Wash., DC 372 pp.

Lugo, A.E. and Snedaker, S.C. 1974. The ecology of mangroves. Ann. Rev. Ecology and Systematics 5: 39-64.

MacCaughey, V. 1917. The mangrove in the Hawaiian Islands. Hawaiian Forestry Agr. 14: 361-366.

Macnae, W. 1974. Mangrove forests and fisheries. Reports IOFC, International Indian Ocean Fisheries Survey and Development Programme, No. 74/34. 35 pp.

Mallery, C.M. and Teas, H.J. 1979. Sodium and Chloride distribution and root cellular compartments in the mangroves Rhizophora mangle and Avicennia germinans. In press. 13 pp.

Mattox, N.T. 1949. Studies on the biology of the edible oyster, Ostrea rhizophorae Guilding, in Puerto Rico. Ecological Monographs 19: 339-356.

McGill, J.T. 1958. Map of coastal landforms of the world. Geograph. Reviews 48: 402-405.

McMillan, C. 1971. Environmental factors affecting seedling establishment of the black mangrove on the central Texas coast. Ecology 52: 927-930.

McMillan, C. 1975 a. Adaptive differentiation to chilling in mangrove populations. pp. 62-68. In Internat. Symp. on Biology and Management of Mangroves. Univ. Florida Press.

McMillan, C. 1975b. Interaction of soil texture with salinity tolerance of black mangrove (Avicennia) and white mangrove (Laguncularia) from North America. pp. 561-566. In Internat. Symp. on Biology and Management of Mangroves. Univ. Florida Press.

Miller, P.C. 1972. Bioclimate, leaf temperature, and primary production in red mangrove canopies in south Florida. Ecology 53:22-45.

Moorman, F.R. and Pons, L.J. 1975. Characteristics of mangrove soils in relation to their agricultural land use and potential. pp. 529-547. In Internat. Symposium on Biology and Management of Mangroves.

Morton, J.F. 1965. Can the red mangrove provide food, feed and fertilizer? Economic Botany 19: 113-123.

Nestler, J. 1977. Interstitial salinity as a cause of ecophenic variations in Spartina alterniflora. Estuarine and Coastal Marine Science 5: 707-714.

Noakes, D.S.P. 1955. Methods of increasing growth and obtaining regeneration of the mangrove type of Malaya. Malayan Forest. 18:23-30.

Odum, W.E. and Heald, E.J. 1972. Trophic analyses of an estuarine mangrove community. Bulletin Marine Science 22: 671-738.

Pannier, P.F. 1959. El efecto de distintas concentraciones salinas sobre el desarrolo de Rhizophora mangle L. Acta Cient. Venozolana, Caracas 10: 68-78.

Pethick, J.S. 1974. The distribution of salt pans in tidal salt marshes. Journal of Biogeography 1: 57-62.

Pitman, M.G. 1963. The determination of the salt relations of the cytoplasmic phase in cells of beet root tissue. Australian J. of Biological Science 16: 647-668.

Pool, D.J., Lugo, A.E. and Snedaker, S.C. 1975. Litter production in mangrove forests of southern Florida and Puerto Rico. pp. 213-237. In Internat. Symp. on Biology and Management of Mangroves. Univ. Florida Press.

Provost, M.W. 1973. Salt marsh management in Florida. Proc. Tall Timbers Conference on Ecological Control by Habitat Management. pp. 5-17.

Rabinowitz, D. 1975. Planting experiments in mangrove swamps of Panama. pp. 385-393. In Internat. Symp. on Biology and Management of Mangroves. Univ. of Florida Press.

Reark, J.B. 1975. A history of the colonization of mangroves on a tract of land on Biscayne Bay, Florida. pp. 776-804. In Internat. Symp. on Biology and Management of Mangroves. Univ. Florida Press.

Scholander, P.F. 1968. How mangroves desalinate seawater. Physiologia Plantarum 21: 258-268.

Scholander, P.F., Bradstreet, E.D., Hammel, H.T. and Hemmingsen, E.A. 1966. Sap concentrations in halophytes and some other plants. Plant Physiology 41: 529-532.

Scholander, P.F., Dam, L. van, and Scholander, S.I. 1955. Gas exchange in the roots of mangroves. Amer. J. of Botany 42: 92-98.

Scholander, P.F., Hammel, H.T., Bradstreet, E.D. and Hemmingsen, E.A. 1965. Sap pressure in vascular plants. Science 148: 339-346.

Scholander, P.F., Hammel, H.T., Hemmingsen, E. and Garey, W. 1962. Salt balance in mangroves. Plant Physiology 37: 722-729.

Scholl, D.W. 1965. High interstitial water chlorinity in estuarine mangrove swamps, Florida. Nature 207: 284-285.

Schuster, W.H. 1952. Fish culture in brackish water ponds of Java. Indo-Pacific Fisheries Council Spec. Public., No. 1, pp. 1-143.

Shapiro, S. 1958. The role of light in the growth of root primordia in the stem of the lombardy poplar. pp. 445-465. In The Physiology of Forest Trees.

Shaw, G. and Ross, N. 1977. How to build a floating tire breakwater. Information Bull. No. 1. Univ. Maine/Univ. New Hampshire Coop. Inst. Sea Grant Program. 12 pp.

Shisler, J.K. and Jobbins, D.M. 1977. Salt marsh productivity as affected by the selective ditching technique, open marsh water management. Mosquito News 37: 631-636.

Steenis, C.G.G.J. van. 1962. The distribution of mangrove plant genera and its significance for paleogeography. Koninkl. Ned. Akad. Wetensch. Ser. C. 65: 164-169.

Stoddart, D.R., Bryan, G.W. and Gibbs, P.E. 1973. Inland mangroves and water chemistry, Barbuda, West Indies. J. Natural History 7: 33-46.

Teas, H.J. 1974. Mangroves of Biscayne Bay. Mimeo. Dade County. 107pp.

Teas, H.J. 1976. Productivity of Biscayne Bay mangroves. pp. 103-113. In Biscayne Bay; Past, Present and Future. Univ. of Miami Sea Grant Spec. Report 5, University of Miami.

Teas, H.J. 1977. Ecology and restoration of mangrove shorelines in Florida. Environ. Conservation 4: 51-58.

Teas, H.J. 1979. Stresses on the mangroves of Guayanilla Bay. Proc. Puerto Rico Mangrove Conf. In Press. 10 pp.

Teas, H.J. and Handler, S. 1978. Notes on the pollination biology of Rhizophora mangle L. Proc. Internat. Symp. on Marine Biogeography and Evolution in the southern Hemisphere. In press. Dept. Scientific and Industrial Research. New Zealand.

Teas, H.J. and Jurgens, W. 1978. Aerial planting of Rhizophora mangrove propagules in Floria. Proc. Fifth Annual Conf. on Restoration Coastal Vegetation in Florida. Hillsborough Com. College, Tampa, Florida. In Press.

Teas, H.J., Jurgens, W. and Kimball, M.C. 1975. Plantings of red mangrove (Rhizophora mangle L.) in Charlotte and St. Lucie Counties, Florida. pp. 132-161. In. Proc. Second Annual Conf. on Restoration of Coastal Vegetation in Florida. Hillsborough Com. Coll., Tampa, Florida.

Teas, H.J., Wanless, H.R. and Chardon, R. 1976. Effects of man on the
    shore vegetation of Biscayne Bay. pp. 133-156. In Biscayne Bay;
    Past, Present and Future. University of Miami Sea Grant Spec. Re-
    port No. 5, Univ. of Miami, Coral Gables, Florida.

Ungar, I.A. 1974. Inland halophytes in the United States. pp. 235-
    305. In Ecology of Halophytes. Academic Press.

Untawale, A.G., Balasubramian, T. and Wafar, M.V.M. 1977. Structure
    and production in a detritus rich estuarine mangrove swamp.
    Mahasagar-Bulletin of the National Institute of Oceanography 10:
    173-177.

Vu Van Cuong, H. 1964. Flore et vegetation de la mangrove de la
    region de Saigon-Cap Saint Jaques, Sur Viet-Nam. These Doct.,
    Univ. Paris, 199 pp.

Wadsworth, F.H. 1959. Growth and regeneration of white mangrove in
    Puerto Rico. Caribbean Forester 20: 59-70.

Walsh, G.E. 1974. Mangroves: A Review. pp. 51-174. In Ecology of
    Halophytes. Academic Press.

Walter, H. and Steiner, M. 1936. Die Okologie der Ost-Afrikanischen
    Mangroven. Zeit. Botan. 30: 65-193.

Watson, J.D. 1928. Mangrove forests of the Malaya peninsula. Malayan
    Forest. 6: 1-275.

Zahran, M.A. 1975. Biogeography of the mangrove vegetation along the
    Red Sea coasts. pp. 43-51. In Internat. Symp. on the Biology and
    the Management of Mangroves. Univ. of Florida Press.

BIOCHEMICAL CONTROL OF PHOTOSYNTHETIC CARBON ASSIMILATION IN MARINE ORGANISMS FOR FOOD AND FEED PRODUCTION

Clanton C. Black, Jr.

Department of Biochemistry
University of Georgia
Athens, GA 30602

It is widely agreed today among plant biologists that higher plants possess several biochemical pathways for assimilating CO_2 which are related to the production capabilities of plants. However the research which allows such a broad conclusion was conducted almost exclusively with terrestrial plants. The biochemistry of carbon assimilation with marine plants has not been vigorously studied. So, of necessity, we will summarize carbon metabolism and topics related to productivity in other plants and then consider marine plants. The purposes of this manuscript are: to assess the current status of knowledge on photosynthetic carbon biochemistry in marine plants; to identify some research areas where their biochemistry/physiology is poorly understood; and to describe some potentially beneficial uses of marine plants for food, feed, fiber and fuel production.

To assess the current status of photosynthesis research in marine plants we will consider first related work in other plants. From research within the last three decades we know that plants assimilate CO_2 via several pathways. The first pathway discovered was the reductive pentose phosphate cycle (C_3 cycle). More recently the C_4-dicarboxylic acid cycle (C_4 cycle) was discovered and the uniqueness of Crassulacean acid metabolism (CAM) was reconsidered. The biochemistry of these carbon assimilation pathways has been thoroughly reviewed (12). Plants also have been found recently which are intermediate between C_3 and C_4 plants (11) in their biochemistry as well as in other characteristics. Tables 1 and 2 describe some features of these 4 types of plants which can be used to distinguish each plant type. Even within these broad plant groups other biochemical variations in pathways of CO_2 assimilation are known in specific plant species (12). Since most of the topics

163

in Tables 1 and 2 have been considered previously (5,7), these topics will not be analyzed here.

In addition to multiple pathways for assimilating CO_2, plants also exhibit a phenomenon known as photorespiration. Photorespiration is the loss of CO_2 in light primarily due to the metabolism of glycolate (18,26). The loss of CO_2 via photorespiration is very significant in terms of net carbon assimilation since it can result in losses equal to 50% of the carbon fixed during C_3 photosynthesis. However, with C_4, CAM, and C_3-C_4 intermediate plants, the losses of CO_2 via photorespiration are not as severe. Indeed C_4 plants have eliminated the actual loss of CO_2 from leaves by sequestering glycolate synthesis and CO_2 loss reactions in bundle sheath cells and by carboxylating all of the internal leaf CO_2 supply with phosphoenolpyruvate (PEP) carboxylase so that no CO_2 is lost to the atmosphere. Thus in nature the loss of CO_2 from photorespiration has been regulated in specific plants with specific types of carbon biochemistry.

Photorespiration can be essentially eliminated in C_3 plants by lowering the O_2 level to 2% (Table 2) while such treatment has little immediate influence on C_4 or CAM plants photosynthesis (Table 2). Since CO_2 can be lost from C_3 plants, resulting in an apparently detrimental loss of fixed carbon, much research today is aimed at studying the regulation of photorespiration either chemically, genetically, or environmentally. Whether or not any of these efforts will be successful in finally increasing the net carbon fixation of plants is unknown. The author believes that plants in nature have already regulated photorespiration as in C_4 plants and we can hope to use this and other means to regulate CO_2 loss in other plant types.

From studying these tables readers can realize that these groups of plants have little relationship to classical taxonomic classifications since species within the same genus, i.e. Panicum, Euphorbia, or Suaeda have specific anatomical, biochemical, and physiological features. More importantly for our considerations, this grouping of plants has definite relationships to such important economic factors as plant water use efficiency (transpiration ratio, Table 2), maximum plant growth rates (Table 2), plant productivity (Table 2), plant competition (8), and plant nitrogen use efficiency (Table 2; 10). Thus it is quite important from the viewpoints of plant production and economics to understand the relationships between plant biochemistry/physiology/anatomy and the efficiency with which plants utilize water, nutrients, and other environmental resources such as light, CO_2, HCO_3^-, and heat.

In summary we know that terrestrial plants have developed several pathways for CO_2 assimilation and they have a major CO_2 dissimilation process via photorespiration. Specific types of

TABLE I. DESCRIPTION OF SOME FEATURES EMPLOYED BY VARIOUS PLANT GENERA TO ASSIMILATE CARBON DIOXIDE

Primary Pathway(s) of CO_2 Assimilation

Feature	C_3	C_4	CAM	C_3-C_4 Intermediate
Leaf anatomy cross section	Diffuse distribution of organelles in mesophyll or palisade cells with similar or lower organelle concentrations in bundle sheath cells if present.	A definite layer of green bundle sheath cells surrounding the vascular tissue; one layer of green mesophyll cells surrounding the bundle sheath cells; or two distinct green cell types.	Spongy appearance 6-10 green cell layers. Mesophyll cells have large vacuoles with the organelles evenly distributed in the thin cytoplasm. Generally lack a definite layer of palisade cells.	Green bundle sheath cells; 6-10 green mesophyll cells between veins.
Representative Genera	Panicum, Euphorbia, Suaeda, Glycine, Fescue, Atriplex	Panicum, Euphorbia, Suaeda, Zea, Cynodon, Digitaria, Atriplex	Euphorbia, Suaeda, Cactus, Opuntia	Panicum
Major leaf carboxylation during the day	RuDP Carboxylase	PEP carboxylase then RuDP carboxylase	PEP carboxylase at night RuDP carboxylase in day	RuDP carboxylase (PEP carboxylase at T)

TABLE 2. BIOCHEMICAL AND PHYSIOLOGICAL CHARACTERISTICS OF PLANT GROUPS DIFFERING IN THEIR PATHWAY OF PHOTOSYNTHETIC CO_2 METABOLISM

Characteristic	Primary Pathway of CO_2 Metabolism			
	C_3	C_4	CAM	C_3-C_4 Intermediate
Initial Product of PS	3-PGA	Oxaloacetate	Dark, oxaloacetate Light, 3-PGA	3-PGA
Energetics of PS (CO_2:ATP:NADPH)	1:3:2	1:5:2	1:6.5:2	--
Transpiration ratio (gm H_2O/gm of dry wt)	450 to 1000	250-350	25 to 125	--
PS CO_2 Compensation (ppm CO_2)	35 to 85	0 to 10	Dirunal variability 0 to 60	15 to 20
Response of leaf PS to: ~ 2% O_2 ~50% O_2	30-40% increase Inhibition	No effect slight inhibition	No effect in short times ($\leq$ 10 min.)	Intermediate between C_3 & C_4
Plant isotopic ratio $^{13}C/^{12}C$ (o/oo)	-22 to -34	-11 to -19	-13 to -34	-25
Leaf Fraction I Protein (% of soluble)	40 to 50	8 to 25	--	--

Characteristic	C_3	C_4	CAM	C_3-C_4 Intermediate
Leaf N for maximum PS (% DM)	6.5 to 7.5	3 to 4.5	--	--
Requirement for sodium as a micro-nutrient	None detected	Yes	Yes	--
Optimum day temperature for dry matter production	20 to 25°C	30 to 35°C	--	--
Response of net photosynthesis to increasing light intensity at temperature optimum	Saturation reached about ¼ to ½ full sunlight in most plants. Some desert species tend not to saturate.	Either proportional to or only tending to saturate at full sunlight.	Uncertain, but apparently saturation is well below full sunlight.	--
Maximum growth rate: gm of dry wt/dm² of leaf area/day;	.5 to 2	4 to 5	.015 to 0.18	--
gm/m² of land area/day	19.5 $\pm$ 3.9	30.3 $\pm$ 13.8	--	--
Dry matter production (tons/hectare/year)	22.0 $\pm$ 3.3	38.6 $\pm$ 16.9	Extreme variability in literature data.	--

PS = Photosynthesis; a dash indicates no information; DM = dry matter.

photosynthetic plants, e.g. C_4 plants, have developed means for regulating CO_2 metabolism to maximize photosynthesis and in turn productivity. With this brief introduction we now turn to the question of the status of similar knowledge in marine plants.

PHOTOSYNTHETIC CARBON BIOCHEMISTRY IN MARINE PLANTS

What pathways of carbon assimilation are present in marine plants? Is photorespiration an active process in marine plants? How does the productivity of marine plants compare to terrestrial plants? If one attempts to collate a variety of data on marine plants to answer such questions (as in Tables 1 and 2), it is soon evident that similar data are difficult to locate and in general not available. Furthermore there are a variety of marine ecosystems to consider in comparison to terrestrial plants. For organization purposes we will concentrate on four broad categories of marine plants or ecosystems which are common near tropical waters; namely, seagrasses, algae, coral reefs, and salt marsh/swamp/sand dune ecosystems. Table 3 cites some characteristics which have been studied in these four broad categories of marine plants.

In regard to pathways of photosynthetic CO_2 assimilation, as I assess the literature, there is no evidence for CAM in marine plants. The two CAM plants in Table 3 grow in the coastal sand dunes.

In submerged marine plants there is only one report definitively claiming C_4 photosynthesis (3). The seagrass, Thalassia testudinum, fixed 66% of its $^{14}CO_2$ into malic plus aspartic acid in 5 seconds of photosynthesis while only 2% was in 3-phosphoglyceric acid (3). This is similar to C_4 photosynthesis (Table 1; 5), but carbon turnover experiments were not reported. The ^{13}C value was -9%, which also is characteristic of C_4 plants (Table 2). Thalassia and other marine seagrasses, i.e. Cymodocea, Posidonia, and Zostera have an unusual leaf anatomy in that the leaf epidermis has no stomata and the epidermis contains most of the leaf chloroplasts and mitochondria (3, 27). So seagrass leaves are surrounded by a single layer of green cells which are not typical of C_4 plants (Table 1). In the freshwater plants Elodea, Egeria, and Lagarosiphon a similar leaf anatomy occurs and a fixation of $^{14}CO_2$ into organic acids has been reported (9,14) but the labeled organic acids do not turnover at rates commensurate with photosynthesis and the evidence leads one to conclude that C_3 photosynthesis is present. So with seagrasses we must conclude that the photosynthetic pathway of CO_2 fixation is uncertain (Table 3).

In algae and coral it seems clear that C_3 photosynthesis is the pathway for carbon assimilation (Table 3; 21,22,27). There are reports on elevated PEP carboxylase levels and early ^{14}C-labeling

TABLE 3. SOME PHOTOSYNTHETIC CHARACTERISTICS OF MARINE PLANTS

General Types of Marine Photosynthetic organisms	Representative Genera	Primary Pathway of PS CO_2 assimilation	Carbon Isotope Fractionation $^{13}C/^{12}C$ (o/oo)	Photorespiration Detected
Seagrasses	Cymodocea	--	Range from −6 to −13	--
	Zostera	--		--
	Thalassia	C_4(?)		--
	Vallisneria	C_3		Yes
Algae, Plankton Diatoms	Scytonema	C_3	Range of −12 to −32	--
	Sargassum	C_3		--
	Entermorpha	C_3		--
	Laurencia	C_3		--
	Coccaneis	C_3		--
Symbiotic Coral	Soft coral	C_3	−17	--
	Zooxanthallae isolated from Tridacna	C_3	−23	--
Emergent Salt Marsh, Swamp, and Dune Plants	Spartina	C_4	−13	No
	Distichlis	C_4	−15	No
	Mangroves	C_3	--	--
	Juncus	C_3	--	--
	Salicornia	C_3	−25	No
	Uniola	C_4	−15	Yes
	Ammophila	C_3	--	--
	Yucca	CAM	--	--
	Tillaea	CAM	--	--

during photosynthesis with marine algae (1,13,15,19). But these reports show very low enzyme activities and no ^{14}C turnover. The reports on high PEP carboxykinase levels in brown algae indicate a
definite role in dark CO_2 fixation (1,21) but no evidence for a C_4-
type photosynthesis (5). I conclude that there is no supporting
evidence for the presence of the C_4 pathway or CAM in algae or coral.

When we examine the salt marsh plants data we find that C_4
photosynthesis is present in specific plants (Table 3). In the salt
marsh these emergent C_4 plants, such as Spartina and Distichlis, can
be very important as illustrated in the extensive salt marshes along
the Gulf and Atlantic coasts of North America. C_3 plants also are
present in the marsh and the great mangrove swamps of the tropics
seem to be dominated by C_3 plants. The only known CAM plants grow
above high tides.

In regard to photorespiration, it seems clear that glycolate
(a photorespiration substrate) metabolism is involved in carbon flux
in marine environments (27). So we can conclude that photorespiration is present in marine photosynthetic organisms (17,27). However,
its quantitative contribution to net carbon assimilation is unknown.
In freshwater algae, photorespiration may even occur via another
metabolic pathway (4). The C_4 plants in the salt marsh (Table 3)
do in fact have photorespiration but it is not expressed as net
loss of carbon (5).

One of the most widely used tools in current studies on plant
metabolism is the ^{12}C enrichment due to ribulose 1,5-diphosphate
(RuDP) carboxylase activity (5). In brief, RuDP carboxylase uses
[12]C in preference to ^{13}C whereas PEP carboxylase has little preference. If one knows the $\delta^{13}C$ value of the beginning source of CO_2
or HCO_3^-, and then assays the biological material, one can determine
whether carbon is being assimilated via PEP carboxylase or RuDP carboxylase. Therefore a definite range of $\delta^{13}C$ values distinguishes
terrestrial C_3 from C_4 plants (Table 2).

In Figure 1 $\delta^{13}C$ values are plotted for a variety of organisms
and for sources of carbon. The marine plant data only contain values
for submerged marine plants. The salt marsh, mangrove swamp, and
dune plant data are plotted within the C_3, C_4, and CAM data. The
submerged marine plants would have as a carbon source sea water while
the atmosphere is the carbon source for the other plants. It is
noteworthy that marine plants generally show less ^{12}C enrichment
than terretrial plants. But the range -6 to -32% is much greater
than for other plants (5,25). Based on this wide range of $\delta^{13}C$
values we should expect to find some major differences in the biochemistry of carbon assimilation among marine plants.

Sodium is a micronutrient for C_4 and CAM plants (Table 2) but
its role in plant metabolism is unknown. Since marine plants live

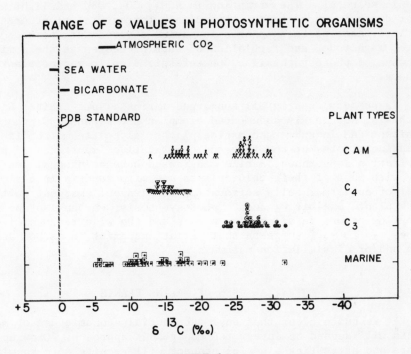

FIGURE 1

Carbon isotope fractionation values of photosynthetic organisms.
The values are relative to a limestone standard (6,25).

in high sodium environments it could be very useful to know the role of sodium in marine, as well as terrestrial plants.

Submerged marine plants live in a "closed system" in their underwater environment which doubtless is very important in understanding their biochemistry and physiology. The available carbon and O_2 are quite different from terrestrial plants. More carbon can be available than in air since seawater can be 1 to 2mM in HCO_3^-. But gas diffusion rates are 4 to 5 orders of magnitude less in water than in air. HCO_3^- rather than CO_2 is the major carbon species available in seawater. The variation in HCO_3^-, CO_2, O_2, and light with water depth is much greater than within terrestrial plant ecosystems. The importance of such changes in marine environments on the occurrence, magnitude, and regulation of such processes as the dominant pathways of photosynthesis or photorespiration in marine plants is unknown.

In summary, nearly all submerged marine plants exhibit C_3 photosynthesis; no clearly documented examples are known of C_4 photosynthesis or CAM in submerged marine plants; glycolate metabolism occurs in marine plants but the role and quantitative importance of photorespiration is unknown; marine seagrasses have an unusual leaf anatomy with most of their chloroplasts and mitochondria in a single layer of epidermal cells surrounding the leaves; emergent salt marsh plants exhibit C_4 and C_3 photosynthesis; carbon isotope fractionation occurs in all marine plants and the wide range of ^{13}C values, −6 to −32% suggests that variations exist in carbon assimilation and/or dissimilation pathways.

POTENTIAL USES OF MARINE PLANTS

It is quite clear that any substantial economic use of marine plants will demand an integrated management approach. Growing the plants and harvesting a useful product will require more knowledge and skill than terrestrial agriculture. Even so it is clear, primarily from aquaculture work (2,23,24), that it is feasible to culture plants as the primary producers and then utilize animals such as carp, mullet, prawns, and shrimp to harvest the plants. The following is one example where it seems particularly feasible to utilize salt, high solar energy, and arid seacoast land as resources.

The salt marshes are very productive ecosystems and they could be manmade on appropriate seacoast lands. We already know many C_4 and C_3 plants which thrive in salt marshes (Table 3). These marsh plants such as _Spartina_ have never been subjected to the intensive scrutiny of plant genetics or plant breeding. I suggest a plant breeding program be founded for developing such marine plants for seed, food, forage, and fiber production and that world plant collections be made of these salt marsh plants and that plant development studies be initiated just as with traditional agricultural crops. Some of the engineering as well as the ecological problems associated with manmade salt marshes already have been studied (2,20,28). In addition the ability of salt marsh plants such as <u>Juncus</u> to fix substantial quantities of nitrogen (16) seems susceptible to fitting into productive salt marsh agricultural systems.

REFERENCES

1. Akagawa, H., T. Ikawa, and K. Nisizawa. The enzyme system for the entrance of $^{14}CO_2$ in the dark CO_2-fixation of brown algae. Plant and Cell Physiol. 13, 999-1016 (1972).

2. Bardach, J.E., J.H. Ryther and W.O. McLarney. Aquaculture. Wiley-Interscience. pp. 868 (1972).

3. Benedict, C.R. and J.R. Scott. Photosynthetic carbon metabolism of a marine grass. Plant Physiol. 57, 876-880 (1976).

4. Bidwell, R.G.S. Photosynthesis and light and dark respiration in freshwater algae. Can. J. Bot. 55, 809-818 (1977).

5. Black, C.C. Photosynthetic carbon fixation in relation to net CO_2 uptake. Ann. Rev. Plant Physiol. 24, 253-286 (1973).

6. Black, C.C. and M.M. Bender. $\delta^{13}C$ values in Marine Organisms from the Great Barrier Reef. Aust. J. Plant Physiol. 3, 25-32 (1976).

7. Black, C.C., R.H. Brown, and R. Moore. Plant Photosynthesis. <u>In</u>: Limitations and Potentials of Biological nitrogen fixation in the Tropics. Eds., J. Dobereiner, R.H. Burris, and A. Hollaender, in Press (1977).

8. Black, C.C., T.M. Chen, and R.H. Brown. Biochemical basis for plant competition. Weed Science 17, 338-344 (1969).

9. Brown, J.M.I., F.I. Dromgoole, M.W. Towsey, and J. Browse. Photosynthesis and photorespiration in aquatic macrophytes. <u>In</u>: R.L. Bieliski, A.R. Ferguson, M.M. Cresswell, eds. Mechanisms of Regulation of Plant Growth. The Royal Society of New Zealand, Wellington. pp. 243-249 (1974).

10. Brown, R.H. A difference in nitrogen use efficiency in C_3 and C_4 plants and its implications in adaptation and evolution. Crop. Sci. 18, 93-98 (1978).

11. Brown, R.H. and W.V. Brown. Photosynthetic characteristics of Panicum milioides, a species with reduced photorespiration. Crop Sci., 15, pp. 681-685 (1975).

12. Burris, R.H. and C.C. Black, eds. CO_2 Metabolism and Plant Productivity. University Park Press, Baltimore–London–Tokyo. p. 431 (1976).

13. Colman, B., K.H. Cheng and R.K. Ingle. The Relative Activities of PEP carboxylase and RuDP carboxylase in blue–green algae. Plant Sci. Letters 6, 123–127 (1976).

14. Degroote, D. and R.A. Kennedy. Photosynthesis in Elodea canadensis Mich. Four carbon Acid Synthesis. Plant Physiol. 59, 1133–1135 (1977).

15. Döhler, G. C_4-Pathway of Photosynthesis in the blue–green algae Anacystis nidulans. Planta 118, 259–269 (1974).

16. Evans, H.J. and L.E. Barber. Biological Nitrogen Fixation for Food and Fiber Production. Science 197, 332–339 (1977).

17. Helder, R.J., H.B.A. Prins, and J. Schuurmans. Photorespiration in the leaves of Vallisneria spiralis. Afeeling voor de Wissen Natuurkundige Wetenschappen. Proceedings, Biological and Medical Sciences 77, 338–344 (1974).

18. Jackson, W.A. and R.J. Volk. Photorespiration. Ann. Rev. Plant Physiol. 21, 385–432 (1970).

19. Joshi, G.V., M.D. Karekar, C.A. Gowda, and L. Bhosale. Photosynthetic carbon metabolism and carboxylating enzymes in algae and mangrove under saline conditions. Photosynthetica 8, 51–52 (1974).

20. Kamps, L.F. Mud distribution and land reclamation in the eastern Wadden shallows. Rijkswat St. Commun. 1–73 (1962).

21. Kremer, B.P. and U. Küppers. Carboxylating Enzymes and Pathway of Photosynthetic Carbon Assimilation in Different Marine Algae-Evidence for the C_4-Pathway? Planta 133, 191–196 (1977).

22. Kremer, B.P. and J. Willenbrink. CO_2-fixation and Translocation in Benthic Marine Algae. Planta 103, 55–64 (1972).

23. Parker, H.A. The culture of the red alga genus Eucheuma in the Philippines. Aquaculture 3, 425–439 (1974).

24. Shang, Y.C. Comparison of the economic potential of aquaculture, land animal husbandry and ocean fisheries: The case of Taiwan. Aquaculture 2, 187–195 (1973).

25. Smith, B.N. and S. Epstein. Two categories of $^{13}C/^{12}C$ ratios for higher plants. Plant Physiol. 22, 45–74 (1971).

26. Tolbert, N.E. Microbodies - peroxisomes and glyoxysomes. Ann. Rev. Plant Physiol. <u>22</u>, 45-74 (1971).

27. Toblert, N.E. and C.B. Osmond, eds. "Photorespiration in Marine Plants." CSIRO, Melbourne and University Park Press, Baltimore, p. 139 (1976).

28. Wheaton, F.W. Aquaculture Engineering. Wiley-Interscience. p 736 (1977).

BIOSALINE RESEARCH: THE USE OF PHOTOSYNTHETIC MARINE ORGANISMS IN

FOOD AND FEED PRODUCTION

Akira Mitsui
Division of Biology and Living Resources
School of Marine and Atmospheric Science
University of Miami
4600 Rickenbacker Causeway
Miami, FL 33149

INTRODUCTION

All food production ultimately depends on photosynthetic organisms and through them, upon solar energy. Since ancient times man has primarily tapped solar energy through land agriculture. Agricultural production is limited by the availability of arable land, fresh water, light and nutrients.

Food requirements are increasing as population increases. Already, demand for food has increased beyond the ability to produce it. This has resulted in malnutrition and starvation that will continue into the foreseeable future. Intensive efforts are being made worldwide to increase productivity of arable land, to provide more fresh water, and to develop new sources of nutrients.

The magnitude and immediacy of the food shortage problem demand that all under-exploited food resources be developed to their fullest potential. New resources will have to be found, and a technology developed to meet future requirements. These resources should be renewable and free of harmful by-products in order to be a long-lasting and stable source of food.

The natural productivity of the oceans has provided food since ancient times. Until recent times, little effort has been spent in increasing its productivity. Examination of this environment reveals a great diversity of organisms with potential as new resources for food and feed. The thrust of this paper is to examine possibilities for new research development in the utilization of the oceans and coastal areas for food and feed production.

LIMITS ON FOOD AND FEED PRODUCTION

FAO estimates that for the next ten years, the percentage ·of
people experiencing malnutrition will decrease, but that the absolute
number of malnourished people will remain the same. The problem can
be defined as one of availability; availability of fresh water, light,
and fertilizers to regions otherwise unproductive or yielding less
than their potential.

Fresh Water

Rainfall of predictable amounts and periodicity is necessary for
advanced agriculture. Low rates of precipitation or irregular occur-
rences call for irrigation as a supplement or back-up. Irrigation on
a vast scale is possible technologically, and should be pursued. But
limits to this must be recognized. The benefits may not justify the
cost, or political considerations may not permit realization of otherwise logical irrigation projects. At this meeting, two alternative
technologies will be presented that are directed at the problem of
fresh water distribution. Dr. Bassham's presentation seeks methods
of circumventing the great dependence of agriculture on fresh water
in areas deficient in this resource. The paper which I am presenting
explores the possibility of growing marine photosynthetic organisms
in the oceans and coastal areas. Figure 1 is a world map of coastal
areas existing under drought or near-drought conditions.

Light

Since we are dealing with technologies directly or indirectly dependent on photosynthesis, the availability of solar energy is a critical factor. Intensity and duration are the major variable. Figure
2 is a schematic map of incident solar radiation of the world. The
successful development of bioconversion technologies will depend to
a great extent on finding or adapting organisms which can optimally
utilize the available light.

Fertilizer

The technology for producing high crop yields depends on large
amounts of fertilizers, especially combined nitrogen compounds such
as ammonia. The synthesis of ammonia by industrial methods spends
non-renewable fossil fuels and requires a considerable input of energy
for the conversion process. These energy costs are also carried over
in the production of animal protein. The study and application of
biological nitrogen fixation and the control of denitrification would
reduce the requirements for fertilizer.

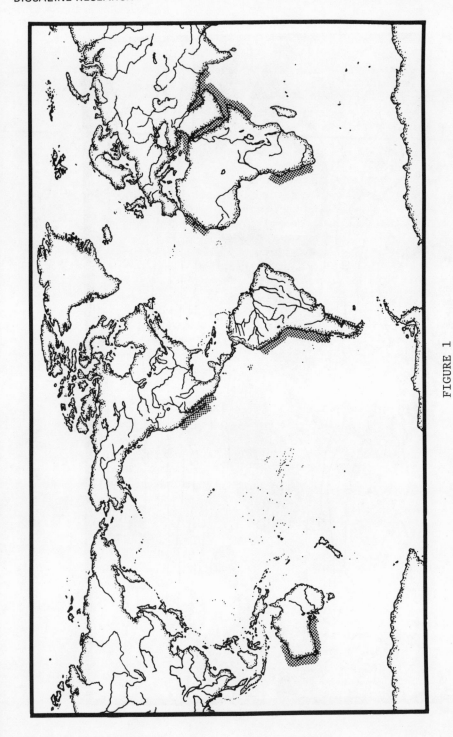

FIGURE 1

Coastal Areas of the World with Low Rainfall

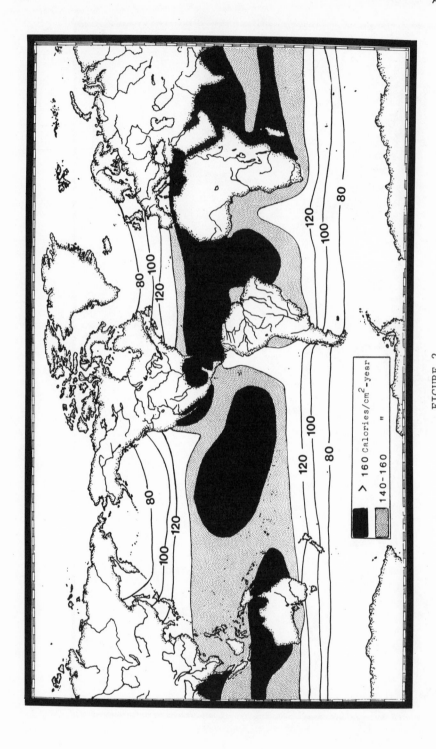

FIGURE 2

Worldwide Distribution of Solar Energy in the Marine Environment

ADVANTAGES OF MARINE RESOURCE UTILIZATION

The factors discussed above indicate a clear mandate for the development of innovative methods for food and feed production. Since 71% of the Earth's surface is covered by ocean, it is reasonable to propose that investigations include marine systems in the search for new food resources and technology. Fisheries will not be discussed, but rather food production systems operating at the primary productivity level. The scope of this paper is to explore the advantages of developing marine food resources using photosynthetic organisms. This would permit the increased exploitation of solar energy. A schematic drawing of the marine system expressed in trophic levels is given in Figure 3. The gains created when bypassing steps in the food chain are enumerated in orders of magnitude.

Physical Aspects

The marine environment offers several unique characteristics when considering food production from the physical point of view. Organisms have a proportionately larger surface area for the absorption of nutrients. The temperature regime is much less prone to strong perturbations, especially to freezing. The flow of water created by tides or currents provides for nutrient input and recycling of metabolic products. In addition, water provides a three dimensional space in which algae grow, and in turn are fed upon by higher trophic levels.

Economic Aspects

Coastal and estuarine areas are largely underexpolited, and as such represent a new resource which can be put into cultivation. It can be seen in Figure 4 that the costal areas of the world are very dissimilar in their offshore nutrient concentrations (represented by productivity measurements). Appropriate cultivation and harvest strategies can be employed based on this knowledge together with other geological, climatological, and vegetation data.

A combined agriculture-mariculture approach will provide a stable and dependable food supply, even in years of adverse weather, such as drought. This will lessen fluctuations in the available food supply.

By establishing coastal marine fish and algal cultures we could expand protein production to levels which would yield a substantial increase in total food supply (see Table 1). For example, the utilization of 10% of the shallow coastal area of El Salvador for fish cultivation could provide for a 50% increase in total protein consumption.

If a similar analysis is extended to the cultivation of algae the resultant increases are even higher. Again, taking the case of

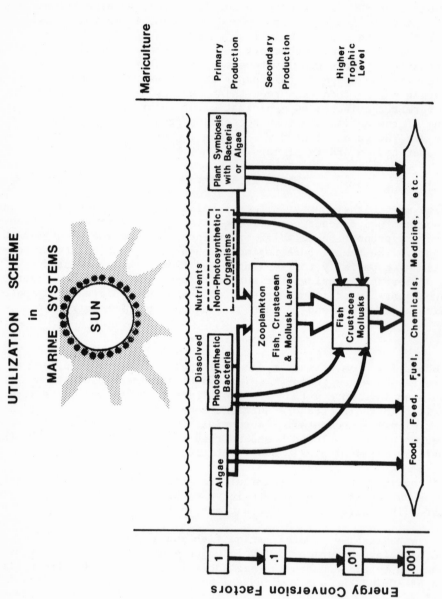

FIGURE 3

Utilization Scheme in Marine Systems

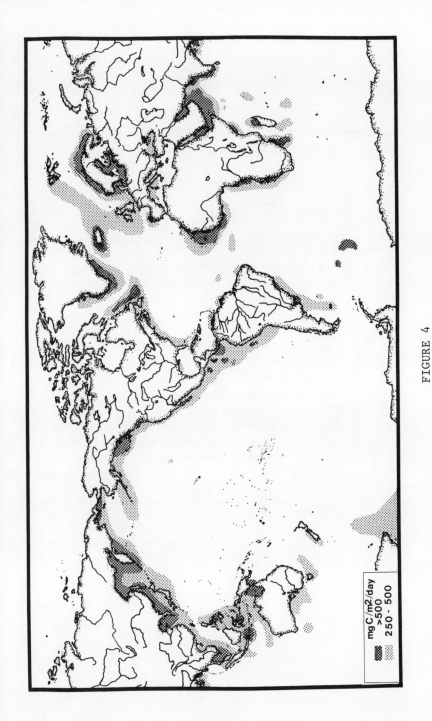

FIGURE 4

Phytoplankton Production (after FAO, 1972, Fisheries Circular No. 126, Rev. 1)

TABLE 1

Algal and Fish Production Potential Based on Utilization of Coastal Area

Country	Population 1975	Total[1] Protein Consumption /Capita/Day (g)	Available[2] Coastal Area (ha)	Potential Culture Production if 10% of Available Coastal Area Is Used				
				Total[3] Fish (tons)	Fish[4] Protein /Capita /Day (g)	OR	Total[5] Algae (tons)	Algal[6] Protein /Capita /Day (g)
Algeria	16,776,000	44.7	777,600	155,520	3.0		1,555,200	25.3
Angola	6,761,000	39.9	777,600	155,520	7.5		1,555,200	63.0
Bangladesh	75,000,000	40.0	1,944,000	388,800	1.7		3,888,000	14.2
Cameroon	6,539,000	58.9	518,400	103,680	5.2		1,036,800	43.4
El Salvador	4,240,000	51.3	583,200	291,600	22.6		1,166,400	75.4
Ecuador	6,733,000	49.0	1,360,800	952,560	46.5		2,721,600	110.7
Ethiopia	27,030,000	68.6	453,600	90,720	1.1		907,200	9.2
Guinea	4,527,000	43.9	583,200	116,640	8.5		1,166,400	70.6
Haiti	5,070,000	38.7	1,296,000	259,200	16.8		2,592,000	140.1
India	610,000,000	52.6	8,164,800	4,082,400	2.2		16,329,500	7.3
Indonesia	138,133,000	42.8	23,587,200	11,793,600	28.1		47,174,400	93.6
Kenya	13,399,000	70.9	1,296,000	259,200	6.4		2,592,000	53.0
Nigeria	72,833,000	59.9	2,073,600	414,720	1.9		4,147,200	15.6
Philippines	41,831,000	44.5	5,832,000	2,916,000	22.9		11,664,000	76.4
Tanzania	15,155,000	42.5	1,555,200	311,040	6.7		3,110,400	56.2
Yeman, Rep.	6,060,000	62.0	103,680	20,736	1.1		207,360	9.4

[1] FAO. 1975

[2] Bell and Canterbery. 1976

[3] Based on technology transfer recommendations by Bell and Canterbery (1976). All countries' production rated at 2 tons/ha/yr except El Salvador, India, Indonesia, and Philippines, which were 5 tons/ha; Ecuador production rated at 7 tons/ha.

[4] All fish protein conversions were 12% of fresh weight.

[5] All algae production rated at 20 tons/ha/yr.

[6] All algal protein conversions were 10% of fresh weight.

El Salvador, the increase in protein available from consumption could
be as high as 150%. For some nations these figures are even higher.
Conversely, countries with less coastal area per capita would not have
such high potentials. These figures are detailed in Table 1. A representative group of coastal countries experiencing protein shortages
has been chosen for evaluation (18). Potential production figures for
these countries assume that further advances in culture technology
will be made.

STATE OF THE ART

Marine photosynthetic resource culture has had a short history
when compared to land agriculture. Most of the technology presently
available emanates from Asian countries. With some exceptions, only
recently have major research efforts been initiated in the Western
Hemisphere.

The Cultivation and Harvesting of Marine Photosynthetic Organisms

The cultivation of marine plants and algae has largely been con-
fined to certain Asian countries. The achievements made by these
countries serve as good examples of the potential harvests to be derived from such mass culturing systems.

Macroalgae - Three classes of macroalgae are cultured, the browns
(Phaeophyceae), the reds (Rhodophyceae), and the greens (Chlorophy-
ceae). The world harvest of each class over the past several years
is given in Table 2. Harvests of natural beds of macroalgae could be
increased on a world basis by severalfold. At the recent FAO technical conference on aquaculture held in Tokyo, it was also estimated
that algae culture would grow at an annual rate of 10% (19). These
increases are attributed to two major factors: increased cultivation
effort and improve cultivation technology.

Red Algae - It is useful to look at the recent advances made in
the culture of the red algae Porphyra in Japan. According to Imada,
Saito, and Teramoto (32), Porphyra tenera production is the most important coastal marine industry in Japan. The increased harvests
over the past 15 years can, to a great extent be attributed to several
biological "breakthroughs" and technological advances. The production
of conchospores through culture of the conchocelis stage has become
possible. The use of nets, cultivation in offshore water by employing
floats, low temperature storage of nurserynets, and the artificial
seeding of conchospores onto the nets have all enhanced production
capability (48).

Today, prefectural and municipal laboratories culture the conchocelis stage in tanks, and for a small fee, farmers dip their culture
nets in these tanks. Thus the conchospores released by the concho-

TABLE 2
WORLD HARVEST OF MARINE PLANTS*
(in metric tons)

Year	Brown Algae	Red Algae	Green Algae	Other	Total
1970	438,000	380,700	1,500	81,200	901,400
1971	483,600	375,500	1,100	80,300	940,500
1972	506,500	322,800	700	89,800	919,800
1973	586,900	450,300	900	95,400	1,133,500
1974	696.567	525,654	2,237	94,834	1,319,297
1975	633,182	422,424	2,487	104,857	1,162,950

*From FAO Yearbook of Fisheries Statistics Vol. 38 and 40

celis stage are seeded directly onto the netting, bypassing the unreliable and much less efficient natural seeding that would occur
along the coast. Almost all Porphyra production today is based on
this artificial production of conchospores (48). The breeding of
Porphyra species is also being studied in order to improve its quality
(49). Production of Porphyra has been hampered because industrial
and municipal pollution is taking its toll (48).

Red algae are also cultured outside of Japan. Korea and the
Republic of China are involved in the culture of Porphyra and are
seeking to expand algal culture to new areas of the Pacific (3). Other
species of red algae are being cultured on an experimental basis in
several areas of the world.

Brown Algae - Brown algae are harvested on the largest scale
world-wide. Since 1970, a 36% increase has been recorded over a four-
year period (Table 2). The major genera of brown algae include Macrocystis (giant kelp), Undaria, and Laminaria. The latter two genera
are cultured in the Orient, again with Japan being responsible for
the greatest total production.

KELP - The culture of kelp is being studied in California by
Wilcox (77), North (54) and their co-workers. Use of oceanic waters,
as opposed to coastal waters, is being investigated for kelp production. The biological problems of seedling production, nutrition, and
environmental requirements have been studied. The idea of bringing
up nutrient rich deep water to the surface culture area (artificial
upwelling) has been successfully applied.

UNDARIA - The macroalgae Undaria represents another success in
the artificial cultivation of a seaweed (61). Young sporophyte plants
are grown to maturity in vats or tanks, often in small operations
quite distant from eventual culture areas. These mature plants release zoospores which in turn are collected on racks strung with
coarse yarn. Once the racks are sufficiently populated with spores,
they are placed in grow-out tanks. After sufficient growth of the
zoospores germling, the racks are removed from the grow-out tanks and
the threads are cut into short pieces. These pieces are then woven
into long lines which are placed out in the open ocean for growth
over the winter months. Disease and fouling are the only major problems encountered by this industry. Genetic crossing experiments to
produce plants more tolerant of warm water are underway, and the hybrids are coming into use.

LAMINARIA - In China, the culture of Laminaria has developed into
a technique rivaling even fish culture. The plants are extremely
sensitive, requiring strict temperature control, nutritional supplements (especially nitrates), and control of light while being cultured
or transported. New genetic work has produced more tolerant species.

The above culturing process requires a large labor input and heavy dosages of nitrogen fertilizers. Less intensive culture of Laminaria occurs in Japan. Seeding methods similar to that described for Undaria are used. Currently, research is being conducted on shortening the tank grow-out time and "planting" the long-lines earlier in the season (28).

Green Algae - Green algae make up only a small portion of the total algal harvest in the world. Korea, Japan, Mexico, and Argentina harvest these algae, usually incidentally to other algal harvests. Enteromorpha and Monostroma are the genera harvested. Culture of Monostroma nitidum is practiced in Japan. It is the most important species after Porphyra in the seaweed cultivation industry there (55). Conditions that are conducive to gamete liberation are being studied. The green alga Ulva is collected in Europe and the Orient, but no culturing of this genus is carried out.

Marine Vascular Plants - Very little information exists on the culture of marine grasses for food and feed. They are considered low in food value.

Marine Microorganisms - The culture of marine microorganisms for direct utilization is a relatively young but expanding effort. Culture work on freshwater species of the green algae Chlorella and Scenedesmus, and the blue-green algae Tolypothrix and Spirulina is much more advanced. The encouraging results with these genera should provide valuable assistance when applied to the culture of marine organisms. Spirulina maxima culture has reached a sophisticated level in the vast pond and processing plant operations in central Mexico (15). This organism is highly nutritious, being 65% protein. The SOSA-TEXCOCO company describes production as averaging better than 10g dry matter/m^2/day, or for the 11 month season, 3 to 4 kg/m^2/year. Chlorella is cultured for human consumption in Japan (74). Their high technology, factory type production, yielding 3 to 8 kg/m^2/year, may eventually bring the costs down. Culture of marine Chlorella is scheduled to begin this year (74). Recently, Chlorella culture has begun in other Asian countries for food production. Both closed and open circulation systems are in use (69).

The growth of microalgae in sewage systems will be covered in the paper to be presented by Dr. Oswald. Dr. Ryther and Dr. Goldman at Woods Hole are also investigating feed production from sewage effluent. Cultivation of microalgae is also practiced by aquaculturists interested in supplying feed to culture animals higher on the food chain. Large scale, highly automated systems of continuous culture and harvest are in wide use. Marine fish and shellfish are also fed microalgae during various stages (i.e. mullet, milkfish, tilapia, penaeid shrimp, blue crab, oysters and mussels) (3).

Many marine species of blue-green algae and photosynthetic

bacteria have been isolated from the tropical environment. These are
being cultured for the study of food and feed value (44).

The Use of Marine Photosynthetic Organisms in Aquaculture

The quantity of algae serving as food sources for fish and shellfish under culture has been estimated to exceed 600 million tons
annually (19). Many organisms, especially mollusks, are efficient at
straining phytoplankton and converting it to higher-quality and more
easily harvestable protein. While harvesting a higher level in the
food chain represents a caloric loss of approximately an order of
magnitude, many situations exist in which the harvest of marine primary productivity is too energetically expensive to be economical.
For this reason, a brief review of the role of marine plants in the
mariculture of fish and shellfish will be given.

The least energy intensive culture method on a protein basis is
carp culture in China. Each level in the water column is occupied by
one of the three cultured species, with plants or zooplankton at each
level serving as the food source. The efficiencies of these ponds
are approached by the milkfish-penaeid shrimp polyculture in the
Philippines. Mats of blue-green algae and diatoms constitute the main
food of the system. Only minor supplementary feedings are necessary
for maintaining either the carp or milkfish ponds.

Two other marine fish species have become objects of intensive
culture. Both the tilapia and the mullet are herbivores. Considerable potential exists in using these species, along with the milkfish,
for expanding coastal aquaculture to underdeveloped nations (5). The
need to exploit natural food chains and recycled nutrients was recognized and practiced in Asia literally thousands of years ago. In the
West, the cultivation of oysters and mussels have evolved to take advantage of the same principles. These organisms live almost exclusively on unicellular algae. Bivalve culture is considered to be perhaps the most successful form of mariculture practiced today (60).

The Supply of Nutrients for Culturing

Primary productivity of the oceans and coastal areas (see Figure
4) may be considered the essential link in developing the biological
resources of these areas. The sun provides the energy for all subsequent levels by means of photosynthesis in marine plants. In polar
and sub-polar regions of the world, the amount of solar energy available is often the limiting factor for primary productivity (see Figure
2). In the tropical and subtropical marine areas, the major rate
limiting factor is nutrient concentrations. The limiting nutrients
are primarily nitrogen and/or phosphorus. Trace elements such as iron
may be limiting to a given species or bloom condition.

Intensive investigations are currently underway on how to supply sufficient nitrogen and phosphorus to marine culture systems deficient in these elements. Other studies are concerned with exploiting those resources such as sewage outfall and upwelling systems that are rich in these nutrients.

Several approaches to supplying nitrogen for plant systems are being investigated. Nitrogen fixation by bacteria and blue-green algae is the most intensively studied. While the vast majority of the work has involved terrestrial systems, a number of papers have been published documenting nitrogen fixation activity in the marine environment. These papers deal largely with rates of productivity and nitrogen assimilation in marine ecosystems.

The emphasis in nitrogen fixation research has been placed on the biochemistry of nitrogenase and associated reactions. Also, genetic aspects of symbiotic relationships between bacteria and host plants have been heavily studied. Other workers have concentrated on the physiology and productivity of nitrogen fixing plants.

In all, the study of biological N_2 fixation and its potential application as a source of nutrients has grown steadily over the past decade. This research is taking place in institutes around the world.

Abiological nitrogen fixation research is also being conducted in several laboratories. The goal of this project is to produce ammonia through a less energy intensive method than the current methods employing the Haber-Bosch process. Supplying nitrogen to the marine environment through artificial upwelling systems is being investigated in California (54, 77) and St. Croix (59b). Such a process·would provide nitrogen, phosphorus, and other trace elements which are required for large-scale culture operations.

The use of urban waste, especially sewage, and agricultural run-off to provide nitrogen and phosphorus are under study at several laboratories. The details of this work will be reviewed by Dr. Oswald in another chapter.

The Problem of Solar Energy Bioconversion Efficiency

The fundamental limiting factors in the solar energy conversion efficiency of photosynthetic organisms are the light absorption and electron transport processes. Based on the two light reaction models for photosynthesis, the maximum attainable efficiency has been calculated to be 12% of the total incident light (4, 38). In addition, the solar conversion efficiency is directly related to the metabolic demands placed on organisms and the nature of the biochemical pathways which handle these needs. As such, the problem of increasing bioconversion efficiency involves a complex of variables associated with photosynthesis and plant metabolism.

On the basis of these considerations recent research into ways of enhancing solar energy conversion efficiency has taken four major directions:

1) Biochemical and physiological regulation of major pathways.

2) Finding ways to reduce photorespiration. The proportion of stored chemical energy lost through photorespiration varies according to the species and the physiological condition of the organisms in question. For many C_3 plants the loss of organic carbon through photorespiration can exceed 50%, whereas photorespiration by C_4 plants is considerably lower (79).

3) Achieving more efficient photosynthetic processes through genetic engineering.

4) Creating cell-free technologies.

None of these four avenues have as yet been extensively explored in marine systems. Therefore, the volume of literature is limited. However, preliminary studies and previous work with land plants (e.g., 9, 47b) indicate hope for future success.

Photorespiration Research - In the area of reducing photorespiration, the suggested methods have included biochemical, environmental and genetic strategies. From a biochemical viewpoint, several researchers have observed that certain chemicals may lower photorespiration in inhibiting the associated pathways (79). From another point of view, it has been suggested that elevated levels of CO_2 may inhibit photorespiration in C_3 plants (4, 7, 79) and thereby enhance net CO_2 fixation. In the long run, however, the most promising alternative appears to be reduction of photorespiration through genetic alteration. It has been suggested that through genetic engineering, photorespiration in C_3 plants could be decreased (79).

Up until recent years, there have been few studies of photorespiration in marine habitats. The work which does exist has been reviewed by Tolbert et al. (72).

Environmental and Physiological Regulation - The problem of solar energy conversion efficiency is not limited to photorespiration. There are many biochemical, physiological and environmental factors (i.e., temperature, salinity, light intensity and quality, pH) which affect the overall efficiency of photosynthesis. These subjects will be discussed in more detail in other chapters of this book.

A considerable body of research exists on the biochemical and physiological regulation of photosyntehtic reactions (4,9,11,13,23,

26,27,30,39,43,44,45,78). Most of this research involves land or freshwater organisms, rather than marine species.

Genetic Techniques - There are many aspects of the solar energy conversion problem which could be effectively approached from a genetic point of view. Research into the alteration of the electron transport system itself, as well as many other crucial metabolic pathways, could lead to breakthroughs in bioconversion technology. Advanced techniques are now being developed in many areas, including: 1) the mutation of higher plant and algal cells, 2) cell or nuclear fusion of different genotypes; 3) DNA transfer and the derepression of genetic loci between species, and 4) plant cell culturing (12). While most of these technologies have not been applied to marine organisms, the potential clearly exists. In fact, the life cycles of many marine algae and higher plants include extensive halpoid stages which could serve as excellent targets for studies of genetic alteration.

Cell-free Research - There is yet another avenue to increasing efficiency: the cell-free approach. It is well known that natural, intact-cell biochemical systems operate under a network of checks and balances. This system helps to maintain organisms within a stage of homeostasis (equilibrium). This precludes excessive production of any single metabolite and thus sets an upper limit to the nitrogen and CO_2 fixation rates (as well as other biochemical products). One of the most efficient ways of removing these restraints would be to isolate the CO_2 and nitrogen fixing systems from their cellular environment. Several laboratories have begun to study the technologies necessary for such cell-free applications (10,11,23,47a), but the practical applications of these methods still appear to be far off in the future.

The Utilization of Marine Photosynthetic Products as Food and Feed

The origins of land agriculture had their basis in the survey and collection of native plants which were palatable and nutritious. By comparison, the survey and collection of marine algae and plants as foodstuffs have received only cursory examination.

For many years algal and marine plant products have received regionally limited use as food items in Asia, Northern Europe, the Pacific Islands, and in some areas of South America. Macroalgae such as Porphyra and Laminaria have a high commerical value in Japan and nearby regions. As discussed above, considerable effort has been devoted to the cultivation of these algae as a food source, especially in Asian countries such as Japan. In the western world brown algae have traditionally been valued as fertilizers and animal feed (33).

Over the past three decades, the cultivation of several types of unicellular green and blue-green algae as a protein source has been undertaken. Fresh water and brackish species of the photosynthetic microorganisms Chlorella, Scenedesmus, Spirulina, and Rhodopseudomonas have also been cultured for use as a protein source in aquaculture, feed additives, and food additives (15,36,37,59a,70, 71,74).

Chemical Composition - In searching for marine photosynthetic organisms which could be cultured as a food source, it will first be necessary to survey a large number of species in order to find those which exhibit rapid and stable growth patterns. Next, chemical analyses of the selected species will have to be performed to predict their value as nutritional resources. These general analyses of protein, carbohydrate, lipid, and mineral content will have to be followed by more detailed analyses of the composition of major components. Species can then be chosen which will give high yields of the desired products, whether they be protein, lipid, carbohydrates, vitamins and minerals which are comparable to those of conventional food sources.

Protein - Numerous dietary factors are necessary for optimal human nutrition. The demand for these factors is constantly growing as the world's population increases. There is no doubt that one of the main limiting nutritional factors is the availability of protein. Several marine photosynthetic organisms are known to have a high protein content.

MACROALGAE - Seaweeds contain from 5% to as high as 59% crude protein on a dry weight basis (2). Among edible seaweeds, the red algae Porphyra has been found to have a protein content of 35%. The brown algae Undaria and Laminaria contain 12% and 6.2% protein respectively (53). Edible algal species from Japan and Iceland have been found to contain considerable quantities of the essential amino acids (2,51). For example, protein from the red algae Chondrus crispus has been demonstrated to have a nutritional value comparable to that of egg albumin (42).

PHOTOSYNTHETIC MICROORGANISMS - Protein content of photosynthetic microorganisms (unicellular algae and photosynthetic bacteria) appears to range from 10% to 65% on a dry weight basis. The green microalgae Chlorella and Scenedesmus are known to contain about 55% crude protein. These have been mass cultured and used both as a food additive and as a fish feed (70,71,74). The blue-green algae Spirulina, which has also been grown in mass culture, contains about 65% protein. Species of photosynthetic bacteria involved in sewage treatment have been found to have a crude protein content of 60%. It has been suggested that these bacteria have considerable potential as a fish feed (36,37).

Feeding experiments with a variety of animals have shown that green microalgal protein is of high nutritional quality. For example, the nutritional value of Scenedesmus protein is also comparable to that of egg white, or skim milk (20,70).

MARINE PLANTS - The protein content of marine seagrasses range from 10 to 20% of dry weight. However, much less is known about marine seagrasses than some species of algae because of the limited use as food or feed in the past.

Carbohydrate - The carbohydrate content of various algae ranges from 10% to 65% of the dry weight (76). Cellulose, mannans, xylans, alginic acid, porphyran, carrageenan, galactan, agar, and other carbohydrate substances have been found in algal cell walls or mucilages (40). In addition, compounds such as polyglucans, fructosans, inulin, trehalose floridoside, mannoglyceric acid, sucrose and polyhydroxy alcohols are found as storage products in algae (80). In some cases, algal carbohydrates consist largely of complex polysaccharides which cannot be digested by humans. This may limit the potential usefulness of some species as a direct source of caloric energy for man.

LIPIDS - Brown algae have been reported to contain 0.16%-6.3% lipid on a dry weight basis, while most red algae contain 0.4%-3.2% lipid(42). These lipid levels are relatively low in comparison with those of green microalgae. A crude fat content of 8-30% has been reported for unicellular green algae (76). This lipid content may become even higher under certain culture conditions.

Lipids from the blue-green algae Spirulina have been characterized and found to consist of unsaturated fatty acids. Gamma linolenic acid, one of the principle lipid compounds, comprised about 20% of the total fat content (15).

VITAMINS AND MINERALS - Many algae contain large quantities of vitamin A, the B group vitamins (B_1, B_2, B_{12}, niacin, pantothenic acid, and folic acid) and vitamins C and D. In certain species of seaweeds and unicellular algae, vitamin content is as rich and varied as that of meat (34,35,42).

Numerous minerals are also provided by algae. Sodium, potassium, calcium, magnesium, and iron may constitute up to 15-25% by dry weight in certain algae (42,53). These inorganic components, play an important role in preventing blood acidosis (66).

Other Nutritional Aspects - After the selection of organisms by predictive chemical analysis, the essential amino acid composition and nutritional value of the proteins must be studied in greater detail. These studies should include determination of the protein efficiency ratio (PER), net protein ratio (NPR), net protein utilization (NPU), and nitrogen balance (57,64).

At this stage, there are two possible methods of enhancing protein quality. One way is to genetically modify the algae so that it produces greater quantity and higher quality protein. The other way is to improve protein quality by the later addition of limiting essential amino acids. All of these subjects have been extensively studied in land agriculture. Much of this technology could be more or less directly applied to marine products.

Acceptability of Marine Algae and Plants as Food - Macroalgae are already commonly eaten in some regions of the world. They are used in several forms, including nori (dried Porphyra), salad vegetables, cooking vegetables, and in soups (53,75). Unicellular photosynthetic microorganisms may also be used in dried or powdered form as a food or feed additive to improve dietary quality. Cultivation programs will surely lead to an increased use of algal food in countries where the consumption of seaweeds is common. However, populations unaccustomed to the use of algae as food may be reluctant to accept it at first. Therefore, once safe and nutritious algae have been cultured, we must study the effect of factors such as odor, flavor, texture, color, and stability during storage. If an algae is unacceptable because of one or more of these factors, several measures may be taken. One possible remedy is to determine the identity of the compounds responsible for the poor acceptability, and to remove them during processing. Another method is to reduce the level of the offending compound, by controlling the biochemical and physiological conditions under which the algae is cultured or harvested. Yet another approach would be to develop a more acceptable species by means of genetic manipulation.

If an algae supplies large quantities of protein, but is not acceptable for direct consumption as food, the protein may be extracted and used as an additive to improve the nutritional quality of other foods. This technique could also be used with other algal products. By mixing algal nutrients with other nutrients, it may be possible to develop new types of food, such as algal "meat." Such new foods could be developed by the application of existing food technology. For example, present technology already includes sophisticated methods for protein concentration by extraction, food texturing, and flavoring. In addition, the use of plastin reactions to make high quality protein could serve as a source of nutrition for special dietary needs (21).

Toxic Substances and Marine Foods - More than a century of technological advancement in the United States has clearly shown that technological progress does not always go hand in hand with the well-being of man and the biosphere. It behooves the developers of new marine resources to take this fact into consideration and to make a concerted effort to avoid unnecessary environmental disruptions. It is, of course, inevitable that the ecosystem will be altered some-

what through the development of new large-scale food technologies.
Careful examination of potential problem areas should minimize detrimental change.

In selecting algae for use as food and animal feed, it is important that they should be screened for toxin production. Several
species of microalgae produce substances which are poisonous to
animals and humans. Among these are the dinoflagellates Gonyaulax
and Gymnodinium, both of which produce several potent toxins. Toxin
production by blue-green algae has also been documented. The blue-
green algal species Microcystis aeruginosa, Anabaena flos-aquae,
Aphanizomenon flos-aquae, and Schizothrix calcicola all produce toxic
substances (63). Several other seaweeds and unicellular algae have
been reported to be toxic to fish and other experimental animals (29).

Additional Considerations - Ideally, an algal food or feed product should be stable during long-term storage. In order to minimize
the deterioration of algal products during storage, studies in this
area will have to be carried out. Existing food technology can also
be applied to preserve algal products.

FUTURE DIRECTIONS IN USING THE SEA FOR FOOD AND FEED PRODUCTION

In the introduction of this paper it was emphasized that the
contemporary system of world agricultural production and distribution could not sustain the dietery needs of tomorrow's world population. This means that major changes will have to be made in the present attitudes towards food and feed production. In order for such
changes to be effective they will have to include the following
elements:

1) An increase in the productivity of existing land crops -
   probably through the increased use of irrigation, fertilizer, and the development of innovative cultivation technology,

2) More efficient redistribution of food and feed between productive and unproductive areas of the world, and

3) The development of new technologies in food and feed production.

It is within the last category "the development of new technology" that the greatest hope lies for meeting the needs of the
future.

Since 71% of the earth's surface is covered by seawater it seems
entirely reasonable to propose that any program for the expansion of
food and feed technology must include a major effort in the direction

of marine resource utilization.

At the present time fisheries play the dominant role in our harvest from the sea. While the improvement of this industry is essential the future outlook, in terms of productivity, is largely limited (50). The emphasis of future research in fisheries will probably shift towards conservation of existing stocks and the reduction of operating costs.

One of the major courses of marine resource development would seem to be in the direction of utilizing organisms low in the trophic structure, i.e., primary (marine photosynthetic organisms) and secondary (zooplankton). The chief advantage to this approach is that the yield per unit of incident solar energy is much higher at this level. This stems from the fact that the passing of energy from one trophic level to another normally operates at only a 10% level of efficiency (Figure 2).

The technology available for using photosynthetic marine organisms in food and feed production is still, for the most part, either experimental in nature or limited in scope. However, as can be seen from the previous section on the state of the art, there are areas of work which exhibit considerable potential for success. The problem which confronts us now is the establishment of a set of priorities and goals for future research and development.

The Survey of Organisms with Utilization Potential

As in the case of conventional agriculture, the key to successfully developing new food or feed production methodologies is finding organisms which exhibit the desired characteristics. This means that one of the research priorities should be a survey of marine photosynthetic organisms. Such a survey should include consideration of all the factors which could affect the eventual success or failure of the venture, as illustrated in Figure 5. These factors include: 1) adaptability for mass culture, 2) nutritional value, 3) marketability, 4) environmental hazards, and 5) economic aspects.

Adaptability for Mass Culture - The first element in such a survey would be testing organisms for their adaptability to artificial mass culture conditions. The future success of the "food from the sea" effort will depend heavily on our ability to find organisms which can be established in stable and highly productive ocean farms.

In addition to productivity, these organisms will have to be selected for their tolerance to environmental stress and for their suitability to the regions of the world in which they will be farmed.

Nutritional Value - The organisms selected for productivity and adaptability must also be tested for nutritional value and market-

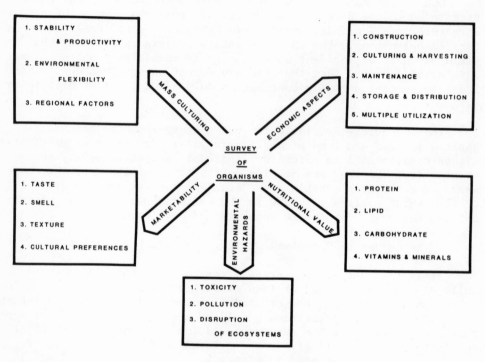

FIGURE 5

Important Areas of Study in the Survey and Application of Marine
Organisms for Food and Feed Production

ability. For testing food value, a rather sophisticated technology
is already available, courtesy of the land based food industry.

   · Marketability - When it comes to marketability, there are numerous special problems to be dealt with. Many marine plants have tex-
tures, smells and tastes unfamiliar and often distasteful to human
consumers. It will be necessary to develop new ways of processing
marine products for the market place.

   Environmental Hazards - An even more serious problem is the
question of toxicity. Relatively little is known about the biological hazards involved in the consumption of many potential marine
products. Before a product is marketed it will be of paramount importance to assure its safety. The question of safety not only includes the toxicity of organisms consumed by man, but also the harm
done to the environment in general, whether it be from biofouling or
the disruption of natural ecosystems.

   Economic Aspects - The final task in this survey will be the
careful consideration of economic limitations. These factors cannot
be underestimated since they can be the downfall of a project irrespective of its scientific merit.

   One consideration which warrants careful study is regional differences in the need for supplementary food technology and the ability to pay for it. It is unfortunate, though understandable, that
many of the nations which seem to suffer from the greatest agricultural inadequacies, also lack the economic strength to actively
search for solutions. At the same time, however, many of these nations do have the natural coastline or estuarine areas necessary to
support shallow marine farming projects (see Table 1). It seems
reasonable to suggest that industrially advanced nations could assist
in the development of marine farming methods suitable for such developing countries (31). The subtropical and tropical regions of
the world have the greatest need for such assistance. Therefore,
future research efforts should be keyed to these areas. Meanwhile,
the potential for developing temperate and cold water areas should
not be ignored.

   Along the lines of developing new technologies, the most important economic factor is the high expense of constructing and maintaining oceanic farms. This has led to pessimism amongst some researchers over the future of such ventures. It must be remembered,
however, that land farming enterprises have often encountered and
successfully overcome such barriers. In order to resolve these problems considerable effort and money will have to be devoted to research on ways to minimize construction and maintenance costs of
saltwater utilization projects, while at the same time maximizing
profit.

Profit is important since the eventual cooperation of private enterprise in these ventures is highly desirable if not indispensable. In many cases the farming of marine plant species for one purpose, such as protein production, will not be profitable enough to justify its existence. It will be necessary to implement a program for the multiple utilization of harvested materials. Many marine plant species would probably lend themselves to the production of several economically valuable substances. Some raw plant materials could be subdivided and refined to yield food, feed, medicine, and fuel (through such processes as fermentation) (Figure 6). These types of multiple utilization procedures have already been successfully applied with several land plant species. There is no reason why they could not be applied to marine systems. This multiple utilization potential includes:

Hydrogen gas photoproduction (45) – The capability of many photosynthetic organisms to produce hydrogen gas has been recognized for many years (22,43,45). The process itself is directly or indirectly linked to the light dependent photosynthetic pathway. However, the solar energy conversion efficiency of the process has always been relatively low. This low efficiency, and the fact that hydrogen production is often sensitive to oxygen inhibition, have largely precluded the use of this theoretical pathway in applied research.

Recent experimental results indicate that there may be hope for resolving some of these problems and thereby enhancing the future application of hydrogen photoproduction. Active research in this area may result in the development of innovative and valuable sources of fuel.

There are numerous advantages to the biological approach to hydrogen photoproduction as opposed to thermo-chemical and physical methods:

1) The biological system could be operated at low (physiological) temperatures (i.e., 10-40°C) as opposed to the high temperatures required for chemical or physical production of hydrogen (400-1000°K).

2) The only major input into the system would be solar energy and a hydrogen donor, probably water (salt water).

3) The production of hydrogen would not involve the formation of pollutants, as in the case of fossil fuel refineries.

Applications to aquaculture (44) – One of the factors which determines the success or failure of ventures in aquaculture is the proper choice of primary food sources. Green algae and diatoms have already been successfully employed, as the base of the food chain,

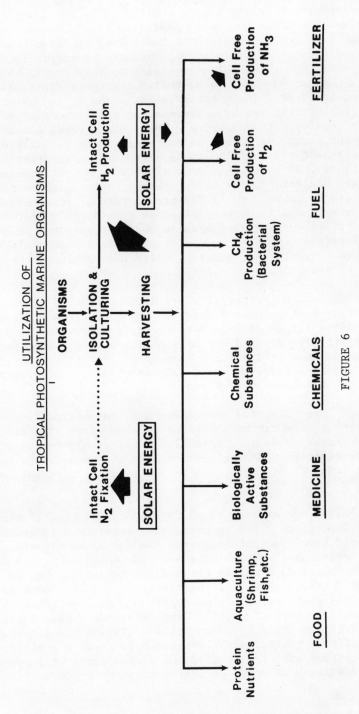

FIGURE 6

Multiple Utilization of Marine Photosynthetic Organisms (after 44, 45)

to feed zooplankton, which in turn are utilized by the larval and
adult stages of the organisms being exploited (i.e., crustaceans,
shellfish and fish). Dr. Kobayashi's laboratory in Japan has successfully used freshwater or soil photosynthetic bacteria for aquaculture purposes. However, cultured marine photosynthetic bacteria
and blue-green algae have yet to be tested for their applicability
as food in aquaculture. The question might arise as to why marine
photosynthetic bacteria and blue-green algae should be used instead
of presently developed food sources. There are several advantages
which favor the former as a food resource: 1) The cell walls of
photosynthetic bacteria and blue-green algae are much softer and
easier to digest than those of green algal species, 2) Some marine
photosynthetic bacteria and blue-green algae have a high nutritional
value as mentioned above, and 3) The growth rate of some marine photosynthetic bacteria and blue-green algae is considerably faster than
many of the presently used food organisms. Mass cultures of H_2-pro-
ducing algal strains could also provide food for the aquaculture
of shrimp, crabs, shellfish, and fish (either directly or through
the culture of zooplankton).

Methane production - Several years ago, Oswald and his co-
workers reported that methane production using sewage and algae
might be economically feasible. More recently Wilcox, North and
their colleagues have studied the possibility of using kelp as a
substrate for methane production. Some other photosynthetic strains
may prove to be an economical source of carbohydrate material for
bacteria-mediated methane production.

Medicine - The surveying and research of metabolically active
substances produced by marine algae has received some attention in
the past few years. However, it remains a relatively little-studied
area for investigation. New metabolically active substances may
therefore be found in many of these as yet relatively unexplored
species.

Several algae are already known to produce substances of potential medical importance. The anthelmintics L-kainic acid and L-
allokainic acid have been obtained from the red algae Digenia simplex
(52); domoic acid has also been obtained from Chondria armata (14,
68). Laminine, obtained from the brown algae Laminaria, has been
shown to be effective as an antihypertensive (56,67). There are
also indications that sodium alginate prevents gastroenteric absorption of radioactive strontium (65). The compounds sodium laminarin
sulfate and fucoidin could be potentially useful as blood coagulants
(76). In addition, several algae are known to produce antibiotic
substances. Survey and further study of marine algae will probably
lead to the discovery of other medically useful compounds.

Chemicals - There is an enormous range of chemical substances
produced by algae and marine plants which are of interest to man.
In a project in which there is a potential for developing systems
which will continuously produce large quantities of material, it is
important to explore ways of exploiting this resource from a chemical
point of view.

Algal polysaccharides such as agar, carrageenan, and furcelleran
from red algae and algin from brown algae are important algal products presently in commercial use. These substances are used in
numerous industries as emulsifying agents, gelling agents, stabiliz-
ers, suspension agents, and thickeners. Cultivation of algae to
provide these substances could be combined with their utilization
for other purposes. For example, mass production of Macrocystis
(Ocean Farm Project) will supply algin as the main product and also
fertilizer and feed as byproducts (41,77).

Sugar alcohols represent another potentially important algal
product. The accumulation of these compounds appears to be a common
feature among algae. Some brown algae are known to accumulate mannitol to levels as high as 50% of the dry weight (80). Halophilic
species of the unicellular green algae Dunaliella have been found to
store glycerol under conditions of high salinity (25). When cells
of this algae are suspended in 1.5 M NaCl solution, an intracellular
glycerol concentration of 2 M can be reached (6,8).

Relatively few surveys of the chemical and biological properties
of marine photosynthetic microorganisms have been conducted. Further
studies on the chemical composition, biochemistry, and physiology of
these organisms will undoubtedly reveal new chemical resources of industrial value.

The Enhancement of Production Capability

The survey for organisms with economic potential and the subsequent development of basic technologies for their application is only
the beginning of a successful saltwater utilization program. As in
the gradual evolution of land agriculture, research in ocean farming
will have to involve work on (Figure 7):

1) Improving solar energy conversion efficiency of the organ-
    isms being utilized,

2) Improving the culturing and harvesting technology,

3) Developing new genetic strains of organisms with more favor-
    able characteristics,

4) Developing renewable sources of nutrients.

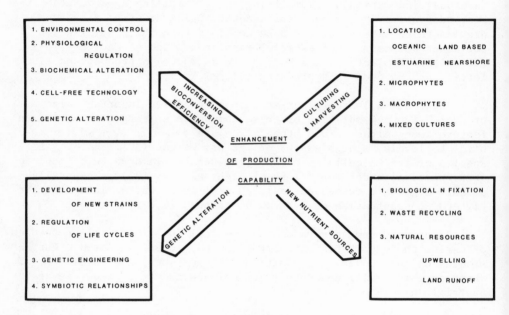

FIGURE 7

Technologies Involved in the Enhancement of Marine Food and
Feed Production Capabilities

Increasing Solar Energy Conversion Efficiency - As discussed in
the "State of the Art" section, research groups around the world
have begun studying many of these problems. In the study of solar
energy conversion efficiency, work is being done on the physiological,
environmental, and biochemical regulation of whole living cells (e.g.
47b). There still remains a large gap in our understanding of the
conditions which regulate the productivity and conversion efficiency
in most marine photosynthetic species. This understanding is fundamental in outlining the course of applied research.

Considerably more work must also be done on the elucidation of
metabolic pathways in photosynthetic marine organisms and the development of methods for their regulation. These areas include photorespiration, membrane transport, uptake of assimilates, dark respiration, electron transport chains, CO_2 fixation and many other metabolically important functions.

In addition, a technology for cell free hydrogen production
could be developed (46). A similar approach could be used for the
cell free nitrogen fixation process (45). Although long term and
continuous research efforts will be required for the development of
these systems, they could provide a new method for the production
of both fuel and food.

Improving Culturing and Harvesting Technology - The potential
success of many future ocean farming projects will be dependent on
the development of economically feasible culturing and harvesting
methods. This fact has been clearly demonstrated and discussed by
several research groups. In the case of photosynthetic microorganisms, harvesting technology may be crucial for their successful
utilization. If economic methods for harvesting phytoplankton could
be invented, they could open up many new avenues for marine resource
utilization, not only in saltwater farms, but in the collection of
natural populations of phytoplankton.

Genetic Alteration - The search for new genetic strains of organisms which exhibit better characteristics for farming has long
been a part of land agriculture. The use of similar techniques in
marine organisms is as yet practically non-existent. As ocean farming grows in importance so undoubtedly must this approach. At the
present time the most active research going on is in the area of
mutational studies with microorganisms. In addition, attempts are
being made to transfer useful gene complexes, such as N_2 fixation
genes, from one species to another. These studies warrant full
support, but there are other areas which remain largely untouched.
Among these are genetic studies with marine macrophytes and the regulation of life cycles. The great success of genetic techniques
in land plant development indicates that similar success with marine
plants could be forthcoming.

Developing Renewable Sources of Nutrients - The final issue of
immediate importance is nutrient availability. There is little doubt
that the achievement of high productivity in ocean farms will demand
considerable input of supplementary nutrients, especially nitrogen,
and to a lesser degree, phosphorus. In the natural environment these
two compounds are utilized by photosynthetic organisms in a 15:1
ration (N:P), and are considered growth limiting.

Several laboratories have already begun to tackle the problem
of extra nutrient supply. Two examples are the artificial upwelling
experiments in the growth of kelp and the use of microorganismal
sewage recycling. As discussed previously, land crop production is
now almost totally dependent on industrially produced nitrogen supplements. Unfortunately, the cost of nitrogen from this source is
escalating rapidly, due to the shortage of fuel. Ocean farming would
be hard pressed to attain economic stability if it had to depend on
industrial fertilizer. The answer may come from the development of
biological means for N_2 fixation, from upwelling research, from
sewage and waste recycling technology, nutrient concentration from
land runoff, or perhaps from genetic alteration. Whatever the source,
it must be renewable and free from the need for fossil fuel input.

International Program for Research Coordination

The achievement of the goals outlined above will require the
implementation of a well organized and extensive program of research.
This task is complicated by the fact the program must be international in scope. An efficient way to deal with these problems would
be to establish several international research centers at various
key locations throughout the world. Such institutes would give researchers from various disciplines an opportunity to work with close
communication and cooperation towards a common goal. Communication
is of primary importance to this entire program. The transfer of information is still one of the major problems in scientific research.
The creation of major institutes and the regular use of workshops
and meetings would greatly reduce such barriers.

Another problem which could be greatly helped by the creation
of central institutes is that of unnecessary duplication, both in
effort and instrumentation. The formation of central institutes,
where major facilities (e.g., computers, instrumentation, etc.)
could be shared, would reduce costs.

The design of such institutes would have to fill the needs of
many areas of research. Each institute should be capable of solving
a wide range of problems with an interdisciplinary strategy. Figure
8 illustrates the form which such an institute could take. Some of
the essential components of such an international marine food and
feed resource development institute would be:

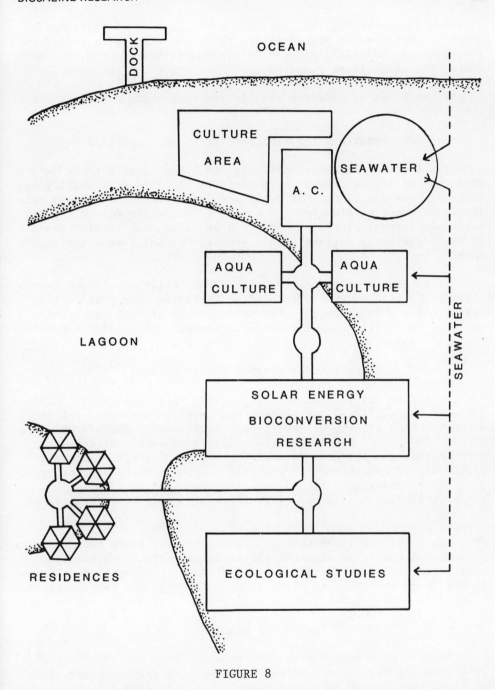

FIGURE 8

Future International Institute for the Development of Bio-Saline
Food and Feed Technology

1.   Structurally, the institute would consist of three major
parts; a solar energy bioconversion research facility, a center for
the study of ecological aspects of bioconversion research, and a
center for aquaculture research,

2.   A fleet of research vessels for open ocean and nearshore
work, and

3.   Accommodations for visiting researchers and students.

The location of these institutes would in large part be determined by regional differences in environment, land availability and
economic considerations.   It would seem most reasonable if separate
institutes were erected for the study of cold water temperate, and
subtropical-tropical utilization.   In addition, the special needs
of the "developing nations" would seem to warrant a separate institute.

It is important to point out that the formation of such institutes would not negate the role of universities, and private laboratories in this research.   The primary role of such institutes
would be to act as a focal point for research and a guide for international policy.

International Program for Education

As marine utilization research grows, there will be an increasing need for qualified and experienced research personnel.   This
means that any program of development will have to include an educational component.   The institutes outlined above could act as the
core of this educational system.   For the sake of efficiency this
system should concentrate on training at the graduate and post graduate level, and should include considerable practical as well as
academic education.   In addition, the system must be international
in scope.

The goal of this educational component must be to equip persons
from around the world with the skills to: 1) direct and maintain
marine food and feed operations, and 2) develop innovative solutions
to the needs and problems faced by both advanced and developing nations.

Funding of Research

Funding of research of marine food and feed development must
come from all levels of government, as well as private enterprise.
At the highest level an international fund should be set up primarily
to aid countries whose economies cannot support major research and
development efforts.   The financial basis for the major institutes
describes above will most likely have to be provided by the govern-

ments of more affluent nations. At the same time, state and local governments should assist in the procurement of land for research facilities and for the support of studies of primarily local interest.

The contribution of private enterprise to the funding effort will also play a vital role in the rapid development of technology. The most substantial function of private funding will probably be in the area of improving upon, expanding, and marketing the projects which have demonstrated potential for success at the research level.

Most of the projects discussed above are presently at the early stage of development. The funding of research in this area must involve a commitment to long-term support. This stable base of support will ensure the proper development of new technology, and will help to guarantee that the final products of this research will meet up to our needs and standards.

ACKNOWLEDGMENTS

The author would like to thank Mr. E. Duerr, Mr. S. Kumazawa, Mr. E. Phlips, Ms. J. Radway, Mr. J. Richard, Ms. D. Rosner, and Mr. S. Trine for the preparation of the manuscript.

REFERENCES

1. Andersen, K., K.T. Shanmugam, and R.C. Valentine. 1977. Genetic derepression of nitrogenase-mediated H_2 evolution by Klebsiella pneumoniae. p. 339-346. In: Biological solar energy conversion. (ed. by Mitsui, A. et al.) Academic Press, New York.

2. Arasaki, T. and N. Mino. 1973. Alkali-soluble proteins in marine algae. Eiyo To Shokuryo 26(2):129-133.

3. Bardach, J.E., J.H. Ryther, and W.O. McLarney. 1972. Aquaculture. Wiley Interscience, New York.

4. Bassham, J.A. 1977. Synthesis of organic compounds from carbon dioxide in land plants. p. 151-165. In: Biological solar energy conversion. (ed. by Mitsui, A. et al.) Academic Press, New York.

5. Bell, F.W. and E.R. Canterbery. 1976. Aquaculture for the developing countries. Ballinger Publ. Co., Cambridge, MA.

6. Ben-Amotz, A. and M. Avron. 1973. The role of glycerol in the osmotic regulation of the halophilic alga Dunaliella parva. Plant Physiol. 51: 875-878.

7. Black, Jr., C.C. 1973. Photosynthetic carbon fixation in relation to net CO_2 uptake. Ann. Rev. Plant Physiol. 24: 253-286.

8. Borowitzka, L.J. and A.D. Brown. 1974. The salt relation of
    marine and halophilic species of the unicellular green alga
    Dunaliella. The role of glycerol as a compatible solute. Arch.
    Microbiol. 96:37-52.

9. Brown, A.W.A., T.C. Byerly, M. Gibbs, and A. San Pietro. (eds.)
    1975. Crop Productivity-Research Imperatives. Michigan Agri.
    Res. Sta. and C.F. Kettering Res. Fdn. and NSF-RANN

10. Butler, W.L. 1973. p. 68-69. In: Proc. of the workshop on bio-
    solar conversion (ed. by Gibbs, M. et al.) NSF-RANN.

11. Calvin, M. 1976. Photosynthesis as a resource for energy and
    materials. Photochem. Photobiol. 23:425-444.

12. Carlson, P.S. and J.C. Polacco. 1975. Plant cell cultures:
    genetic aspects of crop improvement. Science 188:622-625.

13. Cooper, J.P. (ed.). 1973. Photosynthesis and productivity in
    different environments. Cambridge Univ. Press, Cambridge, MA.

14. Daigo, K. 1959. Studies on the constituents of Chondria armata.
    J. Pharm. Soc. Japan 79:350-360.

15. Durand-Chastel, H. and M.D. Silve. 1977. The Spirulina algae.
    Euro. Sem. Biol. Solar Energy Conversion Systm., Grenoble-
    Autrans p. 1-15.

16. Ehrlich, P.R. and A.H. Ehrlich. 1970. Population resources en-
    vironment. W.H. Freeman and Co., San Francisco.

17. FAO. 1975a. Yearbook of fisheries statistics. Food and Agricul-
    ture Organization, Rome. vol. 40.

18. FAO. 1975b. The state of food and agriculture 1974. Food and
    Agriculture Organization, Rome.

19. FAO. 1976. Culture of algae and seaweeds. Food and Agriculture
    Organization, Rome. Fish. Rep. 188:34-35.

20. Fowden, L. 1962. Amino acids and proteins. p. 189-209. In: Physi-
    ology and biochemistry of algae (ed. by Lewin, R.A.) Academic
    Press, New York.

21. Fujimaki, M., S. Arai, and M. Yamashita. 1975. Food protein im-
    provement and plastin reaction. (In Japanese) Protein, Nucleic
    Acid and Enzyem 20:927-935.

22. Gest, H. and M.D. Kamen. 1949. Photoproduction of molecular hy-
    drogen by Rhodospirillum rubrum. Science 109:558.

23. Gibbs, M., A. Hollaender, B. Kok, L.O. Krampitz, and A. San Pietro (editors). 1973. Proc. of the workshop on bio-solar energy conversion. NSF-RANN.

24. Goldman, J.C. and J.H. Ryther. 1977. Mass production of algae: bioengineering aspects. pp. 376-378. In: Biological solar energy conversion. (ed. by Mitsui, A. et al.) Academic Press, New York.

25. Graigie, J.S. and J. McLachlan. 1964. Glycerol as a photosynthetic product in Dunaliella tertiolecta Butcher. Can. J. Bot. 42: 777-778.

26. Hall, D.O. 1976. Photobiological energy conversion. FEBS letters 64:6-16.

27. Hardy, R.W.F. and U.D. Havelka. 1977. Food and feed by legumes: CO_2 and N_2 fixation, foliar fertilization, and assimilate partitioning. P. 299-322. In: Biological solar energy conversion. (ed. by Mitsui, A. et al.) Academic Press, New York.

28. Hasegawa, Y. 1971. Forced cultivation of Laminaria. p. 391-393. In: Proc. 7th Int. Seaweed Symp. (ed. by Nisizawa, K. et al.) John Wiley & Sons, New York.

29. Hashimoto, Y., N. Fusetani, and K. Nozawa. 1971. Screening of the toxic algae on coral reefs. p.569-572. In: Proc. 7th Int. Seaweed Symp. (ed. by Nisizawa, K. et al.) John Wiley & Sons, New York.

30. Hollaender, A., K.J. Monty, R.M. Pearlson, F. Schmidt-Bleek, W.T. Snyder, and E. Volkin (eds.). 1972. An inquiry into biological energy conversion. Gatlinburg. NSF-RANN.

31. Horstmann, U. 1977. Application of solar bioconversion in developing countries. p. 477-436. In: Biological solar energy conversion. (ed. by Mitsui, A. et al.) Academic Press, New York.

32. Imada, O., Y. Saito, and K. Teramoto. 1971. Artificial culture of laver. p. 358-363. In: Proc. 7th Int. Seaweed Symp. (ed. by Nisizawa, K. et al.) John Wiley & Sons, New York.

33. Jensen, A. 1971. The nutritive value of seaweed meal for domestic animals. p. 7-14. In: Proc. 7th Inst. Seaweed Symp. (ed. by Nisizawa, K. et al.) John Wiley & Sons, New York.

34. Kanazawa, A. and D. Kakimoto. 1958. Studies on the vitamins of seaweeds. 1. Folic acid and folinic acid. Bull. Jap. Soc. Sci. Fish. 24:573-577.

35. Kanazawa, A. 1961. Studies on the vitamin B-complex in marine algae. 1. On vitamin contents. Mem. Fac. Fish. Kagashima Univ. 10:38-69.

36. Kobayashi, M. et al. 1969. Sewage purification by photosynthetic bacteria and its use as a fish feed. Bull. Jap. Soc. Sci. Fish. 35:1021-1026.

37. Kobayashi, M., K. Mochida, and A. Okuda. 1967. The amino acid composition of photosynthetic bacterial cells. Bull. Jap. Soc. Sci. Fish. 33:657-660.

38. Kok, B. 1973. Photosynthesis. p. 22. In: Proc. of the workshop on bio-solar conversion (ed. by Gibbs, M. et al.) NSF-RANN.

39. Kok, B., C.F. Fowler, H.H. Hardt, and R.J. Radmer. 1976. Biological solar energy conversion: approaches to overcome yield stability and product limitations. p. 53-54. In: Enzyme technology and renewable resources (ed. by Gainer, J.L.) Univ. of Virginia and NSF-RANN.

40. Kreger, D.R. 1962. Cell walls. p. 315-335. In: Physiology and biochemistry of algae. (ed. by Lewin, R.A.) Academic Press, New York.

41. Leese, T.M. 1975. Ocean food and energy farm kelp product conversion. (manuscript) Presented to 141st Ann. Meeting Amer. Assoc. Adv. Sci., New York.

42. Levring, T., H.A. Hoppe, O.J. Schmid (eds.). 1969. Marine algae. Cram, de Gruyter and Co., Hamburg.

43. Lien, S. and A. San Pietro. 1975. An inquiry into biophotolysis of water to produce hydrogen. Indiana Univ. and NSF.

44. Mitsui, A. 1975. Multiple utilization of tropical and subtropical marine photosynthetic organisms. p. 13-29. In: Proc. 3rd Int. Ocean Dev. Conf. Seino Printing Co., Tokyo. Vol. 3.

45. Mitsui, A. and S. Kumazawa. 1977. Hydrogen production by marine photosynthetic organisms as a potential energy source. p. 23-51. In: Biological solar energy conversion. (ed. by Mitsui, A. et al.) Academic Press, New York.

46. Mitsui, A. 1976. Long-range concepts: Application of photosynthetic hydrogen production and nitrogen fixation research. p. 653-672. In: Proc. Conf. on Capturing the Sun through Bioconversion. Washington Center for Metropolitan Studies.

47a.Mitsui, A. 1975. The utilization of solar energy for hydrogen production by cell free system of photosynthetic organisms. p. 309-316. In: Hydrogen Energy, Part A. (ed. by Veziroglu, T.N.) Plenum Publ. Co., New York.

47b.Mitsui, A., S. Miyachi, A. San Pietro, and S. Tamura (eds.).
    1977. Biological solar energy conversion. Academic Press, New
    York.

48. Miura, A. 1975a. Porphyra cultivation in Japan. p. 273-304 In:
    Advance of phycology in Japan. (ed. by Tokida, J. and H. Hirose)
    Dr. W. Funk Publ., The Hague.

49. Miura, A. 1975b. Studies on the breeding of cultured Porphyra
    (Rhodophyceae). p. 81-93. In: 3rd Int. Ocean Dev. Conf. Seino
    Printing Co., Tokyo. Vol. 3

50. Moiseev, P.A. 1975. Biological resources of the world ocean.
    p. 53-66. In: 3rd Int. Ocean Dev. Conf. Seino Printing Co.,
    Tokyo. Vol. 3.

51. Munda, I.M. and F. Gubensek. 1976. The amino acid composition
    of some common marine algae from Iceland. Bot. Mar. 19:85-92.

52. Murakami, S., T. Takemoto, and Z. Shimzu. 1953. Studies on the
    effective principles of Digenea simplex. J. Pharm. Soc. Japan
    73: 1026-1029.

53. Naylor, J. 1976. Production, trade and utilization of seaweeds
    and seaweed products. FAO Fish. Tech. Paper No. 159.

54. North, W.J. 1977. Possibilities of biomass from the ocean: the
    marine farm project. p. 347-361. In: Biological solar energy
    conversion. (ed. by Mitsui, A. et al.) Academic Press, New York.

55. Ohno, M. 1971. The periodicity of gamete liberation in Monostroma.
    p. 405-409. In: Prox. 7th Int. Seaweed Symp. John Wiley and Sons,
    New York.

56. Ozawa, H., Y, Gomi, and I. Otsuki. 1967 Pharmacological studies
    on laminine monocitrate. J. Pharm. Soc. Japan 87:935-939.

57. Pearson, W.H. and W.J. Darby. 1961. Protein nutrition. Ann. Rev.
    Biochem. 30:325-346.

58. Pimentel, D., W. Dritschild, J. Krummel, and J. Kutzman. 1975.
    Energy and land constraints in food protein production. Science
    190:754-761.

59a.Pirie, N.W. 1975. The Spirulina algae. p. 33-39. In: Food protein
    sources. (ed. by Pirie, N.W.) Cambridge Univ. Press, Cambridge.

59b.Roels, O.A., K.C. Haines, and J.B. Sunderlin. 1975. The potential
    yield of artificial upwelling mariculture. 10th Eur. Symp. Mar.
    Biol. 1:381-390.

60. Ryther, J.H. 1975. Mariculture: how much protein and for whom? Oceanus 18(2):10-22.

61. Saito, Y. 1975. Undaria. p. 304-320. In: Advance of phycology in Japan. (ed. by Tokida, J. and H. Hirose) Dr. W. Funk Publ. Co., The Hague.

62. Sanchez, P.A. and S.W. Buol. 1975. Soils of the tropics and the world food crisis. Science 188:598-604.

63. Schantz, E.J. 1970. Algal toxins. p. 83-96. In: Properties and products of algae. (ed. by Zaijic, J.E.), Plenum Press, New York.

64. Scrimshaw, N.S., G. Arroyave, and R. Bressani. 1958. Nutrition. Ann. Rev. Biochem. 27:403-426.

65. Skoryna, S.C., K.C. Hong, and Y. Tanaka. 1971. The effects of enzymatic degradation products of alginates on intestinal absorption of radiostrontium. p. 605-607. In: Proc. 7th Int. Seaweed Symp. (ed. by Nisizawa, K. et al.), John Wiley and Sons, New York.

66. Takagi, M. 1975. Seaweeds as medicine. p. 321-325. In: Advance of phycology in Japan. (Ed. by Tokida, J. and H. Hirose) Dr. W. Funk Publ., The Hague.

67. Takemoto, T., K. Daigo, and N. Takagi. 1964. Studies on the hypotensive constituents of marine algae. J. Pharm. Soc. Japan 84:1176-1182.

68. Takemoto, T., K. Daigo, Y, Kondo, and Y, Kondo. 1966. Studies on the constituents of Chondria armata. 8. On the structure of domoic acid. J. Pharm. Soc. Japan 86:874-877.

69. Tamiya, H. 1955. Growing Chlorella for food and feed. Proc. World Symp. Appl. Solar Energy. Phoenix.

70. Tamiya, H. 1957. Mass culture of algae. Ann. Rev. Plant Physiol. 8:309-334.

71. Tamiya, H. 1975. Green micro-algae. p. 35-39. In: Food protein sources. (ed. by Pirie, N.W.) Cambridge Univ. Press, Cambridge.

72. Tolbert, N.E. and C.B. Osmond (eds.). 1976. Photorespiration in marine plants. Univ. Park Press, Baltimore.

73. Tolbert, N.E. 1977. Regulation of products of photosynthesis by photorespiration and reduction of carbon. p. 243-264. In: Biological solar energy conversion. (ed. by Mitsui, A. et al.) Academic Press, New York.

74. Tsukada, O., T. Kawahara, and S. Miyachi. 1977. Mass culture of
    Chlorella in Asian countries. p. 363-365. *In*: Biological solar
    energy conversion. (ed. by Mitsui, A. *et al*.) Academic Press,
    New York.

75. Velasquez, G.T. 1971. Studies and utilization of the Philippine
    marine algae. p. 62-65. *In*: Proc. 7th Int. Seaweed Symp. (ed. by
    Nisizawa, K. *et al*.). John Wiley & Sons, New York.

76. Volesky, B., J.E. Zajic, and E. Knettig. 1970. Algal products.
    p. 49-82. *In*: Properties and products of algae. (ed. by Zajic,
    J.E.) Plenum Press, New York.

77. Wilcox, H.A. 1975. The ocean food and energy farm project. p. 43-
    52. *In*: Proc. 3rd Int. Ocean Dev. Conf. Seino Printing Co., Tokyo
    Vol. 3.

78. Zelitch, I. 1975. Improving the efficiency of photosynthesis.
    Science 188:626-632.

79. Zelitch, I., D.J. Oliver, and M.B. Berlyn. 1977. Increasing
    photosynthetic carbon dioxide fixation by the biochemical and
    genetic regulation of photorespiration. p. 231-242. *In*: Biologi-
    cal solar energy conversion. (ed. by Mitsui, A. *et al*.) Academic
    Press, New York.

80. Meeuse, B.J.D. 1962. Storage products. p.289-313. *In*: Physiology
    biochemistry of algae. (ed. by Lewin, R.A.) Academic Press, New
    York.

SALT TOLERANCE IN MICROORGANISMS

Janos K. Lanyi

NASA-Ames Research Center

Moffett Field, CA 94035

SUMMARY

Extremely and moderately halophilic bacteria may be considered models for biological salt tolerance. These organisms have evolved in saline environments and are able to overcome the deleterious effects of salt, up to saturating concentrations. Their intracellular components: enzymes, ribosomes, membranes, etc. have been modified in a variety of ways, which provide the physical chemical basis for the salt tolerance. An excess of acidic residues and a deficiency of hydrophobic residues in the proteins from extremely halophilic bacteria cause these structures to be stabilized by high concentrations of salt. Many proteins and the ribosomes show marked preference for KCl over NaCl. This is consistent with the fact that the halophiles accumulate K^+ and exclude Na^+. The gradients are produced by coupling cation transport to the protonmotive force generated by various membraneous systems. The regulation, as well as the energization, of Na^+ transport by the gradient of protons proceeds so as to suggest that maintaining low internal Na^+ concentration is a desirable goal for the cells. It has not been established, however, whether the preference exhibited by the enzymes, etc. for KCl is a result of the limitations of salt-polyelectrolyte interaction, or a consequence of adaptation to preexisting intracellular conditions.

INTRODUCTION

Highly saline bodies of water often contain microorganisms well adapted to growing at high salt concentrations. Among these are the extremely halophilic bacteria (e.g. the Halobacteriacea) which grow and survive only in brines containing 18-35% NaCl (1), and the moderately halophilic bacteria (e.g. <u>Paracoccus</u> <u>halodenitrificans</u>, <u>Vibrio</u> <u>costicola</u>, etc.) which tolerate NaCl concentrations up to 20%. Of these the halobacteria are of special interest since a considerable amount of information is available about their evolutionary adaptation to salt. The halobacteria are Gram-negative rods, and have a characteristic red appearance due to carotenoids which confer photoprotection against intense sun-light. These bacteria are obligately aerobic, and their primary source of metabolic energy is oxidative phosphorylation. They possess also a unique light-energy transducing system not related to chlorophyll, but one which utilizes membranes containing a purple pigment called bacteriorhodopsin (2-4).

It has been known for many years that a) the intracellular salt concentration in the halobacteria is high enough to be in osmotic balance with the external medium, and b) the intracellular K content is much greater and the Na content smaller than in the growth medium. The first of these postulates implies that the cellular components of the halobacteria must be able to function in the presence of high concentrations of salt, a condition normally deleterious to enzymes, ribosomes, membranes, etc. Indeed, it was found that such cellular structures not only tolerate salt, but function optimally only at several molar NaCl, or preferably KCl (5,6). Furthermore, the enzymes of extremely halophilic bacteria are unstable at low salt concentrations, and some become inactivated within a few seconds unless protected by salt. These salt-dependent properties must reflect physical chemical features which are unique to the proteins of extreme halophiles, and to some extent to those of moderate halophiles as well.

The second postulate, that large gradients of K^+ and Na^+ exist across the cell membranes of the halobacteria (7), implies that considerable energy is expended by the organisms for cation transport in producing and maintaining the gradients. These gradients are found, to be sure, in most living cells, but nowhere are they created by such massive ion fluxes as in the halophiles. It has been argued that the cation gradients reflect the need to keep the intracellular Na^+ concentration at a low value. The reason for this might be that even salt-tolerant proteins seem to be sensitive to NaCl at high concentrations, but not to KCl. If so, the salt tolerance necessarily requires the powerful apparatus of excluding Na^+ and accumulating K^+ that the extreme halophiles possess.

The extremely halophilic bacteria are the best example of adaptation to saline conditions. The existence and properties of these organisms show that concentrated salt solutions are not necessarily inimical to life. The understanding of the mechanisms by which these bacteria deal with the problems presented by saline conditions should contribute to the assessment of the possibilities and the limitations of salt tolerance in higher organisms.

SALT DEPENDENT PROPERTIES OF THE INTRACELLULAR COMPONENTS OF HALO-PHILES

Virtually all biological structures in aqueous solutions will respond to the presence or absence of salt. At low concentrations of salt the response reflects the fact that the structures are invariably polyelectrolytes, and the effect of salt is primarily on electrostatic attraction or repulsion among charged residues. These electrostatic effects may have direct functional roles in substrate binding and catalysis by enzymes, etc., or secondary roles in shifting the pH of ionizable groups or stabilizing specific confirmations or subunit interactions. At high salt concentrations the response of biological structures reflects their capacity for hydrophobic interaction. In the presence of several molar salting-out type salt (e.g. NaCl or KCl) these interactions manifest themselves in aggregation and/or conformational changes resulting in a more collapsed structure. Most components of non-halophilic organisms show optimal functional properties between 0.1 and 0.3 M ionic strength, reflecting activation and inhibition by salt through the two kinds of effects listed above. The salt-dependencies of the proteins, ribosomes and membranes of halophilic bacteria are undoubtedly based on the same principles, but one or another of the features responsible for salt-response must be exaggerated. What follows is a description of the salt effects on these halophilic cellular components, together with some discussion of what their physical chemical bases might be. The arguments advanced would be much more convincing if direct evidence for structural changes at different salt concentrations were available. Lacking such experimental evidence in most cases, the ideas which follow tend to rely on extrapolations from the known effects of salt on non-halophilic macromolecules.

A survey (6) of the enzymes of halophilic bacteria shows that they fall into three categories with respect to salt dependence: a) those for which the salt requirement for enzyme activity is absolute, and which exhibit little or no activity in the absence of salt, b) those which are merely stimulated by salt, showing optimal activity at 1 to 3 M salt, and c) those which are inhibited by salt. The latter are few, and none of them are greatly inhibited even in saturated salt solutions. All of the enzymes studied are unstable at lowered salt concentrations.

For some halophilic enzymes the effects of salt on activity and stability are separable: protection against spontaneous (thermal) inactivation continues until saturating salt concentration, while maximal enzymes activity is obtained at an intermediate concentration. Examples for this type of salt-dependence are the malic de-

hydrogenase of Halobacterium salinarium (8), isocitrate dehydrogenase
(9, 10) and threonine deaminase (11) of H. cutirubrum. The first
two of these enzymes have been shown to unfold during incubation at
decreased salt concentrations, as determined by sedimentation velocity (8,10), gel permeation chromatography (9) and circular dichroism
measurements (12). Protection against such inactivation was obtained
in the presence of the substrates of the enzymes. Slow addition of
salt (dialysis against several molar NaCl) reactivated the enzymes.
The difference between the optimal salt concentration for enzyme
activity in these systems (about 1M) and for maximal stability (up
to 4M) illustrates the difference between specific and non-specific
effects of salt. The former is presumably the consequent of a distinct manner of polypeptide chain folding, the salt effects arising
from the senitivity of various residue interactions to the presence
of cations, while the latter must be due to the collapse of the protein at high salt concentration and the general resistance of the
closely packed polypeptide chains to unfolding by thermal motion.

The situation is much simpler with the membrane-bound NADH-menadione oxidoreductase from H. cutirubrum. As shown in Figure 1, in
the case of this enzyme the salt-dependence of enzyme activity and
the first-order rate constant of thermal inactivation are mirror
images (13), suggesting that an equilibrium exists between the active,
native form of the enzyme and an inactive, unstable form. The effect
of salt is to shift this equilibrium toward the native form of the
enzyme. The chemical nature of such kinetic entities is at present
uncertain, but obviously holds the key to the understanding of the
nature of the salt-response. New techniques of protein isolation
and purification in the presence of salt (14) will make direct physical chemical examination of the enzyme possible. Some insight was
gained already, however, from studying the effects of different
salts and the thermodynamics of the inactivation process, as well as
by the use of specific protein denaturants with well-known effects
on model compounds. These results suggested (15) that in fact three
forms of the enzyme exist, with different kinetic properties: a) a
highly unstable, inactive form in the absence of salt, b) a more
stable form with partial activity, obtained from the first one at
a few tenths molar NaCl or KCl or a few hundredths molar $MgCl_2$ (due
to charge screening effects), and c) a stable form with full activity,
obtained from the second at several molar NaCl or KCl (due to stabilization by salting-out effects). A schematic representation of this
model for the salt dependence of NADH-menadione oxidoreductase is
shown in Figure 2. Separation of charge-screening and salting-out
(hydrophobic) effects, on the basis of the salt concentration and
the type of salt required, has been made also for cytochrome oxis-
dase (16) and citrate synthase (17) of H. cutiribrum, for malate dehydrogenase of H. marismortui (18), and for alkaline phosphatase
of Vibrio alginolyticus (19), among others.

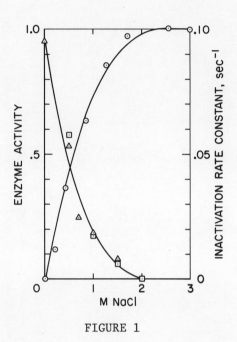

FIGURE 1

Salt dependence of NADH-menadione oxidoreductase from H. cutirubrum. Enzyme activity and stability was determined at different NaCl
concentrations. Symbols: enzyme activity, (☉); apparent first-order
inactivation rate constant calculated from declining enzyme activity,
before (▣) and after (▲) restoring the NaCl concentration to 2M. Reprinted with permission from J.K. Lanyi (1969) J.Biol.Chem.244, 4168-
4173.

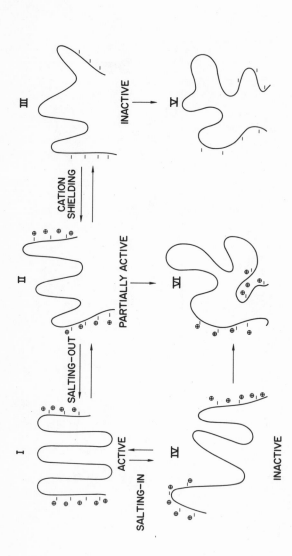

FIGURE 2

Schematic representation of the possible effects of salt on NADH-menadione oxidoreductase from H. cutirubrum. The enzyme is assumed to be optimally folded in State I, which is stabilized by both salting-out (hydrophobic) and charge-screening effects. State I is in equilibrium with State II, which represents a partially active, unstable species, with decreased affinity for menadione. State II, which lacks the salt-dependent hydrophobic folding, is in equilibrium with the inactive State III. The transition III → II is encouraged at lower concentrations (< 1M) of monovalent cations or millimolar concentrations of divalent cations, which will accomplish the screening of negative charges. The transition II → I is encouraged at higher concentrations of salting-out type salts, such as NaCl or KCl, which stabilize weak hydrophobic bonds. Salting-in type salts, such as NaBr, NaNO3 or NaSCN, encourage the formation of State IV, which is in-active and contains less hydrophobic bonding than State II. Reprinted with permission from J.K. Lanyi (1974) Bacteriol. Reviews 38, 272-290.

Although the properties of enzymes will, in general, reflect the amino acid sequences of specific regions of their polypeptide chains, some conclusions may be drawn from the overall amino acid composition of halophilic proteins. The amino acid compositions of these proteins do seem to reflect their unusual physical chemical properties: the proteins tend to contain more acidic amino acids than basic ones (i.e. glutamic and aspartic acid vs. lysine, argi- nine and histidine), and fewer hydrophobic amino acids (i.e. leucine, isoleucine, valine, tryptophan, tyrosine, etc.) than is usual in non-halophilic proteins. A representative group for statistical analysis might be the ribosomal proteins, which were compared for H. cutirubrum and Escherichia coli (Figure 3). The large excess of acidic groups observed (an average of 20 mol%) certainly accounts for the need for charge screening, since these proteins carry a large negative charge. The deficiency in non-polar amino acids, which is statistically significant relative to the group of proteins from E. coli (Figure 3), might be sufficient to destabilize the ter- tiary folding of the halophilic proteins. If so, high concentrations of salt would encourage hydrophobic interactions among residues otherwise not well suited for this purpose (e.g. serine, threonine, alanine, the peptide backbone, etc.), and thereby contribute to the stability of the folding.

Larger cellular structures, such as the ribosome of H. cutiru- brum show the normal 70S particle size only in 3 to 4 M KCl (20), and disassociate at lower salt concentration or aggregate non-speci- fially in NaCl. Since, in contrast with proteins from non-halophilic ribosomes, the proteins from these ribosomes are mostly acidic (Fig- ure 3 and refs. 20-22), the principles which govern the salt-depen- dent assembly of these structures must be similar to those discussed for the simpler case of enzymes. This argument has been applied to the cytoplasmic membranes of extremely halophilic bacteria as well (23), which contain only acidic lipids (24) and acidic proteins (6, 25). The disintegration of these membranes at lowered salt concen- trations proceeded in such a way as to suggest the sequential dis- assembly of the structure (26). The retention of those proteins which are lost above 1M salt could be insured only by adding salting- out type salts, i.e. KCl or NaCl but not $NaNO_3$ or NaSCN. In con- trast, any of the salts tested was able to prevent the disintegra- tion of the membrane below 1M salt. The specificity for different salts above 1M, and the lack of specificity below 1M suggested that both hydrophobic stabilization and charge-screening are involved in the effect of salt on these membranes. The membranes of extreme halo- philes contain very unusual lipids, derived from dihydrophytol (24). Some attempts were made (27,28) to explore the consequences of the presence of these branched lipid chains in the non-polar phase of the membrane bilayer. The structural instability found in this phase, which is associated with the chain-branching, may be one of the rea- sons for the salt requirement, similarly to the case of proteins, where the instability due to the less extensive internal hydrophobic

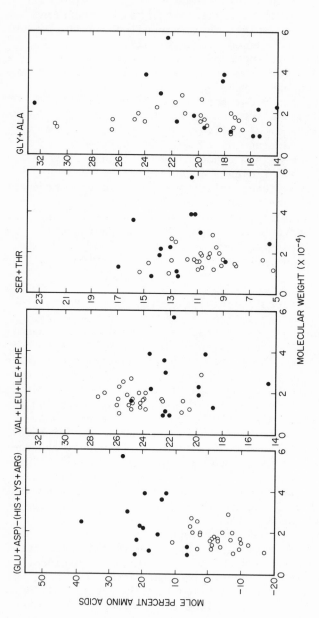

FIGURE 3

Comparison of the amino acid compositions of a group of ribosomal proteins from E. Coli (⊙) and H. cutirubrum (●). From left to right: frequency of acidic minus basic amino acids (glu + asp minus his + lys + arg); frequency of strongly hydrophobic amino acids (val + leu + ile + phe); frequency of weakly hydrophobic amino acids (ser + thre); and frequency of indifferent (short-chain) amino acids (gly + ala). Statistical analysis of the data indicates that the proteins from the extreme halophile contain significantly more acidic groups (probability of co-incidence for means < 1%) than the proteins from E. coli, significantly less strongly hydrophobic groups (probability of coincidence < 1%), significantly more weakly hydrophobic groups (probability of coincidence < 5%), but not significantly different amounts of short-chain groups. Reprinted with permission from J.K. Lanyi (1974) Bacteriol. Reviews 38, 272-290.

bonding is overcome at higher salt concentrations.

MAINTENANCE OF INTRACELLULAR Na^+/K^+ BALANCE

Evidence has been presented that the maintenance of high internal K^+ and low Na^+ concentration in the halophiles will circumvent some of the deleterious effects of Na^+ on intracellular components (29). As discussed above, the preference of ribosomes for
KCl over NaCl is well established. In addition, many halophilic
enzymes are inhibited to some extent as high concentrations of NaCl
but not KCl (6,29). Such inhibitory effects would multiply along
metabolic pathways, and thus multienzyme systems could show marked
preference for K^+.

Given the high concentration of salt in the growth medium,
the production of sodium and potassium gradients across the cytoplasmic membrane of halophiles must involve large cation fluxes.
In principle, the energy for cation transport may be supplied by
phosphate-bond energy (ATP hydrolysis), or by protonmotive force
across the cell membrane (30). The latter possibility has been investigated using cell envelope vesicles, prepared from H. halobium
cells by mechanical breakage. These vesicles are analogous to the
intact cells, but carry out no metabolism since they lack the cytoplasmic contents (31). The vesicles can be energized by either
adding artificial electron donors, such as dimethylphenylene diamine
(32), or illumination (31,33). The latter is possible because of
the presence of bacteriorhodopsin, a light-activated pump for protons (2), in the membranes of the halobacteria. In both cases protons are ejected from the vesicles, giving rise to pH difference
(interior alkaline) and an electrical potential difference (interior
negative) across the membrane. According to the principles of chemi-
osmotic energy coupling, these gradients together constitute the
"protonomotive force", which will yield metabolically useable energy
if the appropriate energy-transducing membrane components for coupling proton flux to other processes are present. Various lines of
evidence have suggested that halobacterial membranes contain a pro-
ton/sodium exchange system, which acts as such a transducing component (34-36). The exchange system, or "antiporter" appears to couple
the transmembrane movements of H^+ and Na^+ in such a way that a single
cycle of translocation involves 2 H^+ for a Na^+. The translocation
thus includes net charge movement, and the antiporter is driven by
pH difference and electrical potential. A scheme for these ion movements in H. halobium cell membranes is given in Figure 4.

Illumination-induced Na^+ efflux is rapid (34-36), and is accompanied by K^+ uptake (37, 38). K^+ accumulation is apparently driven
by the electrical potential, although its mechanism is somewhat un-

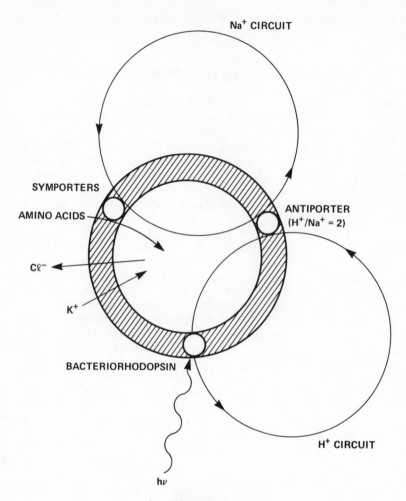

FIGURE 4

Scheme of ion transport in H. halobium cell envelope vesicles. The extrusion of protons is facilitated by bacteriorhodopsin, which is activated by illumination, or by the respiratory chain (not shown). The circuit of protons is completed via a proton/sodium antiporter, which couples the influx of two H^+ to the efflux of one Na^+. Thus, the recirculation of protons results in the removal of Na^+ from the interior. The electrical charge imbalance due to the loss of Na^+ is compensated by either K^+ influx or Cl^- efflux, depending on conditions. In the former case the internal Na^+ is replaced by K^+, in the latter the vesicles lose NaCl and water, and collapse. The concentration gradient of Na^+, in turn, drives the active transport of amino acids via symport systems, which couple the influx of amino acids to that of Na^+.

certain. Recent results indicate that the K^+ uptake in H. halobium
envelope vesicles (38), although not in intact cells (39), is much
slower than the Na^+ efflux. The K^+ influx in the vesicles is linearly dependent on the external K^+ concentration, suggesting that any
specific K^+ permeation mechanism existing in the cells had been lost.
The net result is that at low external K^+ concentrations the Na^+ efflux is accompanied mainly by Cl^- efflux, resulting in the osmotic
loss of water and the collapse of the vesicles, and at high external
K^+ concentrations (above 1 M) or in the presence of the K^+ ionophore,
valinomycin, the Na^+ efflux is accompanied mainly by K^+ influx, resulting in maintenance of volume and the exchange of internal Na^+
for K^+. A suitable control of K^+ permeability in intact H. halobium
cells would allow these organisms to chose between K^+ influx and Cl^-
efflux. No such regulation of K^+ transport has been, in fact observed, although there is some evidence that the cells also show
light-induced volume decrease.

The rate of Na^+ extrusion from H. halobium envelope vesicles is
rapidly increased at proton gradients between -130 and -155 mV (40).
Above this threshold value the rate of Na^+ transport is a steep
linear function of the protonmotive force. Below the threshold the
transport is much slower. The molecular basis for this "gating"
effect is not yet clear, but its physiological consequences must be
far-reaching. It is known (41-43) that anaerobic H. halobium cells
in the dark are poised at or just below the threshold of the K^+
gradient found for Na^+ transport (41-43). Thus, respiration or illumination should result in the rapid depletion of the cells of Na^+,
but when the conditions become less favorable for energy transduction
the process is not reversed and the low internal Na^+ concentration
is maintained.

The internal Na^+ concentration, which will be sustained in cell
envelope vesicles by a given size H^+ gradient, is determined by the
stoichiometry of the antiporter, which was found to be $2H^+/Na^+$ (40).
At saturating light-intensities this results in internal Na^+ concentrations of 3-4 millimolar, while the internal K^+ and the external Na^+
concentrations are kept near 3M. In contrast with these results,
the situation in intact H. halobium cells is not well understood,
and variable internal Na^+ concentrations, up to 1 M, have been reported under different conditions (39, 44, 45).

Evidence has been available for several years that sodium transport is important also in non-halophilic bacteria, and that it also
proceeds by proton/sodium exchange (46, 47). Thus, it appears that
in this respect the difference between salt-sensitive and salt-
tolerant organisms is a quantitative one: the cell membrane of halophilic bacteria can handle cation fluxes which are orders of magnitude greater than in non-halophilic bacteria. The moderate halophiles fall somewhere between the two extremes. In one of these,

$\underline{V}$. $\underline{costicola}$, the internal Na^+/K^+ ratio is also shown to be energy-dependent (48).

THE CASE FOR Na^+ TOXICITY AND K^+ TOLERANCE

From the foregoing it should be evident that most, if not all, of the intracellular components of extreme halophiles have evolved to the point where they can function at very high concentrations of salt. If the salt could be NaCl the need for the elaborate and energy-consuming mechanism for Na^+ removal from the cells would be eliminated. Indeed, it has been argued (49) that the proton/sodium antiport and the K^+ transport system in bacteria and eucaryotes did not arise from the need to keep the internal Na^+ concentration low, but in order to regulate the pH difference across the cytoplasmic membrane and to generate concentration gradients for Na^+ and K^+, which represent long-term energy reservoirs of high capacity. It is in favor of this view that in many bacteria, and particularly in the extreme halophiles (Figure 4), the active transport of at least some of the metabolites is coupled to the concentration gradient of Na^+ (50-54). On the other hand, as tabulated by Lanyi (6) and later elaborated by Brown (29) most enzymes, and certainly the ribosomes, from the extremely halophilic bacteria show marked preference for K^+ relative to Na^+. Thus, it appears that in these cells optimal growth requires that the internal Na^+ be largely replaced by K^+. It may be that there is a limit for molecular adaptation to highly saline conditions, and for reasons which must be rather subtle, even highly salt-tolerant macromolecules will function better in KCl than in NaCl. If so, the K^+ preference of these structures is inherent in the way they interact with cations, rather than a result of adaptation to preexisting conditions. While this issue is not yet resolved, the salt-tolerance of the halophilic bacteria, which evolved to live in saturated brines, is clearly the result of both physiological adaptation (i.e. maintenance of K^+-rich, Na^+-poor internal millieu, favorable for enzymes) and molecular adaptation (i.e. modification of macromolecules and other structures to avoid the deleterious effects associated with salt, even KCl). However, the amount of cellular Na^+ in the moderate halophile, $\underline{V}$. $\underline{costicula}$, was found to be greater than the amount of K^+ at external salt concentrations above 1 M (48). If these cells indeed contain high internal Na^+ concentrations, the argument of Na^+ toxicity is not universally applicable.

Salt tolerance in moderate halophiles is not always associated with Na^+ transport alone. Avi-Dor and coworkers have shown that the response of halotolerant bacterium to increasing NaCl concentration was dependent on the presence of extracellular choline or betaine (55-57). These substances conferred increased salt tolerance on the cells, and were accumulated against high concentration gradients, up to nearly 1M internal concentration, to extents which were dependent on the external NaCl concentration. Although this might be considered simply a means of osmotic regulation, it was found that the rate of respiration increased greatly in the presence of betaine, independently of the transport. Thus, betaine increased the amount of energy available for these cells, and this may have accounted in part for their increased ability to deal with higher salt concentrations.

EVOLUTIONARY CONSIDERATIONS

It might be thought that the phenotypic characteristics of extremely halophilic bacteria are sufficiently unique to offer the possibility of tracing the evolutionary path of these organisms. A beginning toward this goal has been made, although the results have little to do with the acquisition of salt tolerance. Moore and McCarthy (58) tested nucleic acid sequence homologies by hybridization, and found no detectable relationship between extreme and moderate halophiles from DNA sequences. The distinct evolutionary origins of the extreme halophiles have been recently confirmed by comparison of partial sequences from ribosomal RNA (59), which showed that the halobacteria are quite different from all other forms of life except the methanogens (which are obligately anaerobic and not particularly halophilic bacteria), and it was suggested that the halophiles and the methanogens comprise a third kingdom, named archaebacteria (60). The halobacteria, as the methanogens, lack the peptidoglycan cell wall usually found in bacteria (1). Another approach to the study of evolutionary relationships is to compare well-defined proteins from different species. Ferredoxins isolated from H. halobium (61) and H. marismortui (62) show similarities to both plant and bacterial ferredoxins, but are immunologically quite distinct. These are very interesting and relevant findings, and they indicate that it may be a fallacy to ascribe all of the phenotypic variation in the halobacteria to adaptive responses to saline conditions.

REFERENCES

1. Larsen, H. (1962) in "The Bacteria" (Gunsalus, I.C. and Stanier, R.Y., eds.) vol. 4, pp. 297-342, Academic Press, New York.

2. Oesterhelt, D. and Stoeckenius, W. (1973) Proc. Nat. Acad. Sci. U.S.A. 70, 2853-2857.

3. Danon, A. and Stoeckenius, W. (1974) Proc. Nat. Acad. Sci. U.S.A. 71, 1234-1238.

4. Lanyi, J.K. (1979) in "Membrane Proteins in Energy Transduction" (Capaldi, R.A., ed.) Marcel Dekker, New York, pp. 451-483.

5. Larsen, H. (1967) in "Advan. Microbial Physiol." (Rose, A.H. and Wilkinson, J.R., eds.) vol. 1, pp. 97-132, Academic Press, New York.

6. Lanyi, J.K. (1974) Bacteriol. Reviews 38, 272-290.

7. Christian, J.H.B. and Waltho, J.A. (1962) Biochim. Biophys. Acta 65, 506-508.

8. Holmes, P.K. and Halvorson, H.O. (1965) J. Bacteriol. 90, 316-326.

9. Hubbard, J.S. and Miller, A.B. (1969) J. Bacteriol. 99, 161-168.

10. Hubbard, J.S. and Miller, A.B. (1970) J. Bacteriol. 102, 677-681.

11. Liebermann, M.M. and Lanyi, J.K. (1973) Biochemistry 11, 211-216.

12. Wulff, K., Hubbard, J.S. and Miller, A.B. (1972) Arch. Biochem. Biophs. 148, 318-319.

13. Lanyi, J.K. (1969) J. Biol. Chem. 244, 4168-4173.

14. Mevarech, M., Eisenberg, H. and Neumann, E. (1977) Biochemistry 16, 3781-3786.

15. Lanyi, J.K. and Stevenson, J. (1970) J. Biol. Chem. 245, 4074-4080.

16. Liebermann, M.M. and Lanyi, J.K. (1971) Biochim. Biophys. Acta 245, 21-33.

17. Higa, A. and Cazzulo, J.J. (1975) Biochem. J. 147, 267-274.

18. Mevarech, M. and Neumann, E. (1977) Biochemistry 17, 3786-3792.

19. Hayashi, M., Unemoto, T. and Hayashi, M. (1973) Biochim. Biophys. Acta 315, 83-93.

20. Bayley, S.T. and Kushner, D.J. (1964) J. Mol. Biol. 9, 654-669.

21. Bayler, S.T. (1966) J. Mol. Biol. 15, 420-427.

22. Visentin, L.P., Chow, C., Matheson, A.T., Yaguchi, M. and Rollin, F. (1972) Biochem. J. 130, 103-110.

23. Lanyi, J.K. (1975) in "Extreme Environments: Mechanisms of Microbial Adaptation" (Heinrich, M., ed.) pp. 295-303, Academic Press, New York.

24. Kates, M. (1973) in "Ether Lipids; Chemistry and Biology (Snyder, F., ed.) pp. 351-398, Academic Press, New York.

25. Brown, A.D. (1964) Bacteriol. Reviews 28, 296-329.

26. Lanyi, J.K. (1971) J. Biol. Chem. 246, 4552-4559.

27. Plachy, W.A., Lanyi, J.K. and Kates, M. (1974) Biochemistry 13, 4906-4913.

28. Lanyi, J.K., Plachy, W.Z. and Kates, M. (1974) Biochemistry 13, 4914-4920.

29. Brown, A.D. (1976) Bacteriol. Reviews 40, 803-846.

30. Harold, F.M. and Altendorf, K. (1974) in "Current Topics in Membranes and Transport" (Bronner, F. and Kleinzeller, A., eds.) vol. 5, pp. 1-50, Academic Press, New York.

31. MacDonald, R.E. and Lanyi, J.K. (1975) Biochemistry 14, 2882-2889.

32. Belliveau, J.W. and Lanyi, J.K. (1977) Arch. Biochem. Biophys. 178, 308-314.

33. Renthal, R. and Lanyi, J.K. (1976) Biochemistry 15, 2136-2143.

34. Lanyi, J.K., Renthal, R. and MacDonald, R.E. (1976) Biochemistry 15, 1603-1610.

35. Lanyi, J.K. and MacDonald, R.E. (1976) Biochemistry 15, 4608-4614.

36. Eisenbach, M., Cooper, S., Garty, H., Johnstone, R.M., Rottenberg, H. and Caplan, S.R. (1977) Biochim. Biophys. Acta 465, 599-613.

37. Kanner, B.I. and Racker, E. (1975) Biochem. Biophys. Res. Comm. 64, 1054-1061.

38. Lanyi, J.K., Helgerson, S.L. and Silverman, M.P. (1979) Arch. Biochem. Biophys. 193, 329-339.

39. Wagner, G., Hartmann, R. and Oesterhelt, D. (1978) Eur. J. Biochem. 89, 169-179.

40. Lanyi, J.K. and Silverman, M.P. (1979) J. Biol. Chem. (in press).

41. Bakker, E.P., Rottenberg, H. and Caplan, S.R. (1976) Biochim. Biophys. Acta 440, 557-572.

42. Michel, H. and Oesterhelt, D. (1976) FEBS Lett. 65, 175-178.

43. Wagner, G. and Hope, A.B. (1976) Austr. J. Plant Physiol. 3, 665-676.

44. Lanyi, J.K. and Silverman, M.P. (1972) Can. J. Microbiol. 18, 993-995.

45. Ginzburg, M. and Ginzburg, B.Z. (1976) J. Membr. Biol. 26, 153-172.

46. Harold, F.M. and Papineau, D. (1972) J. Membr. Biol. 8, 45-62.

47. West, I.C. and Mitcehll, P. (1974) Biochem J. 144, 87-90.

48. Shindler, D.B., Wydro, R.M. and Kushner, D.J. (1977) J. Bacteriol. 130, 698-703.

49. Skulachev, V.P. (1978) FEBS Lett. 87, 171-179.

50. MacDonald, R.E., Greene, R.V. and Lanyi, J.K. (1977) Biochemistry 16, 3227-3235.

51. Stock, J. and Roseman, S. (1971) Biochem. Biophys, Res. Comm. 44, 132-138.

52. Tsuchiya, T., Raven, J. and Wilson, T.H. (1977) Biochem. Biophys. Res. Comm. 76, 26-31.

53. Tokuda, H. and Kaback, J.R. (1977) Biochemistry 16, 2130-2136.

54. Macdonald R.E., Lanyi, J.K. and Greene, R.V. (1977) Proc. Nat. Acad. Sci. U.S.A. 74, 3167-3170.

55. Rafaeli-Eschkol, D. (1968) Biochem. J. 109, 679-685.

56. Rafaeli-Eschkol, D. and Avi-Dor, Y. (1968) Biochem. J. 109, 687-691.

57. Shkedy-Vinkler, C. and Avi-Dor, Y. (1975) Biochem J. 150, 219-226.

58. Moore, R.L. and McCarthy, B. (1969) J. Bacteriol. 99, 255-262.

59. Woese, C.R. (Personal Communication)

60. Woese, C.R. and Fox, G.E. (1977) Proc. Nat. Acad. Sci. U.S.A. 74, 5088-5090

61. Kerscher, L., Oesterhelt, D., Cammack, R. and Hall, D.O. (1976) Eur. J. Biochem, 71, 101-107.

62. Werber, M.M., Mevarech, M., Leicht, W. and Eisenberg, H. (1978) in "Halophilic Microorganisms" (Caplan, S.R. and Ginzburg, M., eds.) Elsevier, Amsterdam, pp. 427-446.

PROSPECTS FOR FARMING THE OPEN OCEAN

Howard A. Wilcox

Naval Ocean Systems Center

San Diego, CA 92152

INTRODUCTION

Presently visualized prospects for farming the open oceans of
the earth are based mainly on the results of an activity called the
Ocean Food & Energy Farm Project (1). This project, conceived and
initiated by the U.S. Navy in late 1972 and terminated in late 1976,
was composed of scientists and engineers of the Navy, the California
Institute of Technology, the Western Regional Research Center of the
U.S. Department of Agriculture, the Institute of Gas Technology in
Chicago, and a number of other individuals and organizations, all
working together under the joint sponsorship of the Navy, the National
Science Foundation, the American Gas Association, and the United States
Energy Research and Development Administration. It was aimed at developing methods and systems which would hopefully be both technically
and economically successful for using the vast surface waters of the
open oceans to raise, harvest, and convert seaweeds ("kelp"), plus
associated fish and other organisms, into goods for man.

If successful, this project would enable the oceans to become a
huge new source of energy, fixed carbon, and fixed nitrogen for the
benefit of man. Moreover, this great oceanic system would be completely supplemental to (i.e., non-competitive with) the already
existing land-based sources of foods, feeds, fuels, fertilizers, and
related goods for the world's peoples.

Man's supplies of energy and chemicals have been dependent on
the natural gas, crude oil, and coal extraction industries for only
a few short decades, but already the non-renewable deposits of these
commodities are showing signs of serious depletion (2). Before these

235

deposits were invaded by the extractive industries, wood--an abundant type of terrestrial biomass--was both a major fuel and a prime source of man's building materials and chemicals (3). In view of the dwindling character of the earth's coal, natural gas, petroleum, and unexploited farmlands (4), man's eventual return to biomass as the prime generator of energy and fixed carbon chemicals appears virtually inevitable. However, the world's enormous and growing consumption of liquid and gaseous fuels plus petrochemical-like materials each year emphasizes the huge magnitude of the demand for renewable biomass which the future will bring.

Land farming might appear to be the natural answer for supplying this need, but such farming faces many difficult limitations:

. Already evident shortages of suitable land, fresh water, and fertilizers (5).

. Uncertain and destructive climatic cycles and extremes of weather.

. A land-based transportation and distribution system which is relatively expensive as compared to water-borne shipping (6).

All of these difficulties are largely circumvented by the ocean farm concept. Indeed, this concept is based squarely on the following combination of facts:

1. The solar energy received at the surface of the earth is plentiful--2.5×10^{21} Btu* per year (7)--as compared to man's present usage levels of about 2.3×10^{17} Btu per year (8).

2. The flow of solar energy is naturally maintained and highly reliable in both the technical and political sense.

3. Most of the solar energy received by the surface of the earth is absorbed in the upper layers of the oceans because the oceans cover some 71 percent of the earth's surface area (9) and possess relatively low average reflectivities (10).

4. Man's use of the earth's currently received solar energy can avoid upsetting the net balance of the planet's carbon dioxide, oxygen, water, and energy flow cycles as calculated on a global basis over time-spans of a few months (11).

---

*Editorial Note: This chapter uses units of measure common in U.S. engineering practice. For readers unaccustomed to this practice the following conversions are provided: One BTU =(1/3413)KWH; One ton = 2000 lbs. = 0.909 metric tons; One acre = 0.405 hectare; One horsepower = 0.746 KWH; One acre foot = $1233 m^3$; and One Yard = 0.914 m.

5. The sun is the only foreseeable energy source available for the large scale photosynthetic production of vegetation (renewable biomass).

6. Vegetation can be converted by known technology into foods, fertilizers, plastics, synthetic natural gas, synthetic liquid fuels, etc. (12). Indeed, vegetation is the only practically renewable source of such products that is available today.

7. Although the major areas of the surface waters of the oceans are "biological desert" because they are almost devoid of the nitrate and phosphate compounds required for the growth of vegetation, the ocean waters below depths of 300 to 1000 feet generally contain these materials at relatively high average concentrations (13), and these waters stand at a gravitational potential energy deficit relative to the surface layers of less than 3 Btu per ton (14).

8. Plants immersed in the ocean can be highly efficient photosynthesizers (15) and are immune to drought and frost, yet only a very small fraction of the ocean's area has thus far been brought under systematic cultivation by man (16).

9. Present oceanic harvests represent only a small percentage of the ocean's ultimate biomass production potential (17).

10. Much of the biomass producing area of the world's dry land surface has already been exploited (18), and the ocean appears to possess five to ten times more "potentially arable" area than the land.

Two primary problems have thus far prevented farming in the open ocean: the natural bottom is so far down in most places that the sunlight cannot reach it, thus preventing the reproduction and growth of attached plants, plus the earlier-mentioned fact that the surface waters of the open ocean are naturally stable and almost devoid of some needed plant nutrients. This project's answers to these problems are: (1) place a "mesh" some 50-100 feet down from the ocean surface, thus enabling the growth of attached seaweeds irrespective of the depth of the actual seabed, and (2) extend the intake pipes of wave or wind powered pumps vertically down from the surface zone some 300 to 1000 feet in order to create artificial upwelling of the cool, nutrient-rich deep water (see figure 1).

The technical feasibility of many of the processes required by the ocean farm concept has been well established through developments accomplished by numerous investigators and industries in previous decades (19). Therefore, the major question for this concept at present is that of economic feasibility. The issue is more one of "when" rather than "whether" the concept will eventually pay off,

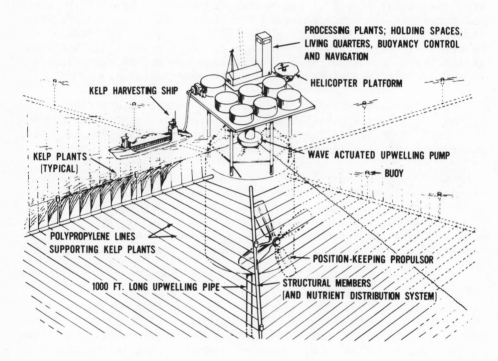

FIGURE 1

Conceptual design: 1000-acre ocean food and energy farm unit.

for as the world's fossil fuel resources and land-based biomass production potentials become increasingly consumed, ocean farming will
necessarily improve in its relative economic feasibility.

Although this project was terminated within the Navy in late
1976, a similar project is understood to be active at the present
time in the General Electric Co. under sponsorship of the American
Gas Association.

THE OCEAN FOOD AND ENERGY FARM CONCEPT

The concept of the Ocean Farm Project is to grow and harvest
suitable seaweeds (kelp) using submerged supporting lines and structures covering thousands of acres and lying 40 to 100 feet below the
surface of the ocean. Figure 1 shows one design which is under
study. The Ocean Farm concept is an advantageous approach to harnessing solar energy because kelp can readily be used as a feedstock
for the production of almost all of man's most desirable materials
and forms of energy, including feeds, foods, electric energy, and
gaseous and liquid fuels. Moreover, seaweeds naturally store the
solar energy, thus avoiding the storage problems inherent in some
other solar energy conversion methods.

The growing seaweeds would utilize the photons from the sun plus
the carbon dioxide and water from the surface layers of the ocean as
their principal "nutrients." The other nutrients necessary for
growth (mainly potassium, nitrogen and phosphorus) would be provided
by recycling process waste to the ocean and bringing up cool, nutrient-rich water from 300 to 1000 feet of depth.

A portion of the growing seaweeds and associated animal communities would be harvested periodically, and these harvests would then
be converted to methane (synthetic natural gas or SNG), liquid fuels,
lubricants, waxes, plastics, fertilizers, fibers, feed supplements,
and human foods at coastal or sea-based processing facilities. Mariculture and livestock operations can also be supported with feeds
derived from the seaweed process streams.

Figure 2 shows the general flow of materials in the Ocean Farm
System. The carbon, hydrogen, nitrogen phosphorus, and other
materials leaving the Ocean Farm as fuel and products would be replaced by the effluent carbon dioxide, water, and other waste streams
from the farm system and from consumers. A large portion of the
fermentation and process wastes can be recycled to the farm as fertilizer, thus reducing the quantity of upwelled nutrients required.

Based on a survey of candidate seaweeds (20) the "giant California Kelp" Macrocystis pyrifera, which grows along the coasts of
California, Mexico, Africa, South America, Australia, and New

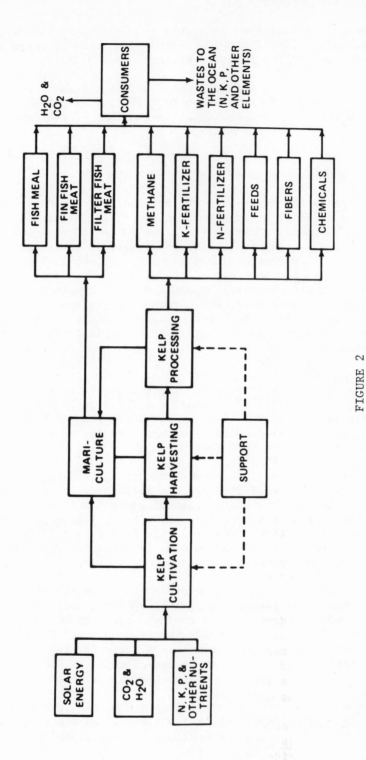

FIGURE 2

Ocean farm project flow chart

Zealand, was selected for initial Ocean Farm feasibility studies. M.
pyrifera--see figure 3-- is one of the world's largest and fastest
growing plants (29). Sargassum species were also under serious consideration as potential crops for ocean farms.

The life cycle of M. pyrifera is well understood. It has been
cultivated in the laboratory and planted and harvested in the coastal
areas of California for several years (22). The basal portion of the
plant is continually producing fronds, the growing fronds produce
numerous blades, and layers of these blades float high in the surface waters of the ocean to absorb the sun's rays. Nutrients are
absorbed from the water by all exposed surfaces of the plants.

In nature, the life expectancy of each frond is about six
months, after which it dies off to make way for the new fronds growing up from below. Therefore, the plant reproduces its own weight
every three to six months or so (23) whether it is harvested or not.
In the Ocean Farm situation, the surface fronds will be harvested
every few months. Replanting of the farm will not be required after
harvesting, and we expect natural replacement of plants torn away
by storms, killed by disease, or excessively damaged by encrustation
or fish grazing (24).

Assuming a spacing of about ten feet between plants (as in a
well-matured natural kelp bed (25), the Farm will support about 400
to 450 plants per acre. Each acre is expected to yield 160 to 320
wet tons of harvested organic material per year (26), which trans-
lates to an energy conversion efficiency of approximately one to
two percent relative to the whole spectrum of the incident solar
energy. This amounts to the harvesting of roughly 200 to 400 million
Btu of stored vegetational energy per acre per year.

M. pyrifera requires water at temperatures below about 20 to
22°C for survival (27). Therefore, the forced upwelling of cool,
nutrient-rich deep water produced by the Farm system should enable
it to be cultivated even in the tropical oceans.

The artifical upwelling system will require relatively little
power, probably less than one horsepower per acre-foot of water
per day (28). We believe that the power needed for upwelling and
distribution of the water can probably be provided mainly by harnessing a small part of the local wave or wind energy.

Because of the high costs of deep ocean mooring lines, most of
the farm systems will probably be "dynamically positioned" by wind,
wave, and/or fuel powered propulsors--see figure 1-- in all areas
where the natural bottom is more than a thousand feet or so below
the surface (more than 90 percent of the ocean is deeper than a
thousand feet). These positioning propulsors will probably need to

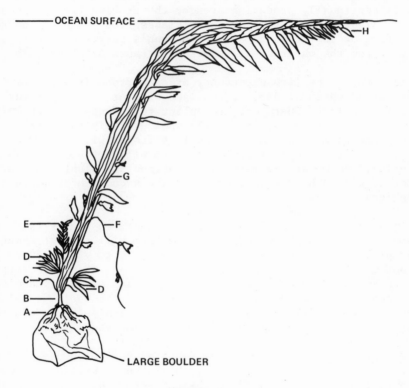

FIGURE 3

Diagram of young adult macrocystis plant. A, holdfast; B, primary
stipe; C, stubs of frond; D, sporophyll cluster; E, juvenile frond;
F, senile frond; G, stipe bundle; H, apical meristem. No root
involved--plant takes all nutrient direct from surrounding water.

FIGURE 4

Kelp harvesting ship in California waters (Photo courtesy of Kelco.,
San Diego).

supply on the average less than two-tenths horsepower per acre of
ocean farm system. It appears likely that the same engines required
for the positioning system can provide the necessary backup power
for the upwelling system.

Some fuel derived from the farm system will be returned to the
farm and stored for use as necessary in powering the farm's positioning and upwelling machinery.

Harvesting of the kelp canopy will be accomplished by special
vessels which will move over the farm and cut off the upper portions
of the fronds. These ships will take in the kelp and transport it to
the processing plants. Some pre-processing, such as removal of water
from the kelp, may be accomplished on the harvesting ships prior to
transferring the kelp to the processing plants. The Kelco Co. of
San Diego currently employs special ships for this type of harvesting
of the natural kelp beds along the coast of California (29)--see
figure 4.

Wastes from the kelp processing plants can also be carried and
dispensed to the farm by the harvesting ships in order to provide
nutrients in addition to the nutrients obtained from deep water by
means of the upwelling system.

Kelp stands naturally shelter and support abundant faunal communities (30). The larvae of small animals attach themselves to and
then grow on the kelp. They obtain nutrition by filtering plankton
out of the water. Fish graze on the kelp, perhaps in order to digest
those encrusting organisms. Blooms of phytoplankton will probably
be enhanced by the upwelling of the deep waters.

The mariculture operations will be based on: (1) the indigenous
low-value and high-value fin fish of the farm, and (2) specialized
culturing of oysters or other organisms. These animals will consume
the sludge and phytoplankton which are inevitably produced by the
farm and which would not otherwise be used so efficiently.

KELP COMPOSITION AND PRODUCTS

The intention of the project is to develop foods, fuels, and a
variety of high-value byproducts utilizing kelp compounds or processed materials. The products would jointly bear the costs of the
Ocean Farm System. The produced quantities of the different products would be responsive to market conditions and the increasing
supply of harvested kelp.

Methane is produced naturally by the anaerobic biodegradation
of organic vegetable mass (31). Many sewage plants produce it as

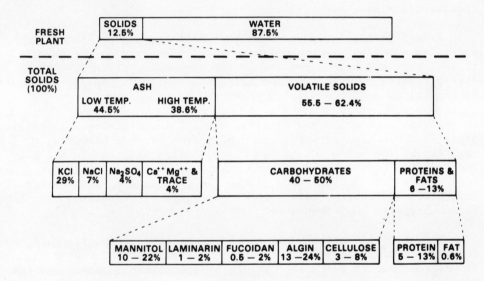

FIGURE 5
Ocean farm project: raw kelp composition

a byproduct to offset their own fuel costs. Methane was selected as
the primary fuel product for this project because of sponsor interest,
the apparent ease and low cost of methane production, its high energy
content plus clean-burning characteristics, and its value as a feedstock in the production of other petrochemical-like materials. Its
production by anaerobic digestion requires relatively low temperature
and low pressure processes.

In order to assess other high-value product potentials from kelp,
a study of the basic composition of M. pyrifera was necessary. The
major compounds and elements in kelp have been analyzed by many
authors (32), and Figure 5 is representative of their data. This project has performed a series of seasonal variation tests at the Naval
Ocean Systems Center, and a rather complete characterization of the
feedstock has been made (33).

From figure 5 it can be seen that kelp has a high ash content
and is an excellent source of potassium. Depending on the characteristics of the brines resulting from processing it is possible
that potassium chloride, sodium sulphate, sodium carbonate, and other
salts will be moderate-value byproducts.

Algin, laminarin, and fucoidan are all industrial gum materials.
Algin is currently extracted for profit from freshly harvested M.
pyrifera by the Kelco Co. of San Diego, California. Algin has many
uses as an additive to foods and industrial compounds. It also
can be made into medium-grade fibers (34), and these can probably
be upgraded into high quality ropes as required for future Ocean
Farm substrates and structures.

All of kelp's carbohydrates except mannitol are polysaccharides.
The sugar monomers may become economically viable products as food
prices continue to increase. Figure 6 shows these monomers and the
energy content of each, as well as the energy content of the overall
kelp.

Mannitol is a soluble carbohydrate (a polyhydric alcohol) and
by appropriate treatment can be converted to a high-value sugar,
d-fructose (35). Fructose syrups have recently been introduced to
the U.S. market.

The digester sludge will be composed of refractory compounds
which only degrade slowly in anaerobic digestion. In order to optimize capital costs the digesters will be sized for economically
efficient gas production, so conversions of fity to sixty percent
will be accepted. This will leave a sludge still rich enough in
energy and nutrients to use as terrestrial or marine animal feed
supplements. Certainly a low to moderate grade nitrogen fertilizer
should result. This fertilizer would not have the virus hazards

KELP COMPOUNDS	SUGAR RESIDUE	FORMULA	HEAT OF COMBUSTION BTU/LB
ALGIN	MANNURONIC ACID GULURONIC ACID	$C_6H_8O_6$	7794
LAMINARIN	D-GLUCOSE	$C_6H_{10}O_5$	6642
D-MANNITOL	D-FRUCTOSE	$C_6H_{14}O_6$	6246
FUCOIDAN	FUCOSE	$C_6H_{10}O_5$	6732
CELLULOSE	D-GLUCOSE	$C_6H_{10}O_5$	6732
PROTEINS	——	$(CH_2)_n(NH_2)_mO$	——
MACROCYSTIS (OVERALL)	——	——	4743

FIGURE 6

Ocean farm project: major kelp constituents

associated with sewage sludges and should be suitable for use on terrestrial farms.

As can be seen, the byproducts projected from the Ocean Farm System will be derived mainly from what are normally thought of as waste sludge and liquid streams.

The conversion of kelp to fuel must eventually compete with the fossil fuels and nuclear energy production on a net cost basis. The conditions for successful competition are drawing closer each year as discovery, extraction, environmental, and societal impact costs increase for the fossil fuel and nuclear energy industries.

The Ocean Farm System obtains photosynthate hydrogen from seawater. The carbon/hydrogen ratio in kelp is approximately 6.28 by weight, so kelp does not require the addition of supplemental hydrogen when being converted to hydrocarbon fuels. Therefore it will not produce competition for fresh water needed in other important areas of civilization.

Entry to the fuels market for kelp-derived methane will probably require some decades of integration and capitalization. The key to successful entry will be low kelp cultivation, harvesting, and processing costs, possibly dependent on use of environmental energy from the ocean site (wave, wind, solar radiation) for aid in the upwelling, positioning, and processing activities. The process approach that will be most successful must integrate well with these constraints. Many candidate processes based on experience with other biomass materials exist in the literature (36).

The relatively low energy content of kelp (590 Btu per fresh pound or 0.33 Kcal per fresh gram; 4,700 Btu per dry pound or 2.6 Kcal per dry gram) indicates that the energy expended in processing must be kept to very low levels. Low temperatures and pressures, low demand for fresh water, and low energy consumption in each step must be stressed. One of the keys to successful conversion is low cost salt removal in order to produce an energy-rich material of high organic polysaccharide and protein content.

Given the properties and potential products of kelp described earlier, an overall process approach was selected and a program was initiated to develop the necessary feasibility and cost data for Phase I of the Ocean Farm project.

Figure 7 shows the process, which is based on: (1) anaerobic digestion as the methane conversion step, (2) successful laboratory tests to pretreat with calcium ion and to remove salt by pressing, and (3) utilization of currently defined byproducts. The heavy lines represent the main material flows and the thin lines represent

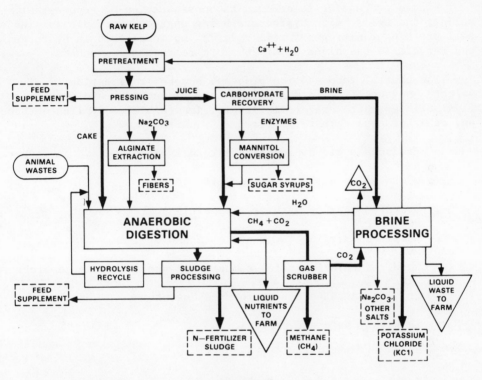

FIGURE 7

Ocean farm project: process chart for production of methane and other products.

lighter material flows or optional product and byproduct process routes.

Steps essential to successful economics are fermentation and brine processing. Of secondary importance (but still critical) is the carbohydrate recovery step, which is currently visualized as an ion-exclusion process. Ion exclusion is presently being used successfully to recover sugar from sugar beet waste brines (37).

Sludge processing is defined to be separation of supernatant, solids, and cells, with byproducts being upgraded for sale as single-cell protein or nitrogen fertilizer. An optional step is hydrolysis and recycle to the digester of the more refractory compounds.

The central fermentation process, anaerobic digestion, was selected because it offers a multitude of advantages and process flexibilities which are, collectively, impossible to duplicate through any other process. That is, it:

. Produces a clean product gas readily absorbed by the market.

. Operates at high conversion efficiency.

. Operates well on varying compositions of feedstock.

. Utilizes existing large-scale technology.

. Requires relatively low temperature and low pressure processes.

. Provides alternate uses for residue energy.

. Is highly safe and reliable.

. Is low in labor intensity.

. Has potential for large future improvements.

The constant dollar cost of the fermentation processes carried out in the relatively energy-rich environment (wave, wind, and solar radiation) of the open ocean may eventually be less than that of current land-based fermentation processes.

The state of the art in natural brine processing has advanced substantially in this century. Modern plants extract and concentrate large tonnages of brine for production of industrial salts and fertilizers. The salt processing step in the Ocean Farm System will require the adaptation of this modern technology because the kelp press brine is potassium-rich. It is intended that this brine provide all in-process treatment salts, such as calcium ion and sodium carbonate, plus enough potassium chloride to be sold as terrestrial fertilizer.

The following table presents some initial results achieved by the Institute of Gas Technology (IGT, Chicago, Illinois) in their studies of anaerobic fermentation of seaweed into methane (38).

Anaerobic Digester Feed type	Conversion Efficiency $\left(\dfrac{CH_4\ Energy}{Feed\ Energy}\right)$	Detention Time (days)
Press Cake	33.0%	10
80/20 PC/Juice	42.4%	10
60/40 PC/Juice	44.0%	10
Juice	31.8%	11.3
Raw Kelp	52.7%	18
Press Cake	42.6%	18
80/20 PC/Juice	49.4%	18
60/40 PC/Juice	47.1%	18

The researchers at IGT further found (38) that ". . . there was no major nutritional deficiency in any of [the kelp] feeds. So, barring any toxicity problems [none was encountered] they might be expected to promote anaerobic digestion without the addition of external nutrients."

Finally, and very importantly, the IGT work showed (38) that "it was possible to acclimate mesophilic [methane generating organisms] to [digester] salt concentrations equivalent to about 29,000 mg/l of NaCl." In fact, the process appeared to work well at salt concentrations up to about 40,000 mg/l of NaCl equivalent. Since seawater has a salt concentration of approximately 35,000 mg/l, this result suggests that the anaerobic digestion system for producing methane will not impose a requirement for large supplies of fresh water.

These results do not yet represent full achievement of the targeted goals of the project in this area, these goals being approximately 50 percent conversion efficiency with an associated detention time of 6 days, but they do strongly suggest that the goals are reasonable and attainable if further research and development effort is pursued.

The feedstock materials used by IGT were shipped to them by the Western Regional Research Center (WRRC), U.S. Department of Agriculture, Berkeley, California, after undergoing preprocessing and processing steps worked out by WRRC under Navy contract. Some data supplied by WRRC (39) are given in the following table:

Harvest Date	11/17/75	12/12/76	2/10/76
Solids conent (% of fresh wgt)	12.5	11.4	10.5
Ash content (% of solid wgt)	41.02	44.40	45.56
N " "	2.59	2.62	2.46
Protein " "	13.62	13.78	12.94
Na " "	4.16	4.45	4.16
P " "	0.64	0.72	0.60
S " "	1.20	1.10	1.00
Cl " "	13.3	13.5	13.6
K " "	10.0	9.3	9.7
Ca " "	1.29	1.05	0.94

The WRRC researchers also found (39) that "processing [of kelp] through chopping and grinding ... would require... approximately 3.6% of the input gross energy ...," while "processing [of kelp] through the [chopping, grinding, cooking, pressing, and carbohydrate recovery seqence] shows from 86-97% [recovery of] volatile solids, depending on ... conditions ... net energy recoveries [would be] 67-70%." Since the "cooking" step mentioned in this latter quotation was the largest single consumer of processing energy, and since this cooking energy was (at this early stage in the work) simply thrown away instead of being conserved by means of appropriately designed heat exchangers, it is believed that the efficiencies of the kelp processing steps occurring ahead of the digestion process can probably reach as high as 80 to 85 percent.

OVERALL DEVELOPMENT PLAN

Tentatively, three phases were originally planned for this project. Broadly stated, the first phase was to answer the question as to whether and when it will be economically feasible to grow seaweed and fish in the open ocean under controlled conditions, harvest them, and process the crops into a mix of profitable foods, methane, and other products. It would therefore include (1) preparation, analysis, and evaluation of design concepts, (2) demonstration of small scale concepts, (3) tests of subsystem models, (4) integration of subsystems, and (5) operation and evaluation of several small farms off the California Coast. The Phase I effort was to last three to five years and was estimated to require between $3 and $6 million to complete.*

Phase II was called "proof of concept" and was visualized as including the scale-up and engineering design of subsystems culminating in two 1,000-acre test farm plus pilot processing facilities.

---

*All dollar values are stated in terms of constant 1974 dollars.

Phase II was projected to last four to five years and cost some $50 to $100 million.

Phase III was identified as the commercial demonstration period, and it was projected to involve the design, development, and operation of a complete 100,000-acre farm system. Phase III was estimated to cost roughly $2 to $4 billion and require six to seven years for completion.

ACCOMPLISHMENTS

<u>Small Experimental Farms and Upwelling Studies</u> - A seven acre farm unit was successfully emplaced in early 1974 in exposed weather and sea-state conditions about 1,000 yards off the north-eastern tip of San Clemente Island, California, where the water is 300 feet deep. This site is some sixty miles west of the California Coast. About 100 <u>M</u>. <u>pyrifera</u> plants were taken from the nearby natural beds and attached to the farm. Preliminary results indicated that the plants survived and fish grazing pressures were low on the farm as compared to those in the natural beds nearby. Also, the natural nutrient levels in the ambient water were low, so both the natural beds and the farm showed low growth rates. The plants reproduced successfully on the farm. No attempts were made to artificially provide nutrients to the plants. A combination of accidents produced total destruction of this farm unit in late January 1975 (40).

Professor Wheeler North of the California Institute of Technology emplaced and studied two small fractional-acre farm units off Corona del Mar and Santa Catalina Island in the late 1974 to mid-1976 time period.

Valuable experience was gained with the three small ocean farm units thus far utilized, and no fundamentally adverse data were obtained in the course of their operations.

An adult kelp plant was exposed to a stream of nearly pure up-welled water drawn from depths of 750 to 1000 feet in early 1975. Experimental difficulties made it impossible to assess meaningful growth rate stimulation in this experiment, but North's final clinical examination of the tissues of the plant revealed them to be in apparently excellent health and without visible signs of starvation or toxicity.

North's work in late 1976 (under grant support from the Energy Research and Development Administration) demonstrated the following growth rate stimulation results in flowing mixtures or surface water and deep water from 870 meters of depth in the Atlantic Ocean near St. Croix Island (41).

Mix Ratio (%Deep/%Surface)	Growth Rate (% per day)
100/0	15.1
50/50	19.3
10/90	17.5
5/95	12.2
0/100	9.5

Control Plant in Flowing Bay Water at Corona Del Mar, California	
	7.9

North also states (42) that "... experiments on Macrocystis held in Atlantic water pumped up from 870 meters deep at ... St. Croix ... verified that previous trends ... described in our September report ... dilution of deep water with equal parts of surface water produced excellent growth rates ... comparable to the highest growth rates measured in Pacific water ..."

Based on these results and other data (43), we believe that the targeted efficiency of 2 percent for converting incident solar energy (whole spectrum) into stored seaweed energy can probably be met or perhaps even exceeded by the ocean farms of the future.

Small Ocean Test Bed - A small "quarter acre module" farm was designed and largely fabricated. The module was to be used to collect data on kelp growth responses and structure responses under various ocean conditions. A shallow area near Corona del Mar, California had been selected as the initial installation site for the farm module. This site had a minimum of logistic problems and was near the California Institute of Technology's Kerckhoff Marine Laboratory so the unit could be monitored frequently. This unit was scheduled to be outfitted with adult plants and operated in shallow water in late 1976, after which it was to be towed to nearby deep ocean water and operated there with appropriate rates of upwelling. Navy activities in connection with this unit were suspended in mid-1976.

Food Studies - Sheep, abalone, fish, and kelp fly larvae were utilized in kelp feeding trials. Results at the University of California at Davis showed sheep digestion efficiencies of 58 percent of the organic matter in the dried kelp -- about the same as for the basal ration composed of alfalfa hay, oat hay, barley, and sodium phosphate (44). Juvenile abalone showed 10 percent conversion efficiency on a fresh-weight-of-kelp to fresh-weight-of-abalone basis (45).

Site Selection Survey - A rather careful study was made (46) of the eastern Pacific ocean from 10°S latitude to 50°N latitude,

and from the U.S. West Coast to Hawaii. Based on this study, it
appears that optimal sites for future ocean farms will be (1) the
latitude band from about 25°N to about 40°N, (2) a correspondingly
similar band south of the equator, and (3) the equatorial band from
about 15°S to about 15°N latitude. These bands encompass approximately 30 percent of the area of the earth and are reasonably free
of the tropical hurricanes that would make ocean farms more expensive in the bands from about 15° to about 25° north and south latitudes.

Economic Analysis - Based on rough estimates using a generalized
farm design, the full-scale Ocean Farm of Phase III, 100,000 acres,
is expected to be able to produce some 22 billion cubic feet (620
million cubic meters) of methane per year at a cost ranging from a
low of about $2.30 or less to a high of about $7.00 per thousand
cubic feet ($0.08 to $0.25 per cubic meter), depending mainly on the
assumptions used for food and byproduct credit values, distance to
coasts, etc. (47). These studies plus another* by the present author
indicate that (1) large systems--8,000 hectares (20,000 acres) and
up--become increasing preferable economically as compared to smaller
units, (2) dynamically positioned farm systems have much lower investment costs than all but the shallowest anchored systems, (3) on-site
structures may cost less than $7,000 per hectare ($3,000 per acre)
of cultivated ocean, (4) harvesting ship costs will possibly amount
to some $3,300 per hectare ($1,300 per acre) and (5) associated on-
shore processing facility investments may amount to about $4,000 per
hectare ($1,600 per acre) of cultivated ocean area.

THE FUTURE

Assuming a two percent efficiency for converting solar radiation
into the stored energy of seaweed compound, five percent efficiency
for the production of human food from the seaweed, and fifty percent
efficiency for the production of fuels and other products from the
seaweed, each square mile ($2.6 \times 10^6 m^2$) of the marine farm is projected to yield enough food to feed 3,000 to 5,000 persons and at
the same time to yield enough energy and other products to support
more than 300 persons at today's U.S. per capita consumption levels.
Since the oceans appear to contain some 80 to 100 million square
miles (2×10^{14} to $2.6 \times 10^{14} m^2$) of "arable surface water," this
means that marine farms could conceivable support a world population
of more than twenty billion persons living at levels of consumption
characteristic to the USA today.

Before any such vast undertaking could reasonably be undertaken,
of course, close and thorough attention and studies will have to be
given to the possible problems of environmental degradation, societal
upset, and legal/political difficulties. Some preliminary studies

*Not yet published.

have already been accomplished (48).

NOTES AND REFERENCES

1. Wilcox, H.A., 1972, "Project Concept for Studying the Utilization of Solar Energy via the Marine Bio-Conversion Techniqe," available from Code 5304, U.S. Naval Ocean Systems Center, San Diego, CA 92152. Also 1975, "The Ocean Food and Energy Farm Project," similarly available.

2. Hubbert, M.K., 1962, Energy Resources, National Academy of Sciences--National Research Council, Washiongton, DC., p. 31. Also, 1969, Chap. 8 of Resources and Man, National Academy of Sciences-National Research Council, Freeman, New York, pp. 157-242.

3. Ayres, E., 1966, "Energy Sources, "Encyclopedia of Science and Technology, McGraw-Hill, Vol. 4, pp. 601-603; also, Saul, E.L., 1966, "Wood Chemicals," ibid, vol. 14, pp. 526-528.

4. In respect to the earth's dwindling farmlands, see Borgstrom, G., 1973, The Food and People Dilemma, Duxbury Press, North Scituate, Mass., p. 39.

5. President's Science Advisory Committee, 1967, The World Food Problem, The White House, Vol. II.

6. Cambel, A.B., et al., 1964, Energy R&D and National Progress, Government Printing Office, p. 161.

7. Sellers, W.D., 1965, Physical Climatology, University of Chicago Press, p. 23.

8. U.N. Statistical Yearbook, 1976.

9. See reference 7, p. 4.

10. See reference 7, p. 21.

11. Wilcox, H.A., 1973, The Energy crunch--Present Trends and Future Prospects for the World and the USA, available from code 5304, U.S. Naval Ocean Systems Center, San Diego, CA., P. 47. Also, Wilcox, H.A., 1975, HOTHOUSE EARTH, Praeger Publishers, New York.

12. Saul, E.L. 1966, "Wood Chemicals", Encyclopedia of Science and Technology, McGraw-Hill, Vol. 14, pp. 526-528; Ziemba, J.V., et al., 1966, "Food Engineering," ibid, vol. 5, pp. 380-405.

13. Sverdrup, H.U., M.W. Johnson, and R.H. Fleming, 1942, The Oceans, Prentice-Hall, New York, pp. 239 ff.

14. Wilcox, H.A., 1975, "Artificial Oceanic Upwelling," available from Code 5304, Naval Ocean Systems Center, San Diego, CA., 92152.

15. Ryther, J.H., 1971, "Seawater Fertility," Encyclopedia of Science and Technology, McGraw-Hill, Vol. 12, p. 157.

16. Bardach, J.E., 1968, The Harvest of the Sea, Harper and Row, New York, pp. 178 ff.

17. Ibid.

18. See reference 5.

19. North, W.J., ed., 1971 "The Biology of Giant Kelp Beds (Macrocystis) in California," Cramer, Lehre, Germany.

20. Jackson, G.S., and W.J. North, 1973, "Concerning the Selection of Seaweeds for Marine Farms," Available from Code 5304, Naval Ocean Systems Center, San Diego, CA 92152.

21. See reference 19.

22. North, W.J., 1973, "Annual Progress Report, Kelp Habitat Improvement Project, 1972-1973," California Institute of Technology, Pasadena, CA pp. 124-136.

23. See reference 19, p. 35.

24. See reference 19, p. 36.

25. See reference 19, p. 46.

26. Mann, K.H., 1973, "Seaweeds: Their Productivity and Strategy for Growth," Science, Vol. 182, p. 975. See also Ryther, J.H., 1959 "Potential Productivity of the Sea," Science, Vol. 130, p. 602.

27. See reference 19, p. 12.

28. See reference 14.

29. Wilcox, H.A., 1973, "Ocean Farm Project Kelp Harvesting Report," available from Code 5304, Naval Ocean Systems Center, San Diego, CA 92152.

30. See reference 19, pp. 52ff.

31. Stadtman, T.C., 1971, "Methanogenesis, bacterial," Encyclopedia of Science and Technology, McGraw-Hill, Vol. 8, p. 397.

32. Vinogradoff, A.P., 1953, "Elementary Composition of Marin Organisms," Sears Foundation for Marine Research, New Haven, CT. Hoagland, D.R., 1915, "Organic Constituents of Pacific Coast Kelps," J. Agricultural Research, Vol. 4, pp. 39-58. Clendenning, K.A., 1962, "Determination of Fresh Weight, Solids, Ash, and Equilibrium Moisture in Macrocystis pyrifera,"Bot. Mar., Vol. 4, pp. 204-218.

33. Lindner, E., C.A. Dooley, R.H. Wade, 1976, "Chemical Variation of Chemical Constituents in Macrocystis pyrifera," available from code 5304, Naval Ocean Systems Center, San Diego, CA 92152.

34. Hall, A.J., 1970, The Standard Handbook of Textiles, Chemical Publishing Co., New York.

35. Legeler, J., 1922, J. Bacteriology, Vol. 112, p. 840.

36. Field, J.H. 1971, "Bergius Process," Encyclopedia of Science and Technology, McGraw-Hill, Vol. 2, p. 165. Also, Field, J.H., 1971, "Fischer-Tropsch Process,"ibid, Vol. 5, p. 300. Also, Hurd, C.C., 1971, "Pyrolysis," ibid, Vol. 11, p. 122.

37. Stark, J.B., 1965, "Ion Exclusion Purification of Molasses," J. of The American Society of Sugar Beet Technologists, Vol. 13, No. 6, p. 493.

38. Ghosh, S., et al.,1976, "Research Study to Determine the Feasibility of Producing Methane Gas from Sea Kelp," Institute of Gas Technology, Chicago, IL, Available from code 5304, Naval Ocean Systems Center, San Diego, CA 92152.

39. Hart, M.R., et al.,1976, "Final Report, Ocean Food & Energy Farm Kelp Pretreatment and Separation Processes, "Western Regional Research Center, U.S. Department of Agriculture, available from code 5304, Naval Ocean Systems Center, San Diego, CA 92152.

40. North, W.J. and H.A. Wilcox, 1975, "History, Status, and Future Prospects regarding the Experimental 7-Acre Farm at San Clemente Island," available from code 5304, Naval Ocean Systems Center, San Diego, CA 92152.

41. North, W.J., 1976, "Marine Farm Studies Progress Report for Sept., 1976," pp. 72-73, available from code 5304, Naval Ocean Systems Center, San Diego, CA 92152.

42. North, W.J., 1976, "Marine Farm Studies Progress Report for Oct. 1976," p. 34, available from code 5304, Naval Ocean Systems Center, San Diego, CA 92152.

43. Ryther, J.H., 1959, "Potential Productivity of the Sea," Science, Vol. 130, 11 Sept, pp. 602-608. Also, Mann, K.H., 1973, "Seaweeds: Their Productivity and Strategy for Growth," Science, vol. 182, 7 Dec., pp.975-981.

44. Garrett, W.N., 1974, "Feeding Value of California Kelp," Final Report on Contract No. N66001-74-C-0376, available from code 5304, Naval Ocean Systems Center, San Diego, CA 92152.

45. Leighton, D.L., 1976, "An Investigation of Feeding, Food conversion, and Growth in the Abalone, with Emphasis on Utilization of the Giant Kelp, Macrocystis pyrifera," available from code 5304, Naval Ocean Systems Center, San Diego, CA 92152.

46. Seligman, P.F., 1977, "Survey of Oceanographic and Meteorological Parameters of Importance to the Site Selection of an Ocean Food and Energy Farm (OFEF) in the Eastern Pacific," Naval Ocean Systems Center Technical Report 121, San Diego, CA 92152.

47. Budhraja, V.A., et al.,1976, "Ocean Food and Energy Farm Project, Overall Economic Analysis, Vol. I," Integrated Science Corporation, Available from code 5304, Naval Ocean Systems Center, San Diego, CA 92152.

48. Wilcox, H.A., 1973, "Expected Thermal Effects of a System of Large, Open-Ocean, Mariculture Facilities for Utilizing Solar Energy," available from Code 5304, Naval Ocean Systems Center, San Diego, CA 92152. Also, 1975, HOTHOUSE EARTH, Praeger Publishers, New York. Also, 1975, "Estimates of the Magnitudes and Time Lags of the Earth's Thermal Response to Man's Global Energy Input Rates," available from code 5304, Naval Ocean Systems Center, San Diego, CA 92152.

AQUACULTURE OF ANIMAL SPECIES

S.I. Doroshov; F.S. Conte and W.H. Clark, Jr.

Aquaculture Program
University of California
Davis, CA 95616

INTRODUCTION

Aquaculture, defined as cultivation of both freshwater and marine aquatic species, is well documented as early as 2000 BC (20). Though aquaculture has been practiced through the ages, it has never received as much attention as it has during the last twenty years. This interest has undoubtedly been promoted by the world's quest for answers to human nutritional problems and the limitations of our natural fisheries.

Fisheries products have long been recognized for their high nutritional quality and the unique health benefits they provide (20). Approximately 13%, a figure which is continually growing, of the animal protein consumed in the world is provided by fisheries products (16). Increasing demand for these products has resulted in increased pressure on our natural fisheries. The bulk of these fisheries, including herring, cod, haddock, flatfishes, shrimp and lobster, are estimated to be at or near their maximum sustained yield. In many instances there has been a disappearance of species (abalone, anchovy and sardine) as existing fisheries in traditional fisheries grounds. These declines and losses have dispelled the common myth that our marine waters offer an unlimited bounty of fisheries products (4,16,43). Traditional marine fisheries that were once thought to be unlimited are now estimated to be capable of producing a maximum sustained level of harvest at 100 to 120 million metric tons (MMT) a year. With expanding fisheries efforts, the harvests are expected to approach this maximum and to reach 94MMT by the year 2000. Current predictions of human population expansion suggest that time demand for fisheries products will far outstrip availability

resulting in higher prices and shortages (16). In addition to being
a source of high-quality protein in great demand, aquatic organisms
have a very attractive attribute when compared to other food animals;
they often devote a higher degree of their food energy for growth
(3).

Recognition of the above has greatly stimulated interest in aquaculture on the part of national and international agencies as well as
private entrepreneurs. In general, many agencies have tended to
devote their attention toward protein-rich, low-cost species, low on
the trophic level; while entrepreneurs have pursued species with high
market demand, high on the trophic level, and requiring supplementary
feeds. In essence, research and development have been promoted in
areas from the culture of primary producers (blue-green and eucaryotic algae) to carnivores (lobsters and salmonids) high on the trophic
level.

Aquaculture has ranged geographically from salmon culture in
coldwaters of the U.S. Pacific Northwest (3) to warmwater fish culture in the coastal deserts along the Gulf of Elat (32). The latter
is typical of large biosaline regions of the world which in the past
have offered little productivity in terms of man's quest for food.
Only recently have the possible advantages of these geographical
regions been seriously considered. For an aquaculturist, the high
solar energy, land availability and unlimited seawater have been
very attractive features.

A discussion of animal aquaculture in biosaline environments
could result in a rather lengthy review of ongoing projects and research, highly varied in their philosophy and approaches. Instead
of such a review, this presentation will focus briefly on the requirements an organism should meet to be a successful aquaculture
candidate and then examine two animals which the authors believe
meet these requirements. An attempt will then be made to show how
these systems can be integrated.

Though the attractiveness of the primary producers is obvious,
one must not lose track of the technical and sociocultural problems
associated with such a food source. The animals discussed in this
chapter are both readily accepted as human food. In the Chapter by
Mitsui, marine algae are reviewed in detail; we will only treat
algae in terms of their usefulness as animal feeds and fertilizers.

AQUACULTURE CRITERIA

In appraising an animal's potential for aquaculture, there are
several questions of a legal, biological, economical, environmental
and sociocultural nature which must be considered. The legal aspects

are usually of little concern during the early exploratory stages of
determining an organism's applicability for culture. Unfortunately,
it is after a species passes the rigors of research and development
and enters the production phase that problems arise. Some of the
more common conflicts are:

- The introduction of a non-indigenous species may be illegal
  since their release could endanger natural fauna;

- The procurement of wild animals for brood or seedstock may
  be viewed with concern;

- The control of protected predators or competitors is often
  a problem;

- The use of or restriction of public waters by aquaculture can
  be a severe obstacle.

There is a myriad of such legal constraints with which the producer can be faced. However, these constraints obviously vary
greatly from one country to another, usually increasing in proportion
to the country's degree of development. Since we are dealing with a
geographic zone encompassing the world and individual countries,
these legal constraints have limited bearing on our discussion in
this chapter.

The biological aspects, on the other hand, are of the utmost importance to this discussion. In our consideration of the biological
criteria vital to aquaculture, it is important that a distinction be
made between extensive and intensive operations. Extensive aquaculture can be defined as the rearing of animals in natural bodies
of water with few, if any, modifications to the environment. Intensive culture is the growing of animals in systems (ponds, tanks or
raceways) where support parameters are carefully maintained. With
these two approaches in mind, we will briefly discuss some biological
factors which are critical for successful culture.

The first of these factors is reproductive control which is
essential for genetic selection and supply of seedstock, alleviating
dependence on wildstock. Intensive operations are totally dependent
on a stable and predictable supply of seedstock which is only provided by reproductive control. Such control is also the first step
towards domestication which is necessary for genetic selection, the
tool needed by culturists for the improvement of such important
traits as growth rate, resistance to disease and increased fecundity.

If reproductive technology is available, one must look carefully
at the larval development of the organism chosen. In many aquaculture systems, the highest loss from mortality occurs during larval

development. Some animals that appear very attractive in their adult forms are unsuitable for culture due to long and complicated larval development periods.

In addition to selecting an organism with a relatively short larval period, an animal exhibiting rapid growth is extremely desirable. The advantages of obtaining a marketable animal in the shortest period possible are obvious. A biosaline environment with extended growing periods, often year-round, is very attractive in view of the above parameter. For example, a catfish farmer in a temperate climate with only five months of optimal temperature may require two years to market his product, whereas a similar farmer with a seven-month growing period can market his animals in one year.

Food requirements are another criteria which must be considered in the selection of aquaculture candidates. In extensive systems animals can graze on natural products; however, under high density intensive systems (tanks and raceways) animals will require complete nutritional supplementation. The degree of intensity will dictate the sophistication of feeds. For example, animals maintained in low density may graze on natural foods in ponds. If the ponds are fertilized, stocking rates and production may be increased; and if supplemental feed is applied, even higher production rates can be achieved. The maximum product may be reached with the same animals in tanks or raceways if total nutritional supplementation is provided. Another item that must be considered is that the higher the trophic level of the animal, the more dependent the animal becomes on costly supplemental feeds.

Even though requiring greater amounts of more sophisticated feeds, increased animal densities are usually desirable. Thus, it is important to select species which will tolerate crowding, as territorial or aggressive behavior often makes species undesirable for use under intensive systems.

Obviously, the site selection must be compatible with the biological requirements of the animal chosen for culture. Animals chosen for a biosaline environment should be flexible in terms of salinity fluctuations. Though marine waters in such localities may have relatively constant salinities, pond and estuarine areas could be expected to occasionally experience great fluctuations in salinities due to evaporation and flooding.

Seasonal high temperatures should be compatible with the upper temperature limits within an animal's normal physiological range. At this high range, the growth rate of poikilothermic animals is more rapid. Thus, high temperatures which often vary little in a biosaline environment can be very advantageous.

In conjunction with the above considerations, the economics of the proposed venture and sociocultural attitudes of the proposed market should be studied. If the product is too costly or for other reasons unacceptable to a local market, its exportation should be a stimulus to the local economy.

The two organisms we will discuss are the grey mullet, Mugil cephalus, and the marine shrimp belonging to the genus Penaeus, mullet being relatively low on the trophic level and shrimp relatively high. These organisms have been chosen since they have been extensively studied for the last decade and appear eminently suited for culture in a biosaline environment. In addition to being appropriate, based on their physiological parameters, they represent food items of universal acceptance. The mullet is a relatively low-cost species while the shrimp is a more highly-prized item with extensive export potential.

MULLET

Market and Economics

In most countries of the world, the main exception being the U.S., there is a high market demand for mullet. This demand usually exceeds the existing supply. In the eastern Mediterranean countries, and in China, Japan and the USSR, mullet is a highly-prized fish. As a food source mullet exceeds many other aquatic products in nutritional value, 21% of the bodyweight being crude protein. It also provides high levels of the B-group of vitamins, particularly B_6 and pantothenic acid (6).

Mullet, long recognized for its high market value, adaptation to great variations in environmental conditions and its low position on the trophic level has been cultured for centuries in the Mediterranean and Far East. This culture has been dependent on wild fry captured in the sea or collected in lagoons. Today, due to pollution and environmental alterations, the number of fry necessary for seeding these operations at optimum levels is no longer available. During the last decade this has resulted in a quest for technology related to reproductive manipulation and hatchery techniques necessary to revitalize mullet culture.

Life History

The majority of the Mugilidae inhabit tropical and subtropical areas; though Crenomugil spp. are abundant in temperate waters. Mugil cephalus has the widest distribution between 42°N and 42°S in all the seas of the world (46). This species is typically an estuarine species which performs periodical migrations along the sea coast depending on the season, temperature changes and freshwater runoff. The characteristic habitat of the grey mullet is shallow coastal lagoons with sandy or muddy bottoms having salinity variations from freshwater to 75 ppt (42). A wide range of salinity tolerance is characteristic for the life stages starting from the fry which can easily resist abrupt changes in salinity from full seawater to freshwater (15). However, the eggs and larvae exhibit a more narrow range (14-35 ppt) which may correspond to the density of seawater required for maintenance of their positive buoyancy.

The temperature ranges which M. cephalus can tolerate are also extremely wide. Developing eggs can resist temperatures of 10-24°C with the optimum being about 22°C (33) and the larval stages exhibit optimal growth at 22-32°C. Subadults and adults may survive temperature variations of 4-35°C, though their normal temperature range in their natural environment is 16-34°C (46).

The ability of mullet to survive in the potentially stressful environments of coastal lagoons and marshes with depleted oxygen content has been observed by McFarland and Moss (28) and is related to their efficient respiratory system which is capable of maintaining a routine rate of respiratory metabolism in hypoxic water (7).

The adult mullet is primarily an iliophagous and herbivorous species characterized by a "grazing" or foraging type of feeding behavior. The main food items of the grey mullet are epiphytic diatoms and unicellular algae incorporated in sand and detritus; filamentous algae and small animals found in the sediments of eutrophic lagoons and marshes (27,36,44). The ability of mullet to withdraw and efficiently utilize the tiny organic materials from sand and mud has been defined by W. Odum (37) as a "telescoping" food chain. Mullet juveniles, 20-55 mm in length, were also found to be herbivorous, however, in contrast to the larger fish the bulk of the organic material they ingest is taken from the water column (11). Fry, less than 20 mm in length are exclusively carnivores and feed on zooplankton (1,2).

Data on growth of grey mullet are relatively scarce and often contradictory which may reflect the wide range of habitat and climatic conditions in which the fish are found (46). M. cephalus reaches sexual maturity within 2-4 years of age at a size of 23-41 cm (weight 1-2 kg) and has a very high fecundity: $1.2 \times 10^6 - 2.8 \times 10^6$ eggs per fish (46). Normally females have a synchronic development of oocytes and spawn once a year. Spawning occurs while large schools of fish migrate offshore during the summer, fall and winter at temperatures of 18-25°C. The eggs are pelagic, possess a large oil globule and range in size from 0.8-1.2 mm in diameter. Larvae of 2-2.4 mm in length hatch 36-48 hours after fertilization and complete metamorphosis within 20-30 days. After metamorphosis the fry migrate inshore and enter saltwater and brackishwater lagoons and estuaries.

Broodstock

Though wild spawners may be used if hatcheries are located near natural spawning sites (25), the maintenance of broodstock is highly desirable. Captive broodstock have been successfully established and used for spawning in Israel (49) and in Hawaii (41). Optimal conditions for ovarian maturation in the grey mullet are a photoperiod of 6 light and 18 dark hours and a temperature of 21°C (24).

Ovarian maturation is monitored by sampling the ovary via catheterization. Once ovarian maturation is complete injections of gonadotrophic hormones are required to promote spawning. After injection, the fish are maintained in 50-gallon aquaria where they undergo natural oviposition or may be stripped and fertilized artificially within 24 hours (22,23,25,49).

Hatchery Technology

Eggs are incubated and hatched in tanks. These systems are designed to allow water circulation and thus maintain a pelagic distribution of the eggs (25,33). Hatchery water is UV irradiated and treated with penicillin and streptomycin to control microbial fouling. A water temperature of 21-24°C and a salinity of 32 ppt is believed optimal for embryonic development (33). Hatching occurs 36 hours after fertilization. The resultant larvae, 2-2.4 mm in length are transferred to large circular tanks at a stocking density of 5 per liter. These tanks contain planktonic algae (Dunaliella sp., Isochrysis sp., Tetraselmis sp. and Chlorella sp.) which favorably affect larvae growth and survival, though not consumed directly as as food item by the larvae.

After 4-6 days from hatch, the larvae begin actively feeding on rotifers (Brachionus plicatilis) and 10-12 days after hatching they begin feeding on brine shrimp nauplii. The larvae continue to feed on the brine shrimp nauplii until they undergo metamorphosis at a size of 20-25 mm at 20 to 30 days after hatching. It has been observed that growth may be promoted by reduced salinities and temperatures of 24-28°C.

Larval survival to metamorphosis is generally poor (10%); though, 33.5% survival has been documented (34). The greatest mortality occurs after the initiation of active feeding and is referred to as a "descent" of the larvae to the bottom of the tank. Strong aeration appears to reduce such mortalities. Development of more efficient larval rearing technology is a subject of intense investigation (25,34).

Growout Systems

Mullet are raised in most countries in extensive culture operations, predominantly, in polyculture with other species. In Italy,
Greece and Egypt mullet are cultured in large estuarine lagoons requiring the voluntary entrance of mullet juveniles from the wild.
Water regimes are often regulated by very sophisticated systems of
canals and sluices (Italian "valli"). Fish are harvested after 2-3
years and production levels are relatively low (100-200 kg/ha). No
fertilization or supplemental feeding regimes are used in these
systems. Additional species entering the lagoons (sea bass, gilthead
bream, eel) are treated as a highly-prized secondary crop.

In Hong Kong mullet are commonly cultured as the primary crop,
with a secondary crop of Chinese carp. A common stocking density
used is 10,000 mullet plus 1,500 Chinese carp per ha. These animals
are reared in 1 to 5 ha ponds requiring a growing period of 300 days
and yielding productions up to 3,500 kg/ha. Such systems require
supplemental feeding with rice bran (2,500 kg/ha), peanut cakes
(3,000 kg/ha) and applications of animal manure every 3 to 5 days.
Mullet fry (24-45 mm) are used for stocking these systems and the
animals are commonly marketed at 300 to 500 grams. The mullet con-
tributes 50 to 90% of the total crop of these polyculture systems
(26).

In Israel mullet are widely used as a secondary crop in brackish
pond polyculture with both _Tilapia_ and carp species. In these systems
extensive supplemental feeding is utilized. These operations are
highly successful and production levels are rapidly increasing. For
example, the total production of mullet was 267 tons in 1964 and 700
tons in 1976 (45). The mullet, though considered the secondary crop,
is highly productive under this system and is marketed at 400 to 700
grams of weight after 120 to 150 days of rearing (48). On the other
hand, in recent experiments in saltwater ponds near the Dead Sea,
mullet and _Tilapia_, receiving no supplemental feeding, have illustrated good production rates--mullet, 500 to 600 kg/ha and _Tilapia_, 100
to 600 kg/ha (12).

Other experiments in the U.S. and Egypt have shown similar success in saltwater ponds without supplemental feeding (5). It is
interesting to note that in some of the systems in the U.S. and Egypt,
mullet are being reared in polyculture situations with penaeid shrimp
(personal communication, Wheeler and Neal, 1978). Thus, the present
utilization of mullet in most aquaculture ventures is directed toward
extensive or semi-intensive culture for which this species is ideally
suited due to its environmental resistance, foraging behavior and low
cost of production. Though mullet can be reared under highly intensive conditions (tanks or raceways), it is not clear whether such
systems are economically viable.

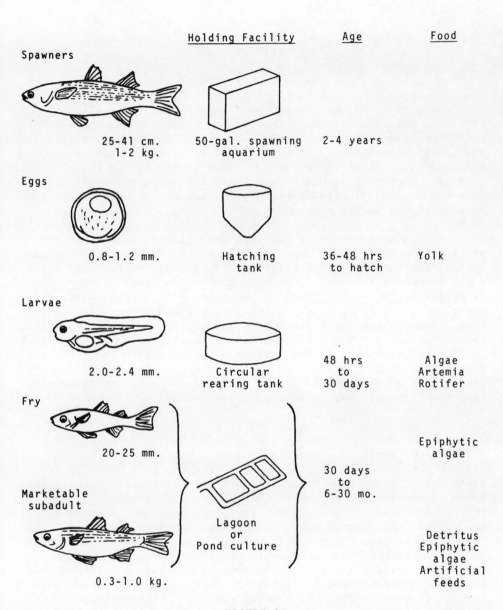

	Holding Facility	Age	Food
Spawners			
25-41 cm. 1-2 kg.	50-gal. spawning aquarium	2-4 years	
Eggs			
0.8-1.2 mm.	Hatching tank	36-48 hrs to hatch	Yolk
Larvae			
2.0-2.4 mm.	Circular rearing tank	48 hrs to 30 days	Algae Artemia Rotifer
Fry			
20-25 mm.			Epiphytic algae
Marketable **subadult**	Lagoon or Pond culture	30 days to 6-30 mo.	
0.3-1.0 kg.			Detritus Epiphytic algae Artificial feeds

FIGURE 1

Life stages and culture technology for _Mugil cephalus_.

Figure 1 diagrammatically illustrates the life stages and culture technology used for mullet.

PENAEID SHRIMP

Market and Economics

The marine shrimp of the genus Penaeus have world-wide recognition as a valuable commodity whose demand seems to have no upper bounds on Asian, European and U.S. markets. On the U.S. market in 1975, while accounting for only 7% of the total landings, shrimp represented 23.3% of the ex-vessel value of all landings (35). With recognition that world-wide demand will soon outstrip availability from the natural fishery, species of marine shrimp have become prime candidates for aquaculture.

Penaeid shrimp culture has been established as a viable and profitable enterprise in Japan and, on a limited basis, in Central America and in the South Pacific. Species under culture and undergoing intensive research include Penaeus japonicus, P. vannamei, P. stylirostris, P. setiferus, P. aztecus, P. californiensis, P. merguiensis and P. monodon. Although labor and operational costs have prevented the establishment of an industry in the United States the technologies being developed to facilitate this aim have direct application to the biosaline environments and many ares of the world. Systems for obtaining spawners, hatchery technology and growout have been developed for penaeid shrimp in areas where the major natural resources are salt water, sand and solar energy.

Life History

Penaeid shrimp are especially suited for aquaculture. The life history of many of the species ranges from the open ocean to estuarine systems and spans large ranges of salinity tolerance. Species that spawn in the open ocean and migrate to bay and estuarine ares during the early stages of their life history are routinely exposed to salinity ranges from 7 to 35 ppt. During periods of increased rainfall penaeids have been collected in near by freshwater areas and under drought conditions have been collected from hypersaline lagoons at 70 ppt salinity (19). This remarkable euryhaline tolerance has made the shrimp extremely adaptive to culture systems where salinity levels fluctuate during the growing season. Although the optimum culture systems operate somewhere between 22 and 35 ppt salinity, pilot facilities have produced commercial production densities with varying ranges of 28 to 50 ppt salinity during a single growing season (38). All penaeid shrimp have a similar life cycle with only slight variations in habitat utilization during specific life stages. Depending on the species, gravid female penaeids capable of spawning

viable eggs, as well as recently spawned females, have been captured
from the edges of the continental chelf to the surf zone. The penaeid
shrimp are characteristically free spawners. The fertilized eggs
when hatched undergo a series of five naupliar stages, three protozoeal stages and three mysis stages before molting into the postlarval
form. The postlarval shrimp exhibits rapid growth through its juvenile form to the adult stage, a portion of the life cycle commonly
found in bays and estuaries.

The combination of characteristics that have made the penaeid
shrimp a viable aquaculture candidate and especially suitable to the
biosaline concept are its tolerance of broad salinity and temperature
ranges, large fecundity, short larval life, and rapid growth. Especially attractive is the shrimp's adaptability to both extensive
culture techniques, where the growout system provides the majority of
the animal's nutritional needs through natural productivity, and
intensive culture,where these requirements are added to the system.

Broodstock

Postlarval shrimp used to stock commercial growout systems have
been obtained by trapping young shrimp in sea-level earthen ponds on
a receding tide, by supplying hatcheries with captured mated females
from coastal waters and by breeding shrimp in captivity. Trapping
postlarvae in sea-level ponds is used in several countries where
low labor costs are compatible with low density growout systems that
require labor intensive means of predator control and harvest. In
more intensive culture systems, requiring more animals and a known
stocking density of postlarval shrimp, gravid females are collected
from coastal waters and spawned in a hatchery. In this process, short
tows are made using an otter trawl and the collected shrimp are examined for ovarian maturation and the presence of a spermatophore.
The shrimp are then maintained at 18-19°C and transported to the
laboratory where they are heat shocked to induce spawning. Reported
spawns range from approximately 9,000 to 750,000 eggs depending on
species, size and maturity of the animal, however, the average spawn
is about 250,000 eggs. The disadvantages of sourcing spawners from
the wild have been the economics of boat operation and the seasonality
and unpredictability of obtaining gravid females. Reproduction of
penaeid shrimp in captivity has recently been accomplished with a
number of species (9,17,21,31). This life cycle control, long considered the highest priority in penaeid research, will not only
eliminate the total dependency on wild spawners but will lead to the
development of a domesticated stock which is a prerequisite for the
optimum development of any animal production industry. There is
little doubt that captive reproduction will completely replace "wild"
broodstock. However, until that time the industry will be supported
by stock from the sea (10,18).

Hatchery Technology

Much of the early production work both in the eastern and western hemisphere was based on stocking ponds with postlarval and juvenile shrimp collected directly from bays and estuaries. Production
figures increased dramatically with the development of hatchery technology by Japanese and U.S. scientists. The fundamental differences
between the Japanese and American hatchery systems are the tank size,
density of larvae and methods for providing feed for the larvae. The
Japanese system using $60m^2$ to $25m^2$ unlined concrete tanks rely on the
promotion of natural blooms of algae and zooplankton as a food source
for young penaeid larvae, supplemented by the addition of rotifers,
oyster eggs and brine shrimp nauplii. In contrast the U.S. system
using 1000-3000 l conical-bottom fiberglass tanks maintains a given
density of food based on a function of water volume rather than an
absolute number of larvae. The predetermined number of larvae,
usually 250/l graze on a given density of prepared algae and freshly
hatched Artemia nauplii. Commercial hatcheries throughout the world
have adopted both methods along with a wide variety of modifications
of each (13,14,30,47). Although both systems could be employed in
a biosaline environment, most laboratories in arid and semi-arid
regions employ the more compact, intensive U.S. system. Briefly,
these systems are maintained at 28°C and a salinity of 30 ppt. Within 14-18 hours the eggs hatch releasing typical crustacean naupliar
larvae. In the next 28-35 hours the naupliar larvae will undergo
four additional molts until they reach fifth stage naupliar larvae.
During this period no external feed is required as the larvae are
sustained on an internal supply of yolk.

The next molt transforms the larval penaeids into the first of
three protozoeal stages, each equipped with a functional digestive
system. For approximately 6 days protozoea are fed a variety of
marine phytoplankton including Skeletonema, Thalassiasira, Phaeo-
dactylum, Tetraselmis, and Cyclotella. Approximately eight days
after spawning the larvae molt into the first of three mysis stages
which clean up the remaining algae and begin feeding on freshly
hatched Artemia nauplii. Within 3-4 days the larval penaeids molt
into the post larval form, continue feeding on Artemia nauplii and
are gradually switched to a suitable commercially prepared flake
diet. Within 8-12 days the post larvae are easily transferable to
any type of growout system (30,47).

Growout Systems

Successful growout systems for the production of penaeid shrimp
have been demonstrated in Japan, the Philippines and in Central
America (14,18), however, in keeping with the philosophy of this
chapter the authors have chosen two pilot systems that have direct
application to the biosaline environment. Two of the most successful

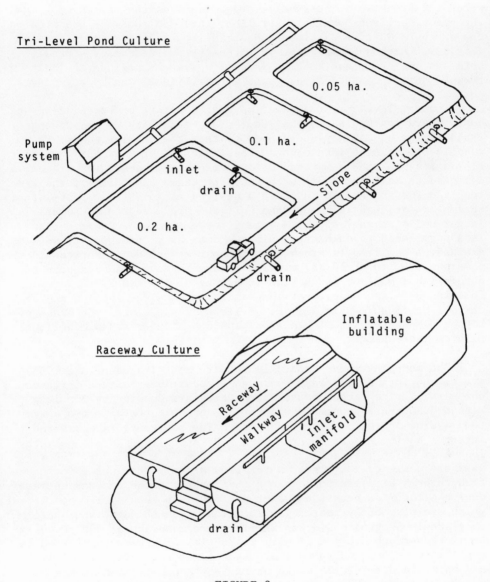

FIGURE 2

Intensive culture system for _Penaeid_ Shrimp.

designs are the pilot production system employed in Corpus Christi,
Texas, U.S.A. consisting of a tri-level pond system capable of moving
water and stock to successively larger ponds as the shrimp grow, and
the pilot, environmentally controlled, enclosed raceways operated on
the northern Gulf of California near Puerto Penasco, Sonora, Mexico
(Fig. 2). Although different in intensity and degree of control,
both systems are classified as intensive systems because of their increased animal densities that require nutritional and energy input.

The tri-level pilot production unit consists of three adjacent
ponds measuring 0.05ha, 0.1ha, and 0.2ha, each with water-exchange
capabilities of 25%/day and sloped so that the smaller ponds are
capable of draining into each successively larger unit (38). Commercial units would be comprised of larger pond systems. The system is
designed to hold shrimp in each pond for about one-third of the time
required for the shrimp to reach a desired market size. Post-larval
shrimp are stocked in the 0.05ha unit and are moved to successively
larger units as they grow. Movement between ponds and harvesting
is accopmlished through drain pipes. Soon after the nursery pond is
drained, a new crop is stocked in the nursery pond and rotated in the
same manner. In a 210-day growing season three crops have been produced with a maximum production of 921 kg of 12 gram P. vannamei
followed by a second crop of 12 gram P. vannamei at 1065 kg and a
third crop of 80mm bait size P. stylirostris at 218kg. In areas of
higher temperature growing seasons would be longer and production
figures increased.

The more inventive controlled environmental aquaculture (CEA)
system consists of raceways enclosed in positive-pressure inflated
domes (49). The "green-house" strucutres are 7m wide by 30m long
with a 0.9m center concrete walkway. Each unit contains two 4x24m
raceways of approximately 0.6m depth. The polyvinyl raceways and
internal equipment are covered with a UV-stabilized 10 mil polyethylene film which when inflated forms a 7x30m half cylinder. The raceways are designed for linear water flow with lateral water introduction through side manifolds along the length of the raceway and are
designed for 7 exchanges of water per day. In this system postlarvae,
which are previously reared in fiberglass flow-through tanks until
they are 10-50mg, are stocked in miniature raceways at 300 to $600/m^2$
until they reach 0.6-1.2 grams. The shrimp are then transferred to
a CEA growout unit where they are stocked at approximately $170/m^2$
and maintained for 20-25 weeks until they are harvested at approximately 21 grams. Using these methods data indicate production units
are capable of producing a 1.25g/week growth rate with a harvest
density of $3.3kg/m^2$.

Both the tri-level pond system and the CEA units are designed
for a biosaline environment and represente alternative approaches to
shrimp culture. The application of either system must be measured
in terms of available energy, space, capital investment and level of

environmental control necessary for profitable returns. The tri-
level unit while requiring more space and exhibiting less environmental control requires less energy input into the system. In turn,
if the location proposed for facility siting had no access to soils
capable of retaining water and liners were prohibitive, the CEA system would be more applicable. The CEA concept also requires a lining
substance in sand, however, the increased density per square meter
and higher production rates may offset the capital investment. In
the tri-level system if liners were a prerequisite it is recommended
that a substrate capable of supporting a benthic infauna be employed
to retain the nutritional advantage the system enjoys. This nutritional advantage may be responsible for the extremely low disease
rate in pond systems as compared to the animal health problems
common to CEA systems. There are obvious advantages and disadvantages inherent in all systems and it should be recognized that this
is not a discussion to define the optimum system. As in all other
aspects of protein production the individual characteristics of each
location and the available levels of energy will dictate system design. The remarkable adaptive capacity of marine shrimp is sometimes very forgiving of the stress parameters we initiate through
our lack of understanding of the "optimum" commercial design.

Figure 3 diagrammatically illustrates the life stages and culture technology used for Penaeid shrimp.

SYSTEMS INTEGRATION

Aquaculture in a biosaline environment ideally should be based
on an integrated system in order to utilize the total benefits provided by climatic conditions and high solar energy. Such a system
should include species located in different positions on the food
chain and utilize either intensive or extensive culture, or a combination of both depending on the geographic location and the capital
potential of the area. A hypothetical example of such a system is
shown in Figure 4.

Unicellular algae as primary producers utilizing solar energy
are the first step in this system. This step requires a significant
nutrient input, particularly N_2 fixation (cyanobacters); the potential application of which has been discussed by Mitsui (29) and
Rains (39). The primary producers have application as:

- a food source for rotifers and brine shrimp;
- a live food source for larval shrimp;
- a component in the preparation of artificial feeds for
  numerous animals including penaeid shrimp and mullet;
- a protein source for human food;
- chemical components for industrial use.

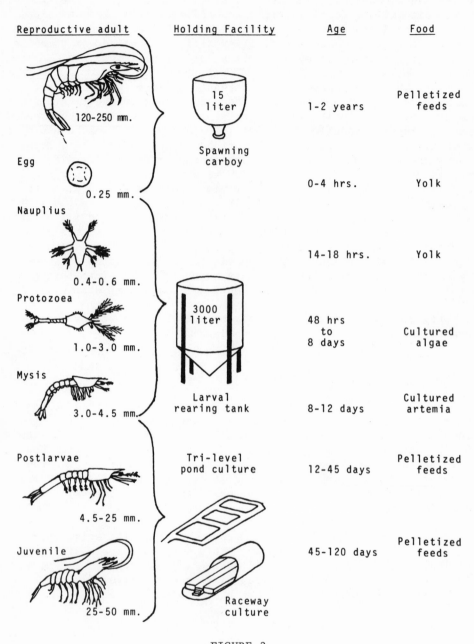

Reproductive adult	Holding Facility	Age	Food
120-250 mm.	15 liter Spawning carboy	1-2 years	Pelletized feeds
Egg 0.25 mm.		0-4 hrs.	Yolk
Nauplius 0.4-0.6 mm.		14-18 hrs.	Yolk
Protozoea 1.0-3.0 mm.	3000 liter	48 hrs to 8 days	Cultured algae
Mysis 3.0-4.5 mm.	Larval rearing tank	8-12 days	Cultured artemia
Postlarvae 4.5-25 mm.	Tri-level pond culture	12-45 days	Pelletized feeds
Juvenile 25-50 mm.	Raceway culture	45-120 days	Pelletized feeds

FIGURE 3

Life stages and culture technology for Penaeid Shrimp.

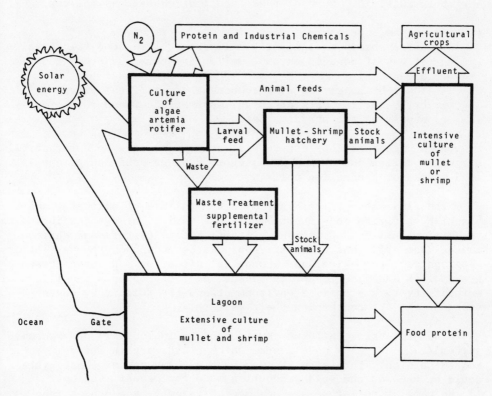

FIGURE 4

Integrated aquaculture system for a biosaline environment.

The effluent from the algal systems can be utilized as a fertilizer in the extensive (lagoons) production systems for penaeid shrimp
and mullet.

The second step in this integrated unit is the production of
zooplankers. In our example we are using rotifers and brine shrimp.
While the use of rotifers in this system is limited to the feeding of
the early larval stages of mullet, the potential application of brine
shrimp is multiple. Some of these applications are as follows:

- use as a human food source;
- brine shrimp nauplii are an important live food for larval
  stages of penaeid shrimp, mullet and numerous other aquatic
  organisms;
- adult brine shrimp as a component of artifical feeds;
- brine shrimp cysts are highly sought after and in short
  supply on the world market.

Both algae and zooplankers will be used as live food for larval
stages of mullet and shrimp in hatchery systems designed to supply
seed stock for production units.

The final step in this sytem is the culture of saltwater and
brackishwater fish and invertebrates. A variety of species including
tilapia, milkfish, rabbitfish, gilthead bream, oysters and shrimp
may be used efficiently for this purpose depending on the particular
geographical, ecological and economical situation. We have chosen
grey mullet and penaeid shrimp because of their obvious physiological
suitability to a biosaline environment and the relatively advanced
state of the art concerning their culture.

The production units we have described include: 1) an extensive
system (lagoon); 2) a pond system which could vary in intensity;
and, 3) a highly intensive raceway system. Any one of these systems
could be used as the sole production mode for penaeid shrimp; however,
in the case of mullet we are unaware of raceway-type production.
Polyculture with mullet and shrimp may be applicable in the ponds and
lagoon systems, though published data on such systems is not available.

An additional benefit of the intensive pond and raceway units is
their high rate of water exchange resulting in effluents rich in inorganic phosphates and nitrogenous compounds. The use of such effluents for salt tolerant crop irrigation as proposed in the Chapter by
Epstein and Norlyn is very attractive.

REFERENCES

1. Albertini-Berhaut, J. 1973. Biologie des stades juvenile de
    Teleosteens Mugilidae (Mugil auratus Risso 1810, Mutil capito
    Cuvier 1829 et Mugil saliens Risso 1810). I. Regime alimentaire.
    Aquaculture, 2:251-266.

2. Albertini-Berhaut, J. 1974. Biologies des stades juvenile de
    Teleosteens Mugilidae (Mugil auratus Risso 1810, Mugil capito
    Cuvier 1829 et Mutil saliens Risso 1810). II. Modifications du
    regime alimentaire en relation avec la taille. Aquaculture,
    4:13-27.

3. Bardach, J.E., J.H. Ryther, and W.O. McLarney. 1972. Aquaculture
    The Farming and Husbandry of Freshwater and Marine Organisms.
    John Wiley and Sons, Inc.

4. Bell, T. 1978. Ed. Fisheries of the United States: Current Fish-
    eries Statistics No. 7500. U.S. Dept. Commerce. NOAA--S/T78-237.
    112 pp.

5. Bishara N.F. 1978. Fertilizing fish ponds. II Growth of Mugil
    cephalus in Egypt by pond fertilization and feeding. Aquaculture,
    13: 361-367.

6. Braekkan O.R. 1962. B-vitamins in fish and shellfish. In: E.
    Heen (Ed.) "Fish in Nutrition", Fishing News (Books) Ltd., pp.
    132-140.

7. Cech, J.J., Jr. and D.E. Wohlschlag. 1973. Respiratory responses
    of the striped mullet, Mugil cephalus (L.) to hypoxic conditions.
    J. Fish. Biol. 5:421-428.

8. Conte, F.S. 1975. Penaeid shrimp culture and the utilization of
    waste heat effluent. Power plant waste heat utilization in aqua-
    culture, Workshop 1. Trenton, New Jersey, PSE7G, NSF, FANN,
    Trenton State College, Rutgers University 1:23-47.

9. Conte, F.S., M.J. Duronslet, W.H. Clark, Jr., J.C. Parker. 1977.
    Maturation of Penaeus stylirostris (Stempson) and P. setiferus
    (Linn) in hypersaline water near Corpus Christi, Texas. Proc.
    World Mariculture Society 8:327-334.

10. Conte, F.S. 1978, Penaeid Shrimp Culture: Current Status and
    Direction of Research. In: Drugs and Food from the Sea: myth
    or reality? Ed. P.N. Kaul and C.J. Sindermann. Univ. Oklahoma
    Press. pp. 333-343.

11. DeSylva, S.S. and Wijearatne, M.J.S. 1977. Studies of the biology
     of young grey mullet, Mugil cephalus L. II. Food and Feeding.
     Aquaculture, 12:157-167.

12. Fishelson, L. and Popper, D. 1968. Experiments on rearing fish
     in salt waters near the Dead Sea, Israel. In: "Proceedings of
     the World Symposium on warm-water pond fish culture". FAO Fish.
     Rep., 44(5): 244-245.

13. Fujinaga, G. 1961. Techniques of shrimp farming during zoea
     mysis and postlarval stages. Tokyo Koho, May 11, 1961. Patent
     application 34-27796:1-3.

14. Fuginaga, M. 1963. Culture of Kuruma shrimp (Penaeus japonicus)
     Curr. Appl. Bull. Indo-Pac. Fish Council (36):10-11.

15. Ganapati, S.V. and Alikuni K.H. 1952. Experiments on the ac-
     climatization of salt-water fish seeds to fresh water. Proc.
     Indian Acad. Sci. 39:93-102.

16. Glude, J.G. 1978. The contribution of fisheries and aquaculture
     to world and U.S. food supplies. In: Drugs and food from the
     sea: myth or reality? Eds. P.N. Kaul and C.J. Sindermann.
     Univ. Okla. Press. pp. 235-247.

17. Griessinger, J.M. 1975. Maturation and spawning in captivity
     of Penaeid prawns Penaeus merguiensis de Man, P. japonicus Bate.
     P. aztecus Ives and Metapenaeus ensis (de Mann). Proc. World
     Mariculture Society 6:123-132.

18. Hanson, J.A. and H.L. Goodwin. 1977. Shrimp and prawn farming
     in the Western Hemisphere: State of the art reviews and status
     assessment. Dowden, Hutchinson and Ross, Inc. Penn, p. 439.

19. Hedgpeth, J.W. 1967. Ecological aspects of the Laguna Madre, A
     hypersaline Estuary. In: Estuaries. Ed. G.H. Lauff. Pub. No.
     83 AAAS. pp. 408-419.

20. Iversen, E.S. 1968. Farming the edge of the sea. Fishing News
     (Books) Ltd. London, E.C.4. pp. 301.

21. Johnson, M.C. and J.R. Fielding. 1956. Propagation of the white
     shrimp, Penaeus setiferus (Linn.) in captivity. Tulane Studies
     Zoology 4(6):175-190.

22. Kuo, C.M., Z.H. Shehadeh and C.E. Nash. 1973. Induced spawning
     of captive grey mullet (Mugil cephalus) females by injection of
     human chorionic gonadotropin (HCG). Aquaculture, I:429-432.

23. Kuo, C.M., C.E. Nash, and Z.H. Shehadeh. 1974. A procedural
     guide to induce spawning in grey mullet (Mugil cephalus L.).
     Aquaculture, 3:1-14.

24. Kuo, C.M., C.E. Nash and Z.H. Shehadeh. 1974a. The effects of
     temperature and photoperiod on ovarian development in captive
     grey mullet (Mugil cephalus L.). Aquaculture 3:25-43.

25. Liao, I.C., Y.J. Lu, T.L. Huang and H.C. Lin. 1972. Experiments
     on induced breeding of the grey mullet, Mugil cephalus Linnaeus.
     In: T.V.R. Pillay (Ed.) "Coastal Aquaculture in Indo-Pacific
     Region." Fishing News (Books) Ltd. pp. 213-243.

26. Ling, S.W. 1967. Feeds and feeding of warm-water fishes in
     ponds in Asia and the Far East. In: Proceedings of the World
     Symposium on warm-water pond fish culture." FAO Fish. Rep. 44(3):
     291-309.

27. Luther, G. 1962. The food habits of Liza macrolepis (Smith) and
     Mugil cephalus Linnaeus (Mugilidae). Indian J. Fish. 9(2):604-
     626.

28. McFarland, W.N. and S.A. Moss. 1967. Internal behavior in fish
     schools. Science, NY 156:260-262.

29. Mitsui, A. 1977. Marine algae and aquatic plants as food compo-
     nents. In: Microbial Conversion Systems for Food and Fodder
     Production and Waste Management. Ed. T. Overmire. Kuwait Insti-
     tute for Scientific Research and Kuwait University. pp. 3-31.

30. Mock, C. and R.A. Neal. 1974. Penaeid shrimp hatchery systems.
     FAO United Nations. CARPAS 6-74-SE29,pp.9.

31. Moore, D.W., R.W. Sherry and F. Montanez. 1974. Maturation of
     Penaeus californiensis in captivity. Proc. World Mariculture
     Society 5:445-449.

32. Myers, S.P. 1977. Marine fish farming in Israel. Feedstuffs.
     49(34):23-25.

33. Nash, C.E., C.M. Kuo and S.C. McConnel. 1974. Operational pro-
     cedures for rearing larvae of the grey mullet (Mugil cephalus L.)
     Aquaculture, 3:15-24.

34. Nash, C.E., C.M. Kuo, W.D. Madden and C.L. Paulsen. 1977. Swim
     bladder inflation and survival of Mugil cephalus to 50 days.
     Aquaculture, 12:89-94.

35. Nichols, J.P., W.L. Griffin and V. Blomo. 1978. Economic and production aspects of the Gulf of Mexico shrimp fishery. In: Drugs and food from the sea: myth or reality? Ed. P.N. Kaul and C.J. Sindermann. Univ. Okla. Press. pp. 301-315.

36. Odum, W.E. 1978. The ecological significance of fine particle selection by the striped mullet Mugil cephalus. Limnol. Oceanogr. 13:92-98.

37. Odum, W.E. 1970. Utilization of the direct grazing and plant detritus food chains by the Striped mullet Mugil cephalus. In: J.H. Steel (Ed.) "Marine Food Chains." Oliver and Boyd. pp. 222-240.

38. Parker, J.C., F.S. Conte, W. McGrath and B. Miller. 1974. An intensive culture system for penaeid shrimp. Proc. World Mariculture Society 5:65-80.

39. Rains, D.W. 1977. Adaptations of Biological Systems to enhance use of saline waters in arid environments. In: Microbial Conversion Systems for Food and Fodder production and waste management. Ed. T.C. Overmire. Kuwait Institute for Scientific Research and Kuwait University. pp. 55-79.

40. Salser, B., L. Mahler, D. Lightner, J. Ure, D. Donald, C. Brand, N. Stamp, D. Moore and B. Colvin. 1978. Controlled environmental aquaculture of penaeids. In: Drugs and food from the sea: myth or reality? Ed. P.N. Kaul and C.J. Sindermann. Univ. Okla. Press. pp. 345-355.

41. Shehadeh, Z.H., C.M. Kuo, and C.E. Nash. 1973. Establishing brood stock of grey mullet (Mugil cephalus L.) in small ponds. Aquaculture, 2:379-384.

42. Simmons, E.G. 1957. An ecological survey of the upper Laguna Madre of Texas. Publ. Inst. Mar. Sci. Univ. Tex. 4:156-200.

43. Sindermann, C.J. 1978. Food production from the sea. In: Drugs and food from the sea: myth or reality? Ed. P.N. Kaul and C.J. Sindermann. Univ. Okla. Press. pp. 233-234.

44. Suzuki, K. 1965. Biology of striped mullet Mugil cephalus Linne. I. Food contents of young. Rep. Fac. Fish. Prefect. Univ. Mie 5(2): 295-305.

45. Tal, S. and Ziv, I. 1978. Culture of exotic species in Israel. Bamidgeh, 30(1):3-11.

46. Thomson, J.M. 1966. The grey mullets. Oceanogr. Mar. Biol. Ann. Rev. 4:301-335.

47. Yang, W.T. 1975. A manual for large-tank culture of penaeid
     shrimp to the postlarval stages. Univ. of Miami sea grant Tech.
     Pub. No. 31, pp. 94.

48. Yashouv, A. 1968. Mixed fish culture. In: "Proceedings of the
     World Symposium on warm-water pond fish culture" FAO Fish.
     Rep. 44(4):258-273.

49. Yashouv, A. 1969. Preliminary report on induced spawning of
     Mugil cephalus (1.) reared in captivity in fresh water ponds.
     Bamidgeh 21:19-24.

BIOLOGICAL WASTE TREATMENT AT ELEVATED TEMPERATURES AND SALINITIES

William J. Oswald and John R. Benemann

Sanitary Engineering Research Laboratory
University of California
Berkeley, CA 94720

INTRODUCTION

The subject of waste treatment at elevated temperatures and
salinities is, perhaps, best introduced by briefly reviewing the history, technology, and fundamentals of waste treatment under ordinary
conditions of climate and water quality. The biological treatment of
liquified human wastes as an essential municipal activity has been a
development of the last 100 years, initiated primarily in Northwestern
Europe and Northeastern United States in response to stream pollution
and epidemics of cholera and other water-borne diseases brought about
by indoor plumbing and the discharge of wastewaters into drainage
channels, streams, rivers, lakes, and the sea. The waste treatment
processes were designed first to remove large floatable and settleable
solids from waste streams by screening and sedimentation and to dis-
pose of these solids in separate systems by burial, drying, incinera-
tion, or fermentation. In the late 1800's, it became evident that
physical removal of solids from waste streams was often not sufficient
and that colloidal and soluble organic materials in waste streams also
had to be removed if the depletion of life-giving dissolved oxygen in
waters receiving the wastes was to be avoided. During this period,
it became evident that by aerating the wastes under controlled conditions of time and temperature, aerobic microbial growth would occur
which would remove most of the biologically degradable organic matter
from the waste and incorporate it in particulate biomass subject to
physical removal. Such removal of soluble organics is usually referred to as secondary sewage treatment to distinguish it from the
simple physical solid removal processes of primary treatment.

Two general forms of secondary treatment emerged from these
early experiences--the first called "biofiltration," (BF) involved

285

passing wastes in thin layers over inert surfaces exposed to the air.
Films of microbes grew on these surfaces at the expense of nutrients
and oxygen from the passing stream. As these films thickened, they
sloughed from the surfaces and were subject to physical removal from
the waste stream and subsequent disposal. The second process, often
called "activated sludge," (AS) involved the growth of particulate
aerobic microbial suspensions under conditions of intense aeration
and mixing followed by a stage of quiescent sedimentation. A fraction of the settleable material, returned to mix with the influent,
maintains the correct soluble organic matter to microbe ratio for high
organic removal, and the balance of the settleable material is dis-
posed with primary sludge. These two methods, and their numerous
variations, of secondary biological waste treatment have come to be
regarded by the sanitary engineering profession as conventional technology to be applied where ever secondary waste treatment may be required.

In recent years, it has become evident that, while these two
methods may be applied worldwide, they are not necessarily the best
alternative for secondary treatment under conditions of high temperature and salinity prevalent in low latitude regions. It should be
emphasized that such treatment was developed in northern climates--
geographical areas characterized by low temperatures and low salini-
ties--and these technologies have predominated the field since their
inception. Such conventional energy-intensive systems involve many
complex high-technology elements and assemblages; they are difficult
to apply in developing countries. They exhibit strong economy of
scale, often leading to the conclusion that large area-wide collection
and treatment systems have lower unit costs than smaller decentralized
systems. Such a conclusion may be erroneous for new kinds of treatment more relevant to high temperature and high salinity areas. Also
conventional secondary treatment has been found to lead to severe
environmental problems under many high-temperature and high-salinity
conditions. Finally, in conventional secondary treatment, most water,
fertilizer, and energy values are lost so that reclamation opportunities are minimized. Non-conventional methods of waste treatment,
based on microalgae and other aquatic plants, are now available and
of greater applicability to arid areas.

HIGH TEMPERATURE AND SALINITIES: CAUSES AND EFFECTS

Before further discussing these subjects, it is worthwhile to
first review the basic underlying factors related to high temperatures
and high salinities and to relate them to problems of waste management. These conditions, of course, stem from geography and climate.
It is obvious that high temperature is related to a high flux of
solar energy reaching from the earth's surface. An examination of a
solar energy table, Table 1, shows that the greatest average amounts

TABLE 1

Solar Radiation[a]

Probable Average Values of Insolation—Direct and Diffuse—on a Horizontal Surface at Sea Level, in Langleys[b] per Day

NORTH LATITUDE Degree	Range	Jan		Feb		Mar		Apr		May		Jun		Jul		Aug		Sep		Oct		Nov		Dec	
		vis[c]	tot[d]	vis	tot	vis	tot	vis	tot	vis	tot	vis	tot	vis	tot	vis	tot	vis	tot	vis	tot	vis	tot	vis	tot
0	max[e]	255	685	266	700	271	708	266	690	249	645	236	626	238	630	252	666	269	690	265	694	256	683	253	667
	min[f]	210	580	219	583	206	536	188	462	182	480	103	274	137	368	167	432	207	533	203	530	202	543	195	527
2	max	250	670	263	693	271	706	267	697	253	655	241	642	244	646	255	673	269	693	262	688	251	666	249	646
	min	206	560	213	560	204	534	188	464	184	484	108	288	141	375	169	442	206	531	200	523	198	526	189	505
4	max	244	650	259	688	270	704	268	701	258	665	247	656	250	657	258	678	269	695	260	680	246	650	244	628
	min	200	540	206	543	202	532	187	466	187	492	113	300	146	385	171	448	204	529	196	513	194	510	183	480
6	max	238	630	254	675	268	702	270	705	262	675	252	668	255	669	261	683	269	697	256	670	240	634	238	610
	min	193	520	199	530	200	530	186	467	189	500	118	310	150	395	172	452	202	524	191	500	188	494	176	460
8	max	230	610	249	665	267	700	270	709	266	685	258	678	260	680	263	688	267	695	252	660	234	616	231	590
	min	187	495	192	510	196	523	185	467	191	506	124	320	154	405	174	456	200	518	186	486	182	478	169	440
10	max	223	595	244	655	264	694	271	711	270	694	262	688	265	690	266	693	266	693	248	650	228	600	225	570
	min	179	475	184	490	193	513	183	464	192	512	129	330	158	414	176	460	196	510	181	474	176	462	162	420
12	max	216	572	239	645	262	690	271	710	273	702	267	700	269	700	267	697	264	691	244	640	221	585	217	550
	min	172	455	176	470	189	500	181	462	193	518	133	343	161	421	176	464	193	502	176	462	169	446	154	400
14	max	208	555	233	630	258	680	271	709	276	710	272	710	273	708	269	700	262	688	240	627	214	567	209	536
	min	163	430	167	450	184	487	179	460	194	524	137	354	164	429	177	467	189	496	170	449	162	430	146	380
16	max	200	530	226	610	255	670	272	707	279	718	276	720	277	715	270	703	259	684	234	615	206	554	200	520
	min	154	400	159	430	180	473	177	456	194	528	141	363	167	435	177	469	185	489	164	434	154	410	138	360
18	max	192	515	220	590	250	664	272	705	282	723	280	728	280	723	272	705	256	680	229	605	198	538	192	500
	min	144	380	150	410	174	459	174	452	194	530	145	375	170	442	177	471	180	479	157	418	146	390	129	340
20	max	183	500	213	575	246	652	271	703	284	730	284	738	282	729	272	706	252	674	224	596	190	520	182	480
	min	134	360	140	390	168	440	170	447	194	532	148	383	172	450	177	472	176	467	150	400	138	370	120	320
22	max	174	480	206	560	241	644	270	701	286	734	286	747	285	736	273	707	248	668	218	582	183	500	172	460
	min	123	335	132	370	162	426	167	440	193	530	152	392	173	454	176	472	170	455	143	380	128	350	110	300
24	max	166	460	200	545	236	625	268	697	288	738	290	753	287	742	273	708	244	659	212	568	175	480	161	440
	min	111	310	123	340	156	410	164	433	191	525	155	403	176	459	174	471	165	443	136	360	119	326	101	280
26	max	156	440	192	530	230	615	266	690	288	741	292	760	288	749	273	706	240	652	205	552	166	460	149	420
	min	99	280	114	310	149	390	160	425	189	518	158	409	177	463	172	469	160	429	128	332	109	300	90	260
28	max	146	420	184	510	224	603	264	683	289	743	294	764	288	755	272	704	236	635	199	537	157	440	138	400
	min	87	250	106	290	142	373	156	415	187	506	161	418	178	467	169	466	154	415	120	310	99	278	80	236
30	max	136	400	176	490	218	587	261	675	290	744	296	768	289	759	271	702	231	625	192	524	148	420	126	380
	min	76	220	96	260	134	362	151	405	184	490	163	425	178	469	166	462	147	399	113	290	90	256	70	210
32	max	126	380	169	470	212	570	258	663	290	744	296	772	289	761	269	700	226	615	185	510	138	400	114	360
	min	63	180	87	240	126	340	146	395	181	475	166	431	178	472	163	458	140	385	104	270	80	224	60	184
34	max	114	360	160	450	204	553	254	657	290	743	297	775	289	763	267	696	221	602	178	490	128	380	101	338
	min	53	155	78	215	118	320	141	385	176	462	168	439	178	472	159	448	134	368	96	250	70	202	47	158
36	max	103	335	150	430	196	538	250	650	288	741	298	776	289	765	264	690	215	590	170	470	118	360	88	314
	min	44	133	70	200	111	300	136	375	172	444	170	443	177	470	155	438	127	350	88	230	60	180	39	134
38	max	90	310	140	415	189	520	246	640	287	738	298	778	288	766	262	684	210	576	162	450	106	336	77	290
	min	36	120	62	180	103	280	131	365	166	428	171	448	175	464	152	429	120	330	80	216	50	158	30	111
40	max	80	280	130	390	181	500	241	630	286	732	298	778	288	765	258	680	203	562	152	430	95	313	66	270
	min	30	105	53	160	95	270	125	355	162	415	173	450	172	455	147	416	112	310	72	202	42	134	24	94
42	max	68	255	119	370	172	485	236	618	283	728	298	777	287	761	254	670	196	547	144	410	84	289	56	244
	min	24	90	45	140	88	250	120	344	157	405	174	451	167	442	143	403	105	290	65	187	34	112	19	78
44	max	55	228	106	340	165	470	230	607	280	722	298	777	285	755	250	660	189	530	132	390	72	263	47	218
	min	20	80	37	130	80	230	114	325	153	395	175	453	164	430	139	389	98	270	58	173	28	98	15	62
46	max	45	200	94	315	156	450	224	598	278	716	298	776	284	749	245	650	181	512	122	370	61	238	39	194
	min	16	74	30	110	72	210	108	315	150	385	175	455	161	420	134	374	90	250	52	158	23	86	11	48
48	max	35	180	82	290	149	430	218	582	274	710	297	776	282	740	241	640	174	496	111	350	50	210	32	170
	min	12	64	25	99	64	190	102	307	146	378	176	458	158	410	129	358	81	230	45	144	18	70	9	37
50	max	28	164	70	265	141	410	210	568	271	703	297	776	280	733	236	625	166	480	100	329	40	183	26	144
	min	10	54	19	80	58	173	97	300	144	371	176	458	155	403	125	342	73	210	40	130	15	60	7	30
52	max	22	140	60	240	134	390	202	555	267	695	296	776	278	725	232	615	158	460	87	307	32	160	21	121
	min	8	45	14	62	51	158	92	295	141	366	176	460	153	398	120	326	65	190	34	120	12	53	4	27
54	max	16	120	50	215	126	370	194	542	263	687	296	776	276	720	224	602	150	440	76	285	25	140	16	99
	min	6	40	11	45	46	145	88	289	139	360	176	460	150	394	116	312	58	170	29	106	9	43	3	23
56	max	12	102	43	200	120	350	188	528	258	680	295	775	273	714	218	587	141	420	64	261	20	120	12	78
	min	4	35	8	35	41	132	85	283	136	352	175	460	148	390	110	297	51	150	24	95	7	36	2	18
58	max	9	80	37	170	113	330	182	516	254	670	294	774	270	710	212	575	134	402						
	min	3	28	6	28	37	118	82	277	134	346	175	460	146	385	106	285	44	132						
60	max	7	64	32	150	107	310	176	500	249	660	294	773	268	708	205	556	126	386						
	min	2	20	4	20	33	105	79	270	132	340	174	460	144	380	100	270	38	116						

Algae Research Project
Sanitary Engineering Research Laboratory
Department of Engineering
University of California, Berkeley

[a] Calculated from data published by the United States Weather Bureau.

[b] Gram calories per square centimeter.

[c] "Visible" = radiation of wave lengths of 4000A° to 7000A° penetrating a smooth water surface.

[d] "Total" = radiation of all wave lengths in the solar spectrum.

[e] Value which will not normally be exceeded.

[f] Value based on, or extrapolated from, lowest values observed for indicated month and latitude during 10 years of record.

Approximate corrections for elevation up to 10,000 ft.:

1. Total radiation: tot (1 + 0.0185 El.)
2. Visible radiation: vis (1 + 0.00925 El.)

Where El. is in thousands of feet.

Correction for cloudiness (approximate):

Min + [(max − min) cl.]

Where cl. is fraction of time weather is clear.

FIGURE 1

Although the hydrology of the Earth is complex in this view from a manned space craft above central Africa, the five major hydrologic zones of the Earth are quite visible. Above and to the right of the Mediterranean Sea is the cloud cover of the Arctic wet zone. The north arid zone is distinquishable by the lack of cloud cover over north Africa, the Arabian Peninsula in Iran and West Pakistan. The equatorial wet zone is the cloud cover over Zaire, Kenya, Tanzania, etc. In the south arid zone, a cyclonic storm hovers off the coast of Angola but there is little cloud cover over the remainder of South Africa and the South Indian Ocean. At the bottom of the photo, the cloud cover of the Antarctic wet zone is apparent. Local geographical features such as the Alps, the Himalayas and indeed the continental masses themselves modify the symmetry of the massive movement of water through the atmosphere in the arid to the wet zones.

of solar energy, about 778 cal/sq cm/day, reach the northern latitude
of 40° in June, decreasing to a minimum in December of 270 cal/sq cm/
day. At lower latitudes insolation is more evenly distributed, at a
latitude of 10° total solar energy carries only between 570 and 694
cal/sq cm/day from December to June. These are the maximum values,
reduced by local amounts of cloudiness. In the case of "visible"
radiation, Table 1 refers only to that which penetrates a smooth
water surface, of interest to the microalgae systems described later.

Although the available solar energy is predictable, one cannot
generalize such factors as cloud cover, atmospheric clarity, the presence or absence of wind, the relative humidity, and the presence of
large bodies of water, sand dunes, glaciers, forest, and similar surface features, all of which may influence local temperature and sunlight.

Just as the world can be divided into five geographic zones on
the basis of temperature, one can also conveniently divide the world
into five hydrological zones based on net evaporation--that is, average annual evaporation minus average annual precipitation. One can
then define an arid zone as any locality where annual evaporation
greatly exceeds precipitation. One can usually detect these zones as
permanently visible bands in the earth's atmosphere as seen from space
(Figure 1). Although the borders of these bands are variable depending on season and local disturbances, it is almost always possible
to find the north, south, and tropical wet zones and the north and
south arid zones. It is of great importance that many of the people
of the world reside on the north arid zone. It is within these arid
zones, marked by net evaporation levels from 100 to as much as 300
cm/yr, that most of the deserts and ground water salinity problems
occur. Mountainous areas in arid zones may receive substantial precipitation, but often runoff does not reach the sea but rather
spreads over low lands and evaporates, leaving a residue of minerals
and creating salt beds, salt lakes, and salt-impregnated surface and
ground waters. In California and elsewhere this problem has been
partly overcome by impounding excessive stream runoff, delivering
the water to crops through canals and leaching the accumulated salts
away in separate drains.

Water quality specialists refer to the salts in water as total
dissolved solids (TDS) and recognize a spectrum of water quality
based on TDS much as indicated in Figure 2 in which one can see that
water quality is inversely proportional to the TDS and is related
to use. For example, studies indicate that for greatest comfort and
economy, TDS of domestic water should not exceed 1 gm/L.

In addition to TDS, water quality experts recognize a large
number of use-related water quality criteria including minerals, organic compounds, and dissolved gasses. Most natural waters contain

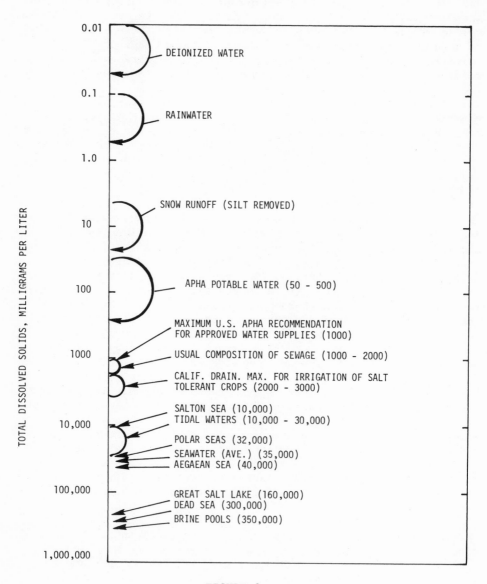

FIGURE 2

Typical total dissolved solids values for various natural waters.

the major cations: sodium, calcium, and magnesium; and the major anions: chloride, sulfate, and bicarbonate. Cations which are usually
minor in an ionic balance but important to water quality effects on
the biota and waste treatment are: nitrate, iodine, phosphate, carbonate, and hydroxide. In the geochemical classification of minerals
in natural waters, carbonate and hydroxide are called alkalinity.
Chloride, sulfate, and nitrate are termed salinity; calcium, magnesium
and iron are termed hardness. Chlorides contribute most of the salinity in natural waters and in the oceans. For example, the salinity
of California's agricultural drainage water is about 2,000 mg/L; the
salinity of sea water is on the order of 17,000 mg/L; and the salinity
of the Dead Sea is about 150,000 mg/L.

With respect to mineral effects on the production of crops, in
addition to TDS, the sodium ratio, defined as the fraction or percentage of the cations represented by sodium, is extremely important.
Only halophylic plants thrive when sodium ratios are greater than
60%; whereas, most plants tolerate sodium ratios of less than 40%.
Higher sodium ratios are tolerated by most plants when the TDS is low
but not when the TDS is high. Bacteria and microalgae are generally
much more tolerant of high sodium ratios and TDS than higher plants,
although many algae and bacteria appear to be inhibited by high
sodium chloride concentrations. Boron is one of the minerals that
appears to adversely affect most higher plants, but there is no
established relationship between boron and the growth of microalgae.

WASTE OXIDATION

Organic substances which have been identified in wastewaters now
number in the thousands and are tabulated in large books such as Water
Quality Criteria (1). Organics in wastes are often measured in terms
of total organic carbon (TOC), chemical oxygen demand (COD), biochemical oxygen demand (BOD), or total volatile solids (TVS). In the BOD
test, which is a bioassay to determine the dissolved oxygen required
to microbiologically oxidize the biodegradable component of the organic matter in a waste, is by far the most useful parameter. In the
BOD test a waste is diluted in water adequately provided with trace
minerals, dissolved oxygen, and microbial flora. The assay is usually
conducted for five days at 20°C. Most sanitary engineering texts
show the derivation of a formula:

$$Y_t = La (1 - 10^{-Kt}) \tag{1}$$

to express the time course of the BOD test. In the equation, Y_t is
the BOD exerted after time t; La is the ultimzte carbonaceous BOD to
be exerted; K is the temperature dependent rate constant per day; and
t is the time in days. The expected value of K at 20°C is approximately 0.1 and K varies with temperature T °C as follows:

$$K_T = K_{20} (1.047^{T-20}) \qquad\qquad (2)$$

in which K_T is the K value at the temperature T °C; T is the temperature in °C; and K_{20} is the rate constant at 20°C. Inasmuchas both
K and La may vary with the type of waste, they are normally determined
experimentally utilizing a sequence of identical assays evaluated over
a period of five to seven days.

With respect to high temperatures and salinities, the BOD is
exerted much more rapidly under conditions of elevated temperatures
and is somewhat inhibited by the presence of sea water. On the other
hand, adapted cultures of microbes have been operated at TDS levels
in excess of 100,000 mg/L with evidence of substantial oxidation.
Most workers feel that for BOD tests to be most meaningful at high
temperatures and salinities, they should be conducted with microbes
that have been adapted to such high temperatures and salinities.
Gases dissolved in water greatly influence the BOD test and the salinity and temperature of water greatly influences the content of dissolved gases. Oxygen, nitrogen, methane, carbon dioxide, ammonia,
hydrogen sulfide, and hydrogen are the more common gases of concern.
Dissolved oxygen is, of course, the requirement for waste oxidation.
Carbon dioxide is essential to microalgae growth, and methane and
hydrogen sulfide are produced during anaerobic decomposition of
wastes. Concerning the influence of water quality and temperature
on the solubility of gases, oxygen is a good example. In Figure 3
the solubility of moleculare oxygen (O_2) is seen to decline with
both temperature and salinity. Thus, both factors tend to adversely
effect the oxygen resources available for waste oxidation. According
to Fair, Guyer, and Okun, at 20°C the difference in dissolved oxygen
per 1,000 mg/L of chlorides is about .088 (2). Extrapolation of this
value to the Dead Sea's 150,000 mg/L of chlorides indicates that for
such salinities little dissolved oxygen will be available at high
temperatures.

PHOTOSYNTHETIC OXYGENATION

An algal assay procedure (1) has come into use in the past ten
years as an endeavor to determine the probability that a given waste
or water will induce nuisance blooms of algae when discharged into
natural water courses. Ammonium, nitrate, organic nitrogen, phosphorus, iron, and, of course, carbon dioxide, stimulate algal growth
under proper conditions of illumination, temperature, seeding, and
mixing. Although chemical analyses of plant nutrients can be highly
accurate, it is not possible to predict algal blooms since toxic substances and environmental conditions may prevent blooms.

In a widely used method, Selenastrum capricornutum, a green alga,
is used to assay for total algal growth potential (AGP). The nitrogen

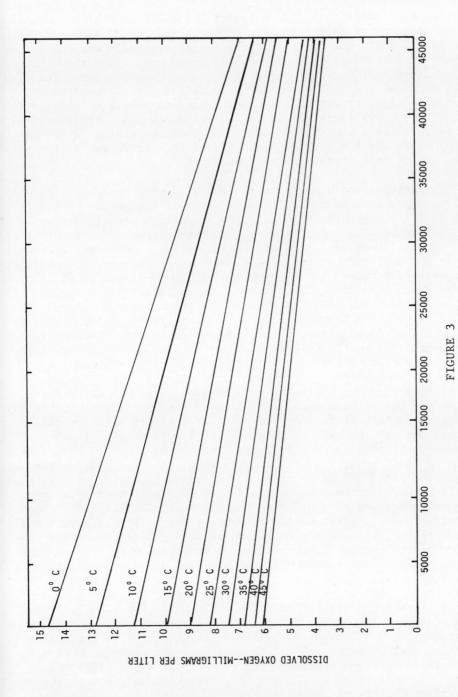

FIGURE 3

Approximate solubility of oxygen in water exposed to an atmosphere containing 20.9% O_2 at 760mm Mercury for displayed temperatures and salinities after Fair Guyer and Okum (2).

fixing blue-green alga, <u>Anabaena flos-aquae</u>, is used to determine the propensity of a water to induce nuisance blooms of blue-green algae. Modified versions of this test have been applied for 25 years in our laboratory to guide work with algal-bacterial symbiosis and the treatment of wastewater in open ponds (4). We have found through this procedure that many types of algae can develop tolerance for toxic substances and, as in the BOD test, algal assays required adaptation. Algal growth in the assay procedure indicates the capacity to use the available nutrients and produce oxygen. This oxygen produced by algae allows the growth of bacteria resulting in BOD reduction. Such algal bacterial symbiosis is applied to waste treatment by use of open "stabilization" ponds.

This photosynthetic oxygenation (PO) of wastes is a practical method of waste treatment used in many areas of the world. Where it can be used it is generally the most economical form of intensive waste treatment, often costing less than one-half as much as activated sludge or biofiltration. However, stabilization type ponds, because of their depth of over one meter and unmixed nature, are not optimized for oxygen production. The high-rate pond was developed to minimize land and energy use for waste treatment by optimization of oxygen and algal production. High-rate ponds are channelized and mixed with a linear velocity of about 5-15 cm/sec with increased mixing velocity of up to 30 cm/sec during short periods, if necessary. The short fast-mix suspends settled sludge that may otherwise becomes sour and malodorous, whereas the slow-mix maintains the algae suspended and in gentle motion and prevents thermal stratification which would deprive surface algae of nutrients and, on warm days, may cause settling of the entire biomass. The important feature of photosynthetic oxygenation is that it is independent of atmospheric aeration and only dependent on the rate of photosynthesis. This rate, in turn, is sensitive to temperature, salinity, and the presence of nutrients. But, under ideal conditions for algal growth, the amount of oxygen in high-rate ponds may be as much as four times the saturation levels shown in Figure 3. Thus, the oxidation of wastes in high-rate ponds can be comparable to that attained with mechanical aeration. Normally, ponds are designed so that the oxygen produced by photosynthesis is about twice that actually required for waste oxidation. This provides sufficient oxygen to carry through the night when photosynthetic oxygenation cannot occur and also provides dissolved oxygen in the pond effluent.

Stabilization ponds, often known as "facultative ponds", have depths of 100 to 200 cm and can provide sufficient oxygen and a stable environment for waste oxidation while avoiding excessive algal growth. Algae from such cultures will often settle in the ponds or when discharged in an isolated settling pond (5). To maximize algal and oxygen production, depths of 20 to 50 cm are required, but provision for algal removal is necessary to avoid accumulations of algae in receiving bodies. Both bioflocculation or rotary screening with 25μ screens could effect inexpensive harvesting of most of the algae, if the ponds are properly operated (6,7,8). Although these methods are not yet well developed a few generalizations can be made. Optimum detention times for production of algae that are harvestable are longer than required for waste treatment and are inversely proportional to the solar flux. At a flux of 250 calories of photosynthetically available radiation (PAR) per day, three days of detention give harvestable algae, whereas at a flux of 100 calories per day PAR, about ten days are required to obtain harvestable algae. Low algae productivity results when the PAR is less than 80 cal/sq cm/day. In fact, photosynthetic oxygenation is not a recommended practice where minimum PAR is less than 100 cal/sq cm/day. Although deep, unmixed ponds are used widely as far north as the Arctic Circle, such ponds obviously do not optimize the use of land and depend only partially on photosynthetic oxygenation, being frozen in winter. They must be sufficiently isolated so that any odors produced go unnoticed.

PATHOGENS AND DISINFECTION

Bacterial content is another imporant biological criteria of water quality that requires some discussion. The coliform bacteria which inhibit the intestines of all warm blooded animals can be detected and roughly quantified in a simple, highly sensitive, assay based on the fermentation of lactose. E. coli organisms are presumed to be present if lactose and brilliant green bile is reduced and if eosine methylene blue is reduced. The presence of E. coli is regarded as an indication of pollution. Assays for other organisms such as fecal E. coli and fecal Streptococcus are more definitely of human origin. At any rate, the philosophy which generally errs on the safe side is that numbers of E. coli and related organisms in human wastes exceed by four or five orders of magnitude the numbers of typhoid, dysentary, and cholera organisms. The tendency is to reject as unsafe any water which consistently has more than one confirmed coliform bacteria per 100 ml. This test is expressed as the "most probable number (MPN) of bacteria." This sensitive microbiological test has doubtlessly saved millions of human lives since its first use, but it is continuously criticized by those adversely affected by its rigid application.

Only little coliform removal is effected in activated sludge or biofiltration, and, thus, effluents from such processes must be disinfected if they are to go into waters to be contacted by people or used for livestock or crop production. On the other hand, disinfection in ponding systems and, particularly, high-rate ponding systems is high as will be described below.

Disinfection of water and wastes to kill pathogens with chlorine has been applied for about 75 years but is now being severly criticized because chloramines and other chlorinated organic compounds are found in chlorinated wastes and have been shown to be mutagenic for Salmonella typhimurium, the major organism used for assays in the Ames test for carcinogenicity (9). Inasmuch as all mutagens appear to be carcinogens, there is now great concern that our water disinfection procedures may have saved us from typhoid and other enteric diseases but exposed us to a long-term hazard of cancer. It is also evident that chlorine is relatively ineffective against enteric viruses which may be responsible for such water-borne infections as infectious hepatitis (jaundice) and related ills. The substitution of ozone for chlorine seems to aid in the viral disinfection attained. The Ames test has not been extensively applied to ozonated water, but there is little doubt that mutants will also be detected to result from this procedure.

The disinfection attained in stabilization ponds without chemical additions and their related dangers is shown in Figure 4. As indicated in the figure, high temperature acts to accelerate the death of E. coli and presumably pathogens. An added factor is pH. The pH of water in a five to ten day detention period pond is usually above 9 for 10 to 12 hours per day and, as shown by Parhad (10), such pH levels tend to cause rapid die-away of E. coli.

It has also been pointed out by Marais and Shaw (11) that a series of ponds gives disinfection superior to that of a single pond. At St. Helena (California) effluents from an integrated series of four ponds are occasionally as low as 10^2 E. coli per 100 ml.

Generally, speaking, all waste treatment systems are designed to modify water quality in order to meet specified levels of such water quality criteria as suspended solids, BOD, AGP, MPN, and TDS. In past years, levels for water quality criteria were established on the assumption that natural waters have a capacity to assimilate a given amount of waste, but more recent legislation such as that contained in U.S. Public Law 92-500 opposes the discharge of any pollutants to navigable or contiguous waters and required not only avoidance of degradation, but also the enhancement of quality in natural waters. This law has been difficult to implement and enforce because the great variations in local conditions do not permit economical application of blanket quality criteria. New legislation is under consideration to ammend this fault and greater consideration of local conditions of climate and water quality should evolve. Certainly, water quality criteria and resultant waste treatment methods for high temperatures and high salinity areas could differ greatly from those now used for low temperature and low salinity conditions.

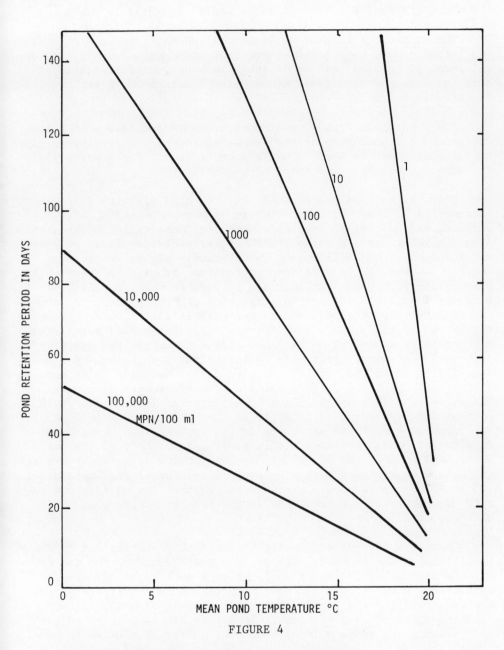

FIGURE 4

Observed MPN/L) ml in a single facultative pond as a function of time and temperature, initial MPN 10^8/ml.

HIGH-TEMPERATURE, HIGH-SALINITY WASTE TREATMENT IN PONDS

The unique combination of resources and needs in arid coastal
and inland zones, together with the fact that much of the future development of mankind must be in the arid zones, makes research in
high salinity and high temperature waste treatment of great significance to our future.

The predominant resource of such areas--an abundance of solar
energy and the accompanying high temperatures--are currently going to
waste. These can be used in wastewater treatment for growing algae
for oxygen production and nutrient recycling.

Evaporation from ponds can be high in arid climates. If water
consumption is low, which it often is in these countries, high
strength wastes will be produced whose treatment will require significantly lower loading rates (depth of wastewater applied per unit
time) and therefore will result in relatively higher evaporation
ratios. As the TDS in arid land wastewaters if high to start with,
evaporation from ponds can increase it even further, resulting in it
becoming unsuitable for use in irrigation of crop plants. A second
factor resulting in water loss is the percolation of water through
the bottom of the ponds. Both factors, evaporation and percolation,
can result in a severely diminished effluent quality and quantity of
reclaimed wastewater applicable to agricultural purposes. A simple
calculation shows that at 100 L/capita/day of domestic water consumption (about one-third of U.S. rate) a sewage strength of 600 mg/L BOD_5
(about twice U.S. averages), and an evaporation loss of 2 m/yr (a
minimum figure under arid conditions) then a loading of 1000 people
Ha (corresponding to 60 kg BOD/Ha/day, a typical loading rate for
facultative ponds) would result in a net loss of half of the wastewater, and an increase in TDS of two-fold. This is a yearly average;
during summer months a much high evaporative losses and increases in
TDS would be experienced, lowering even further the quality and quantity of wastewater applicable to agriculture at a time when it is
most required.

There are several strategies by which this problem can be minimized, though not avoided. The principal one would be the use of
high-rate pond systems which would allow a much higher loading per
unit area. High-rate ponds could be preceded by primary treatment
systems, resulting in significant (33%) removal of BOD_5, further
limiting the pond size required. It may be estimated that in the
arid subtropical zones a loading of over 200 kg BOD_5/Ha would be
possible using high-rate ponds, to which may be added the BOD_5 removed by primary treatment. As greater area is needed in winter time
than summer, not the whole area would need to be operated in the
summer time, again saving on water loss through evaporation. By such
a strategy the actual evaporative loss could be kept to a minimum and

TDS increase would not be large. Percolation would need to be minimized; low cost methods include use of a clay layer, incorporation of
organic matter into the soil, and spraying with sealants. These
methods still need further development.

Another alternative to prevent evaporation may be to cover pond
systems with transparent plastic covers. Although greenhouse covers
have been repeatedly suggested, they should only be considered in
extreme cases. First their cost is quite large, at about $10-$30/sq
m. They would cost about $100,000-$300,000/Ha, making such systems
at least as expensive as mechanical aeration. Secondly, the covers
reflect a significant amount of sunlight, reducing the effectiveness
of the system. Finally, evaporation cannot be completely avoided any-
way, as air must be circulated to prevent oxygen accumulation. It
must be pointed out, however, that the particular requirements of
algal pond systems may allow design of much cheaper plastic covers
than conventional greenhouse covers (e.g. no head room is required)
and that the use of covered ponds in waste treatment requires further
research, particularly for arid regions.

The specific design of wastewater pond system will depend on
local conditions and requirements. In some cases, water reclamation
for agricultural purposes may not be desireable or necessary as prevention of ground water contamination may be the most important consideration. In some cases the treated wastewater would be discharged
into local rivers or lakes, with the least deterioration in receiving
water quality being the main objective. The specific local micro-
climates, topography, soils and other factors would further affect
the design of such systems. Present experience has shown that a
series of at least four ponds is optimal. The concept of "integrated
ponding" has been developed to optimize various aspects of wastewater
treatment in the successive stages. Considerably more research and
development work is required to translate present design experience
(12,13) into an optimal pond system applicable to arid-high temperature conditions. However, it is already possible to design integrated
ponding systems for many situations and requirements which, although
not optimized, can be built and operated at considerably lower costs
than the conventional mechanical systems.

It must also be remembered that waste treatment ponds carry out
not only a disposal (e.g. destructive) process but also perform a
synthetic (productive) function in terms of algal biomass that is a
byproduct of the process. Methods for the harvesting of the algal
biomass have been briefly discussed above: bioflocculation and/or
screening appear to be both cost-effective and simple enough for most
applications (5-8). Although these methods are still in the process
of development, they may already be applied in various situations and
they raise the question of what practical uses may be made of the
algal biomass harvested from such ponds.

The likelihood of Human sewage-grown algae being used as a direct human food is at best remote. However, their use as an animal feed is much more immediate. Indeed, use of fish ponds for treatment of human (and animal) wastes is long established and widely practiced in many countries. Such systems are not intensive; a much high productivity could be achieved by using microalgae biomass harvested from high-rate ponds at a fish food, directly or indirectly, in intensive cultures. The use of waste-grown microalgae as an animal feed has been studied extensively and there appears to be significant potential for such algal single cell protein as animal feeds (14). This is particularly true where industrial effluents do not contaminate the wastewaters, e.g. in the rural and non-industrial regions of the world. In this connection it should be pointed out that the water efficiency of protein production with microalgae (protein content about 30 to 40%) is higher than that of the typical feed grains, which exhibit higher, or at least similar, water requirements per area and lower yields. Although waste-grown algae appear quite suitable for animal feeds, and experimental experience already exists, considerably more animal feeding tests are required, public acceptability must be realized, and, perhaps most important, processing techniques (drying, cell wall breakage) must be developed which allow storage and efficient use of the microalgal protein.

A more immediate alternative use of the microalgal biomass is its utilization, together with primary sludge, as a substrate for anaerobic digestion to produce methane. This is already a well developed technology in wastewater treatment, and recent developments suggest that it can be significantly reduced in both cost and operational complexity. The anaerobic digestion process has the further advantage that it preserves the nutrients concentrated and recovered from the wastewaters by the microalgae. The effluents from the anaerobic digesters can be used as an organic fertilizer. Indeed, it is possible to cultivate nitrogen-fixing blue-green algae on nitrogen depleted effluents from oxidation ponds, thus increasing the total biomass and fertilizer produced (15). The high salinity and temperature tolerances of blue-green algae make them particularly applicable to arid regions.

In conclusion, strategies of wastewater management and treatment through pond systems can be devised which minimize water evaporation while conserving vital nutrients which can be used either for agriculture or animal feeds. Fossil fuel energy is thereby conserved, relative to the conventional mechanical waste treatment processes, and biofuels could be produced through anaerobic digestion of the algal biomass produced. The main advantage of such systems is, however, their capability of enhancing human health and environmental quality at a cost that is affordable in areas where capital and raw materials are scarce. However, as detailed below, present experience is limited and further research and development is required.

EXPERIENCE IN HIGH-SALINITY HIGH-TEMPERATURE WASTE TREATMENT

Although (as evidenced in Figure 3) high temperature and high
salinity in waste systems have not been ignored by sanitary engineers
and microbiologists, efforts to develop special waste management systems for such conditions have been limited. This is, in part, due
to the fact that in tropical or arid regions wastes go uncollected
in many areas and hence are less common then they are in temperate
high precipitation countries. Direct applications of processes such
as activated sludge to waste treatment in arid zones has been carried
out, however, experience indicates that such processes are generally
less preferable under high temperature salinity conditions. The
subjects of wastewater infiltration and aquifer recharge, as a method
of treatment and water reclamation in arid zones has recently been
discussed at the UNITAR Conference on Alternative Strategies for
Desert Development and Management (16).

In some localities seawater is used as the flushing and carrying
agent for domestic sewage and some work has been done on conventional
seawater sewage treatment systems (17). High TDS wastewater reutili-
zation would require expensive secondary plus tertiary treatments
(18), making development of less expensive methods desireable. Systems based on salt tolerant plants appear particularly advantageous.

The following discussion of the state of science and technology
in high temperature and high-salinity waste treatment is limited
mainly to the authors personal experiences and is consequently not
a comprehensive review based on an extensive literature search. This
discussion thus emphasizes the applications of microalgae to waste
treatment problems. However, other types of plants (e.g. salt marsh
plants) could well be applied to waste treatment under saline conditions and process such anaerobic digestion can also be adapted to
function in a high salt environment.

One of the major efforts to develop a special waste treatment
for mildly saline waters began with studies carried out jointly by
the United States Department of the Interior, the U.S. Water Quality
Control Administration (now EPA), and the California State Department
of Water Resources to remove nitrate from the saline drainage waters
resulting from irrigation of salt-laden lands in the Western San
Joaquin Valley. The entire project has been described in detail by
Brown (19). Nitrate concentrations in these drainage waters are as
high as 30 mg/L, whereas an allowable nitrate of 2 mg/L has been set
by water quality authorities for discharges of the drain to the San
Francisco Bay. The study involved two competing microbiological pro-
cesses--growth of algae in shallow high-rate ponds and the use of
nitrate-reducing bacteria in enclosed columns. In the former case it
was not possible to consistently remove nitrate below a 3 mg/L level
with a single-stage algal system. In the case of columns, it was

possible to completely denitrify the waste, but large amounts of methanol were required and the effluent tended to contain both organic nitrogen and ammonium in amounts that increased with run duration. It was also found that backwashing the columns to rid them of bacterial accumulations required a complex and expensive backwash system. Both processes were estimated to cost about $100 per million gallons of waste processed in 1969. Recent work in our laboratories indicates that it is possible to reduce ammonia to 1 mg/L with a two-stage algal process. It thus now appears that an algae process will be more advantageous in nitrate removal from agricultural drainage waters, particularly in view of escalating methanol costs. The microalgae removed during this process can provide an additional valuable crop from the same original irrigation water. The microalgae can be sold for their protein content or alternatively digested to provide methane, with the residue used as agricultural fertilizer. The cultivation of the microalgae in such systems, as in all other applications, will require development of pond operations which allow maintenance of desireable algal strains, in this case algae or diatoms which are acceptable to the high level animals in the food chain.

In extensive studies Goldman and Ryther (20) have grown the diatom Phaedoctylum tricornutum in mixtures of seawater and activated sludge effluents. These cultures were subsequently fed to brine shrimp, oysters, and other shellfish in a controlled food chain. The growth units which were used were not optimized for algal growth, but substantial nutrient removals were attained and large yields of higher marine organisms were produced. This approach to aquaculture, if the technology can be perfected, has great potential for arid regions.

In Elat, Israel, Shelef (21) has grown algae on the effluents from a water supply made up from desalinized seawater and brackish wells. Elat has one of the highest solar energy fluxes in the world and algal cultures there have attained yields (new photosynthate and some carried-over sewage organic matter) up to 60 grams per square meter per day (90 tons per hectare per year) which are a prodigious rate of biomass production. However, the proteins produced on human wastes would at best only be suitable for fish or animal feed, and perhaps only for fertilizer. In a pilot plant project Shelef has studied protein production from algae harvested from high-rate ponds. McGarry and others have studied a similar process in Thailand (22).

One of the most promising microbiological processes for the production of protein suitable for direct human consumption under high salinity and high temperature conditions is the production of microalgae single cell protein, particularly the culturation of the filamentous blue-green algae Spirulina (23). A great deal of work has been done in microalgae SCP (24,14), but much more research remains

to be done. Spirulina production by the Sosa Texcoco Co. near Mexico
City is a natural byproduct of salt and sodium bicarbonate production.
Spirulina growth is enhanced both because of natural concentration in
in the salt-evaporation beds and because limiting nutrients (nitrate,
iron) are added. Maximum Spirulina production at Sosa Texcoco in the
current system is about 2 dry tons/day, and productivity is estimated
at 10 grams/sq m/day average for a 10 month growing season.

Recently, the French Petroleum Institute (FPI) which, according
to Durand-Chastel and G. Clement (23) first isolated and cultivated
Spirulina from the saline lakes of Central Africa, has constructed a
major pilot plant installation for the growth of Spirulina on a mesa
near Cairo, Egypt. Their culture units which are constructed of
concrete are being built in conjunction with a major laboratory
facility for long-term studies of Spirulina production and use under
the high-temperature conditions found near Cairo. The FPI media for
Spirulina has a TDS approximately one-half that of seawater, but is
low in chlorides. A major ingredient is about 17,000 mg/L of sodium
carbonate, making it economically unattractive. A large number of
research projects are being initiated around the world on Spirulina
production. The high price of Spirulina and other microalgae as a
specialty product, the potential high productivities and low production costs, and its large market for animal feed make Spirulina and
other microalgae particularly attractive. If present research efforts were somewhat coordinated and benefited from each others' findings, no doubt rapid progress would be possible.

In Tunisia, the senior author, together with Mr. A.W. Plummer
of the U.S. AID, undertook a study of the influence of waste discharges from the City of Tunis into the Lake of Tunis, a shallow (one
meter) saline bay that has long been a major source of food fish for
Tunis. A number of studies had revealed the detrimental effect of
waste discharge upon the bay as evidenced by massive, recurring uncontrolled algal blooms and fish kills. A reef-building worm also
resides in Lake Tunis and its reefs have increasingly interfered with
natural circulation of water in the lake and its use in boat transportation. Problems with the lake are compounded by a power plant
discharge which increases the temperature of the lake by as much as
5°C. The introduction of partially treated sewage at the edge of
the lake has also produced vast sludge beds in which the sulfate of
seawater is microbiologically reduced to hydrogen sulfide which waffs
over the resort hotels of Tunis and occasionally causes large areas
of the lake to turn red due to photosynthetic sulfur bacteria (probably Thiopedia rosia). This activity is accomplished by the production of large amounts of elemental sulfur which is deposited along
the windward banks of the lake. These findings led to a recommenda-
tion that the wastes of Tunis be diverted into stabilization ponds
built in nearby Sebket Ariana (a dry lake bed) and that pilot plant
studies be made of the production of algae in Tunisian sewage. The

recommended stabilization ponds are now under construction and the
algal production pilot plant has been constructed in conjunction with
waste treatment at a nearby community of Maxula Rades, but no funds
have become available for this study.

A similar situation exists in Manila, Philippine Islands, where,
due to advers tidal conditions, much wastewater and seawater enters
Laguna de Bay which is a major fishing source for Manila and its
only economical supplementary source of municipal water in the 21st
century. Not only does much of Manila's sewage find its way into
Laguna de Bay, but industries, power plants, feed lots, and many
small communities dump their untreated wastes into Laguna de Bay.
For this reason, the United Nations Development Program (UNDP), the
World Health Organization (WHO), the Asian Development Bank (ADB),
and the Laguna Lake Development Administration (LLDA) have jointly
undertaken an extensive study of water quality in the lake, the influence of seawater intrusion, and the development of techniques of
waste treatment compatible with the high temperature conditions found
in Manila. This pilot plant was started up in April, 1977 and currently is being used for an intensive preplanned experimental program
in which depth, loading, and mixing regimes are being varied to determine their influence on waste treatment and algal productivity.

RESEARCH NEEDS

The limiting aspects of the technology of microalgae production
from wastewater ponds, or fertilized growth units, is the frequent
instability of the cultures (due to invasion and dominance by un-
wanted algal types, predation by zooplankton, etc.) and the costs of
microalgae harvesting and processing. Microalgae, as their name
indicates, are microscopic in size and thus difficult to collect
except by expensive centrifugation, filtration, or chemical floccula-
tion processes. These problems may be overcome in two different ways;
either by cultivation of larger forms of microalgae, such as the
filamentous blue-greens that can be readily and cheaply harvested
by screens, or through development of processes which result in auto
bioflocculation and settling (or flotation) of the microalgae. We
have recently investigated both approaches and obtained encouraging
results: By adjusting hydraulic and organismal detention times in
experimental sewage ponds, large colonial algae types were selected
which could be harvested by microstrainers (rotating screens with
backwash) (25). By growing algal cultures to the limits of available
nutrients conditions are induced which favor algal settling (8).
Both processes are applicable to waste treatment processes inasmuch
as low cost algae removal technology is an important prerequisite
for achieving advanced waste treatment standards and allowing energy
and fertilizer reclamation, or protein production.

The technology required to optimize microalgae yield and maintain the desirable algal species under conditions allowing economical production must, however, still be perfected. In this effort conditions of high temperature and salinity are particularly appropriate; they are selective conditions under which only a limited number of algae (or predators) can propser, simplifying the problems of algal species control. The example of Spirulina has already been given: it grows naturally in specific limits of salinity and alkalinity, with few or no other algae being able to compete. Another interesting example is that of the microalgae Dunaliella which grows under the extremely high salt conditions of the Dead Sea. It produces high quantities of glycerol, which allow it to resist the great osmotic pressure exerted by the salt. Industrial glycerol production by Dunaliella is of great interest.

Microalgae systems are relatively well developed and demonstrated for wastewater treatment, however, other plant systems may be also considered: water hyacinths, other floating aquatic plants, marsh systems, and terrestrial plants grown on sewage or sewage treatment plant effluents. Such plant systems may be integrated with conventional treatment or microalgae ponds. They cannot readily remove BOD5 except at very low loadings. They are, however, capable of performing a polishing or tertiary treatment function. Under conditions of high TDS special plant systems will have to be developed. The possibility exists of TDS removal by microalgae (26); it needs to be explored.

Although pond systems can be desiged for most locations, where climate is not limiting, optimization of the systems, particularly in arid settings, will require pilot and demonstration projects to establish specific design and operation parameters. This should become a high priority of future research and development. Another aspect that requires considerable research is the development of microalgae biomass conversion and utilization processes. Although much basic research in the anaerobic digestion and animal feeding trials have been carried out, further work and demonstration projects are required. In particular the use of residues after anaerobic digestion of the microalgae for fertilizer and low cost protein extraction or algal drying methods should be pursued. Finally, the production of specialty or feedstock chemicals from microalgae biomass is an important future area of research.

In conclusion, high salinity and high temperature regions present unique needs and opportunities for wastewater treatment and reclamation. Processes powered by solar energy are most desireable and applicable under these circumstances. Microalgae are particularly adaptable to high and variable salinities and temperatures, and are already widely used in wastewater treatment. Research and demonstration projects are required to develop and apply existing experience and to perfect this technology. The study of other, tempera-

ture and saline resistant plants in wastewater treatment should also
be encouraged. Such systems should be integrated with food-feed-fiber
and fuel production to allow water and nutrient recycling.

REFERENCES

1. Environmental Studies Board, National Academy of Sciences/ National Academy of Engineering, Water Quality Criteria 1972, Supt. of
   Documents, U.S. Govt. Printing Office, Washington, DC 20402 (1972).

2. Fair, G.M., J.C. Geyer, and D.A. Okun, "Water and Wastewater Engineering," Vol. 2, Water Purification and Wastewater Disposal,
   John Wiley and Sons, Inc., New York (1968).

3. Torien, D.F., C.H. Huang, J. Radimsky, E.A. Pearson, and J.
   Sherfig, Provisional Algal Assay Procedures, Final Report,
   Univ. of Calif., San Eng. Res. Lab. 71-4 (1971).

4. Oswald, W.J., H.B. Gotaas, H.F. Ludwig, and V. Lynch, "Algae
   Symbiosis in Oxidation Ponds III Photosynehtic Oxygenation,"
   Sewage and Industrial Wastes, 25, 6 (1953); Also Oswald, W.J.,
   "Metropolitan Wastes and Algal Nutrition," Proc. Algae and
   Metropolitan Wastes, Robert A. Taft San Eng. Ctr, Cincinnati,
   Ohio (1960).

5. Koopman, B.L., R. Thomson, R. Yackzan, J.R. Benemann, and W.J.
   Oswald, Investigation of the Pond Isolation Process for Micro-
   algae Separation from Woodland's Waste Pond Effluents, Final
   Report, San. Eng. Res. Lab., Univ. of Calif., (1978).

6. Benemann, J.R., B.L. Koopman, J.C. Weissman, D.M. Eisenberg, and
   W.J. Oswald, Species Control in Large-Scale Algal Biomass Production, Final Report, Univ. of Calif. San Eng. Res. Lab. 77-5,(1977).

7. Benemann, J.R., J.C. Weissman, D.M. Eisenberg, B.L. Koopman, R.P.
   Goebel, P. Caskey, R. Thomson, and W.J. Oswald, An Integrated Sys-
   tem for Solar Energy Conversion Using Sewage-grown Algae, Final
   Report, Univ. of Calif. San. Eng. Res. Lab. 78-5 (1978).

8. Benemann, J.R., J.C. Weissman, D.M. Eisenberg, R.P. Goebel and
   W.J. Oswald, Large-Scale Freshwater Microalgae Biomass Produc-
   tion for Fuel and Fertilizer, Final Report, Univ. of Calif., San
   Eng. Res. Lab. (1978).

9. Ames, B.N., H.O. Kammen, and E. Yamasaki, Proc. Natl. Acad. Sci.
   USA 72, 2423-2427 (1975).

10. Parhad, N.M. Studies of Microbial Flora in Oxidation Ponds, Ph.D.
    Dissertation, Central Public Health Engineering Red. Inst.,
    Nagpur, India (1970).

11. Marais, G.R. and V.A. Shaw, "A Rational Theory for the Design of
    Sewage Stabilization Ponds in Central and South Africa," Trans.
    South African Institute of Civil Engineers 3, 205 (1961).

12. Gloyna, E.G., Waste Stabilization Ponds, Monograph No. 60, World
    Health Organization, Geneva, Switzerland (1971).

13. Oswald, W.J., "Experiences with New Pond Designs in California."
    in Ponds as a Wastewater Treatment Alternative, edited by E.F.
    Gloyna, J.F. Malina, Jr., and E.M. Davis, Water Resources Sym-
    posium No. 9, Center for Research in Water Resources, College
    of Engineering,Univ. of Texas, pp. 257-274 (1976).

14. Benemann, J.R., J.C. Weissman, and W.J. Oswald, "Algal Single-
    Cell Protein" in Economic Microbiology, (A.H. Rose ed.) Academic
    Press, London (in press).

15. Benemann, J.R., J.C. Weissman, and M.A. Murry, P.C. Hallenbeck,
    and W.J. Oswald, Fertilizer Production with Nitrogen Fixing Heter-
    ocystous Blue-Green Algae, Final Report, Univ. of Calif. San.
    Eng. Res. Lab. 78-3 (1978).

16. Bouwer, H., and R.C. Rice, "Reclamation of Municipal Wastewater
    by Soil Filtration in Arid Regions," Proc. Conf. on Alternative
    Strategies for Desert Development and Management, UNITAR (in
    press).

17. Kessich, M.A., and K.L. Machen, "Sea Water Domestic Wast Treat-
    ment," J. Water Poll. Control Fed. 48, 2131-2136 (1977).

18. Cook de, K.H., "Reuse of Wastewater in the Desert Regions,"
    Proc. Conf. on Alternative Strategies for Desert Development and
    Management, UNITAR (in press).

19. Brown, R.L., "The Occurence and Removal of Nitrogen in Subsurface
    Agricultural Drainage from San Joaquin Valley, Cali., Water Re-
    search 9, 529 (1975). Also Removal of Nitrate by an Algal System,
    Calif. Dept. of Water, Res., (1971).

20. Goldman, J.C. and J.H. Ryther, "Nutrient Transformation in Mass
    Cultures of Marine Algae," J. Environ. Eng. Div. ASCE, 101 EE3
    Paper 11358, (1975).

21. Shelef, G., Interim Report on Elat Algae Production Project,
    Civil Engr. Dept., Technion, Israel (1976).

22. McGarry M.G. and N.L. Ackermann, "Resource Conservation Through Water and Protein Reclamation," Session 9, Proc. Regional Conf. on Water Res. Develop., Asian Inst. Tech., Thailand (1970).

23. Durant-Chastel, H., and G. Clement, "Spirulina Algae: Food for Tomorrow." Proc. 9th Intl. Cong. Nutrition, Mexico (1972).

24. Soeder, C.J., and W. Pabst, Berichte Deutsch Botanische Gesselschaft 83, 607 (1970).

25. Benemann, J.R., J.C. Weissman, B.L. Koopman, and W.J. Oswald, "Energy Production by Microbial Photosynthesis," Nature, 268, 19-23 (1977).

26. Oswald, W.J., A.G. Beattie, and C.G. Golueke, An Engineering Evaluation of Some Currently Proposed Methods for Biological Conversion of Saline Waters, San. Eng. Res. Lab., Univ. of Calif. (1960).

CHEMICALS AND FUELS FROM SEMI-ARID ZONE BIOMASS

John R. Benemann
Ecoenergetics
5619 Van Fleet Avenue
Richmond, California 94804

INTRODUCTION

Biofuels, fuels from biomass, are being recognized as a significant future energy source for the U.S. and the world (1). Present
uses of wood and dung rival or exceed those of fossil fuels in rural
regions of many underdeveloped countries. In the U.S., wood wastes,
sugar cane bagasse, and domestic firewood presently contribute as
much as 2% of total fuel usage. Worldwide, biofuel production could
be expanded considerably by better management and greater utilization
of existing forest resources, by collection and use of agricultural
residues, by gas production from animal wastes, and through municipal
and industrial waste utilization. Besides such readily available
and economically attractive biofuel resources, biomass farming specifically for fuel production has been proposed.

Many types of plants have been recommended or considered for
biomass "energy farming": intensively cultivated short rotation (3
to 7 years) hardwood trees (2-5), microscopic freshwater algae (6,7)
seaweeds (8-10), water hyacinths (11), sugar crops (12), and all
kinds of grains and grasses (13). However, those proposals are presently based on technical and economic extrapolations of doubtful
validity. They are also limited because of severe competition with
conventional food-feed-fiber production. Indeed, most of these proposals only look attractive because of very optimistic assumptions
made about yields (ash-free dry weights) of up to 45 tons/hectare/
year for short rotation trees (4) and over 100 tons/hectare/year for
kelp (10), for example. In addition, input costs for fertilizers,
pest control and management are often unrealistically low in these
proposals, thus calling into question their economic feasibility.

Thus, biomass energy farming where the sole output is fuel, must be
considered at present a subject for long term research and with near
term applications restricted to a few special situations and on relatively small scale. However, the better management of forests to
allow both increased timber/pulp and fuel (firewood, charcoal) production is of both priority and significant potential. Indeed, re-
forestation must become a worldwide effort to counteract the destructive effects of the recent and ongoing wholesale elimination of for-
ested ecosystems.

The potential of biomass fuels derived from wastes and residues
is quite significant. For example, collectable forest and agricultural residues may be capable of providing about 5% of current U.S.
energy requirements using existing or near term technologies (both
collection and conversion) and marginal economics of fossil fuels.
A major obstacle is the lack of a sufficient commercial experience
with biomass fuel markets and the present incompatability of local
biofuel resources with conventional large energy industries. These
problems could be solved through increased governmental-private
efforts at overcoming institutional barriers to commercialization.
Related efforts are urgently needed in many under-developed countries that have significant biomass resources, in relation to their
energy uses.

Another approach to biofuel production, intermediate between the
utilization of available wastes/residues and the concepts for single
purpose energy farming, is the development of plant cultivation systems in which food-fiber are produced alongside with higher value
chemicals and/or biofuels. It may be possible to optimize to some
extent residue production with food-fiber production in conventional
agriculture-silviculture. However, it does not appear likely that
our major grain crops will allow a much increased production of residues usable for fuels. Restricting factors are soil erosion and fertility losses associated with residue removal and the possible decrease in grain yields if the residue fraction is increased cultur-
ally or genetically. Although better forest management will increase
biofuel resources from tree residues and thinnings, it is uncertain
whether the ratio of wood fuels to fiber-timber will increase greatly
in the future due to such practices.

Novel plan systems geared specifically to the production of high
energy and high value chemical products, along with biofuels, appear
to be the most likely near term appraoch to biomass energy farming.
Examples of such systems may be short rotation tree farms integrating
production of "naval stores" (e.g. extractable resins and turpentines)
and fuels, or the cultivation of various "hydrocarbon" or latex
plants. In any such scheme competition with established agricultural-
silvicultural systems for land, water and fertilizers must be minimized for both economic and political reasons. During the initial
stages in the development of such systems, utilization of some com-

mercial agricultural and forest lands will be feasible. However, in
the longer term, concepts that use available land and water resources
not used, or underutilized, by conventional agriculture are the most
attractive. Such lands are primarily found in semi-arid regions and
the applicable water resources are either the limited rainfall in
those areas or waters too saline for conventional agriculture. Thus
it is likely that the development of biomass energy systems will em-
phasize novel plant systems in saline-semi-arid environments geared
to the production of higher value chemicals, with biofuels as byproducts.

The semi-arid zones include those with average annual production
rates of about 0.2 to 2.0g biomass/m^2/day (dry weight ash-free biomass gains of whole, below and above ground, plants). Depending on
the latitude, these areas receive at least 100mm and as much as 500mm
of precipitation yearly. Much of this biomass is harvested through
grazing, although dryland cropping is also very extensive. Over one-
third of the world's potentially arable land is found in semi-arid
regions where crop production is possible, though erratic and with
short growing seasons. These croplands include all those where
actual crop transpiration and yields are only 20% to 50% of irrigated
crop transpiration during the growing season (14). The severe erosion and salinity problems experienced in many of these agricultural
grazing regions is a major factor in desertification. Gentler approaches to the utilization and exploitation of such lands are required. Use of desert plants for the production of novel foods or
feeds and, as discussed in this review, chemicals and fuels, is of
significant potential.

Salinity problems affect an estimated one-third of the world's
160 million hectares of irrigated croplands (15). In such areas,
water quality rather than quantity is the limiting factor. Halophylic plants could be cultivated on waters unsuitable for agriculture
(above 2000mg/1 in TDS - Total Dissolved Salts, or 15m equivalent/1
Na^+). There are already well established economic uses for salt tolerant plants (15) and they could also become sources of chemicals and
fuels. Some plants which remove salt from soils may even be useful
in salinity control, although this is speculative at present.

Historical and present day experience with water collection and
farming in semi-arid regions (16) suggests that it may be feasible to
develop plant production systems adapted to such conditions. Although many problems need to be solved, including erosion control
during runoff collection, the gross damage to such ecosystems due to
overutilization can be avoided. In this review the potential for the
production of chemicals and fuels in semi-arid and saline regions is
discussed. The production of high value chemicals is already taking
place in, for example, managed plantations of gum arabic trees in
several Sahel countries. Specific newly proposed plant systems,

such as guayule, Euphorbia, and microalgae are described. A brief
introduction to plant growth under semi-arid-saline environments is
presented first.

PLANTS IN SEMI-ARID, SALINE ENVIRONMENTS

In semi-arid areas, water is the key limiting factor to plant
growth. (Very arid deserts will be ignored here as no significant
plant growth occurs). As a general guideline, semi-arid zones are
those where precipitation is at least 100 mm/year and not more than
500 mm/year, and where plant ground cover is not complete, except
during short wet periods. In addition, there is a very large year
to year variability in actual rainfall, making conventional agriculture difficult or erratic without elaborate water collection, storage,
transportation and irrigation systems. Therefore, agriculture in
these areas, although of great importance, cannot fully utilize available water resources and is subject to severe restrictions.

The native plants of semi-arid regions have developed a number
of adaptations to such conditions (14,17,18,19). Key morphological
characteristics are the low surface to volume ratio of the leaves,
large root-storage organs resistant to dry periods, and the adaptations of the C_4 and CAM (crassulacean acid) photosynthesis that allow
for high water use efficiency. Of particular importance is the co-
existence of plants with different morphologies and water utilization
mechanisms. This allows full utilization of limited water resources
through occupation of different microenvironments by a variety of
plants.

Although desert ecosystems are characterized by low standing biomass and productivities, many individual desert plants can exhibit
a high photosynthetic conversion efficiency during favorable parts
of the year, or, in favorable microniches, even throughout the year.
(Favorable conditions being synonimous with presence of moisture).
Even the high temperatures commonly associated with desert environments are not necessarily a limiting factor, as shown for at least
one plant capable of active photosynthesis at a leaf temperature of
almost 50°C (20). However, many desert plants, particularly succulents and shrubs, exhibit very low photosynthetic rates with yearly
increments in dry weight as low as 1%. Adaptation to such harsh environments often means restricting growth to the absolute minimum
(19).

Water economy is central to the survival of plants in semi-arid
areas. Basically, the maximization of carbon (biomass) gain is optimized with the water loss. Diffusion of CO_2 through the leaf stomata
will inevitably result in some water loss. Succulent plants have no
leaves, minimizing the area over which water is lost. They store

water in their bodies and carry out CO_2 absorption at night when
water losses are minimal, often allowing year-round photosynthesis.
Drought resistant evergreen shrubs (true xerophytes) are slow growing
perennials with small leaves which can photosynthesize under water
stress and remain essentially dormant when water can no longer be extracted from the soil. Typically such plants carry out a pulse of
CO_2 fixation during and directly after the wet parts of the year followed by a very low rate of photosynthesis during the rest of the
year. Another type of desert plant is the "phreatophytes" which is
characterized by very long, deep roots (commonly 10-20 m but sometimes
as much as 50-100 meters!) which tap subsurface water deposits in
favorable locations (e.g., washes). This group often maintains year
round photosynthesis, although some dormancy and specialization of
leaf structure may be present. The mesquite tree, discussed below,
is an example of such a plant. The phreatophytes can have a severe
effect on local water economy because they can transpire a large
amount of ground water that otherwise might be available to agriculture. Finally, ephermeral annual (often C_4) plants are common in
desert environments, particularly where large year to year variability
in rainfall exists, discouraging perennial plants.

The desert plants of greatest interest for biofuel-chemical production are the true xerophytes and some of the phreatophytes. These
plants are perennial, allowing storage of sufficient biomass to make
eventual harvest worthwhile, and they can overcome a significant year
to year variation in rainfall. Many succulents are too slow growing
for such applications. Of specific interest are those plants which
can produce highly reduced organic compounds such as rubber or latex
compounds, or lower molecular weight hydrocarbons. Such compounds
are of specific economic interest and, in addition may allow increased water economy by producing a greater energy content per dry
weight biomass produced. However, such increased water economy must
still be experimentally demonstrated. Examples of plants that produce reduced organic compounds include guayule, _Euphorbia_, and jojoba,
discussed in the next section.

Other types of plants of interest in biomass energy farming are
those capable of using highly saline waters unsuitable for conventional agriculture. Of course, saline water irrigation has been
practiced with many crop plants with good results as long as sandy
or very coarse, porous soils were used, allowing good drainage and
keeping the root zone aerated and free of salts. On clay soils
without calcium, sodium and magnesium salts are absorbed and clay
particles swell, impeding drainage and exposing roots to high salt
concentrations (21) With the right soil conditions and correct irrigation practices even seawater may be used for growing salt resistant varieties of common crop plants although, thus far, at productivities significantly reduced from conventional yields (22). The
coastal desert areas below about 33 meters in elevation extend for

about 15,000 km, although those with sandy soil suitable for seawater
irrigation have yet to be surveyed (15). A considerable potential
may exist for ocean water farming, particularly for naturally halophylic plants. Vast areas with brackish ground and surface waters
also exist, suggesting these as additional resources for halophylic
plant production. Of course, by necessity, many, but not all, of the
desert plants described above are also salt tolerant. Also, by no
means are all halophilic plants drought resistant (for example,
marsh plants and mangroves). It must also be considered that the
salt concentration and ionic species of inland brackish waters are
quite distinct from those in seawater, being normally much higher in
divalent cations and anions. Thus plants will often respond dif-
ferently, usually better, to such brackish waters than to seawater
ions of the same TDS. Irrigation practices may have to be adjusted
to allow use of such waters.

Finally, the microscopic algae can be considered among those
potentially applicable for chemicals and fuels production in biosaline
environments. Their main useful characteristics are their tolerance
for high and variable salt concentrations and their potentially high
productivities. Their disadvantage is that they require a pond from
which water can evaporate freely. For example, in a semi-desert environment with an annual precipitation of 250 mm/year, and a native
plant community of true xerophytes transpiring 150 mm/year, an algal
pond will exhibit a free evaporation of about 2500 mm/year. (These
values can, of course, vary significantly with latitude, climate,
site, etc.) Thus, an algal pond requires the water resources of up
to ten times the normal precipitation--a very major factor. The
algal pond, however, would produce about 50 tons/hectare/year while
the desert plants produce, at most, 2 tons/hectare/year. Water consumption per unit biomass production could, then, be considerably
higher for higher plants than microalgae. Microalgae are capable of
producing chemicals and other valuable products in high concentrations
using saline waters unsuitable for even the most tolerant plants.
Thus microalgae have a potential role to play in the development of
biosaline chemical-fuel resources. The economic feasibility of microalgae production is discussed later in this review.

GUAYULE: RUBBER FROM SEMI-ARID REGIONS

Because of high demand, favorable economics and relatively advanced technical development, the most immediate feasible plant systems for production of chemicals in semi-arid regions appear to be
those which contain rubber. Of these, the guayule shrub is the one
attracting the most attention as a great deal of information is known
about it. Thus this section will present this option in detail,
based primarily on a recent review of guayule by the National Academy
of Sciences (23) and some other recent sources (24,25).

The world production of rubber is presently divided between natural rubber derived from Heavea brasiliensis (about 5×10^6 tons/year, estimated 1980 production) and synthetic rubber (about 10×10^6 tons/year). Demand for natural rubber, produced mainly on Southeast Asian plantations has increased recently and is likely to do so in the future. Large increases in productivity of Hevea rubber were achieved during the 1950's and 60's. However, this trend is slowing down and increases in Hevea plantations or new natural rubber sources will be required in the future, particularly if synthetic rubber prices increase along with oil prices. About 2000 other plants are known to contain rubber (cis-polyisoprene compounds; other types of "rubber" are also found in plants). However, only one has ever been exploited commercially--the guayule shrub (Parthenium argentum) found in semiarid North American regions, particularly upland plateaus in Mexico. This inconspicuous shrub (less than 1 m tall) was a major source of rubber for the U.S. before World War I. However, only wild stands were harvested and these were soon depleted. The loss of Southeast Asian rubber supplies during World War II led to the "Emergency Rubber Project" (ERP) which, on a crash basis, planted 27,000 acres of guayule in California. At the end of the war, with Hevea rubber again available, this acreage was returned to the farmers, the guayule bushes approaching maturity were plowed under, and research and development efforts wound slowly down until 1953 when the promising genetics program was terminated. Interest in guayule has picked up again recently, both in the U.S. (26) and in Mexico (25).

Guayule has much to recommend it as a new crop plant for the semi-arid regions. It is a perennial bush, surviving 30-40 years in deserts where the average rainfall is 100-400 mm per year. It is heat resistant and can withstand cold moderately well. It's root system is extensive, allowing efficient moisture collection, and its leaves are narrow, helping preserve moisture. Rubber is concentrated in cells of the outer layers of the stems and branches (two-thirds) and roots (one-third). The rubber content of the plant is about 10-25% of dry weight, with the lower number representative of wild stands and the higher values reported for some environmental conditions and genetic isolates. During the "ERP", up to 20% rubber content was achieved. Rubber production in the shrub takes place during periods of water or cold stress (not during active growth) and can also be induced by certain chemicals. Once formed the rubber accumulates and it is not further metabolized. In addition to rubber, the plant also produces a variety of resins (low molecular weight hydrocarbons) constituting up to 10% of dry weight, as well as the waxes that cover its small leaves.

Rubber and resin extraction requires harvesting of the whole plant and processing. The rubber extraction processes which were relatively crude at the beginning of the century, were considerably

improved during and after the ERP in the late 1940's and are the subject of continuing research and development activities. The major
activity is taking place at Saltillo, Coahila, Mexico, where a demonstration plant financed by the Mexican government is operating, harvesting the large natural guayule stands found in the area. The process used at Saltillo is based on a combination of processes adapted
from the pulp-paper and the synthetic-natural rubber industries. The
process uses hot water coagulation, hammer milling, pulping with
caustic soda (to help break open the rubber filled cells), flotation
of the rubber (with the waterlogged residues or "bagasse" sinking),
washing, acetone extraction of the resins, and finally solvent purification. One key factor in the process is that guayule rubber
easily autooxidizes (unlike Hevea rubber) and requires rapid processing. Alternative processing methods are feasible and may be developed to lower costs and/or meet particular market demands. The quality
of the guayule rubber is, or can be, essentially identical with that
of natural Hevea rubber.

    Some of the most important aspects of guayule processing are the
energy and water requirements. The scheme described above requires
a considerable amount of processes heat and power, although a detailed
analysis is not yet available. The ultimate success of any process
will require minimizing these inputs and substituting solar energy
and maximizing utilization of the byproduct resins and bagasse. Some
of the bagasse may also be suitable for paper or cardboard production.
The prospects of solar energy utilization in the chemical processing
of semi-arid (and saline) region plants is considerable. Generation
of low temperature process heat using covered solar ponds or conventional flat plate collectors is presently feasible. Tracking solar
concentrators could be used for higher temperature process heat as
well as shaft and electric power. If breakthroughs in photovoltaic
collectors materialize, these may also be considered. The sites
where guayule processing plants are likely to be located offer particular advantages for the use of solar energy: favorable climate and
remoteness from conventional energy supplies. The guayule residues,
or bagasse remaining after extraction of the rubber has a substantial
fuel value whose utilization could be integrated with that of direct
solar energy conversion processes to meet peak demands, provide back-
up, or increase the overall thermodynamic efficiency of the systems
(for example, by generating high pressure steam from low pressure
solar generated steam). Conversely, several biomass conversion technologies, specifically gasification-pyrolysis, are amenable to a
solar energy "assist", allowing higher fuel recovery from the biomass.
This would be particularly true of the high moisture bagasse derived
from the above described guayule processing. Indeed, a solar-pyroly-
sis unit will be tested at Saltillo, Mexico (27). Although the necessary technologies required for interfacing biomass and solar ener-
gies are only at a beginning stage, and their interfacing should not
be overemphasized at this early stage; they have considerable poten-

tial for the future of semi-arid regions. Thus, for example, guayule
production and processing may not only become fuel-energy self-suffi-
cient but also may allow sufficient extra fuel production to support
ancilliary industrial activities and local community uses.

The agricultural production of guayule will be the main determining factor for commercialization of this new type of crop plant.
Only in Mexico are there sufficient natural resources to support an
industrial scale processing plant, with an estimated 300,000 tons/
year of guayule bush (containing 30,000 tons of rubber) being harvestable on a ten-year rotation. Even in Mexico cultivation would
be desirable as the depletion of natural stands early in this century
demonstrated. For efficient agricultural production high yielding
plant strains must be developed and the most cost effective cultivation conditions worked out. Improvements in plant genetics and plantation management have increased <u>Hevea</u> rubber production rates (per
acre) ten-fold; large increases in guayule rubber production should
also be possible in the future. Thus, projections based on present
data, derived mainly from the ERP, are not a good guideline of the
potential of this plant.

The agricultural production of guayule will require coarse to
medium textured soil, permeable and well drained. Guayule cannot
withstand waterlogging as can occur even in the desert with clay
soils. It is not a salt tolerant plant, being able to withstand only
salt concentrations below 3,000 ppm TDS. The proper amount of moisture at the proper time of the year is the key determining, and uncertain, factor in guayule production. In its native habitats it
grows even at average rainfalls below 200 mm/year, although the ERP
project concluded that commercial rubber production would require
280-640 mm/year, with about 500 mm/year optimal. Higher moisture
would increase vegetative growth at the expense of rubber production
in response to water stress. This may, however, be also controlled
by chemical and/or genetic means. Supplemental irrigation is expensive, both in terms of extra water consumption and the capital investment required in land preparation, water conveyances and operating
costs. Thus, unirrigated agricultural production on sites unsuitable
for dry farming of conventional crops appears to be particularly
favorable. Proper plant spacings and crop establishment procedures
would also need to be worked out. However, once established, a
guayule plantation is likely to require relatively little attention
or pest control. If irrigation is used it must be carefully managed
to minimize weed growth, prevent water logging, and avoid excessive
vegetative growth.

One of the unique features of guayule is its ability to store
the rubber synthesized for several years. This would allow, on the
one hand, accumulation of a reasonably high amount of rubber-biomass
before necessitating harvesting and, on the other hand, would allow

the guayule farmer some flexibility on when to harvest, depending on market prices. Although such flexibility does require increased investment in the guayule processing plant capacity, this is likely to be a worthwhile tradeoff. Harvesting is easily mechanized, a major factor in favor of guayule rubber production. Harvesting involves cutting at about 5 cm above the soil, allowing rapid regrowth of the plant and minimizing costs for replanting. The proper rotation period and number of allowable regrowths is not yet determined; estimates range from 4 to 8 years and from 2 to 4 crops.

Rubber yields are, of course, dependent on cultivation conditions, with common irrigated yields being in the range of 0.5 tons/hectare/year of rubber. The best yields (irrigated) reported by the ERP were in the Salinas Valley at 0.6 to 0.8 tons/hectare/year of rubber. Higher yields were found in small test plots, and, as mentioned above, improvements in genetic stocks and cultural practices are likely to result in increased yields in the future. If we assume that a 20% dry matter rubber content can be achieved commercially, that 250 mm of annual rainfall and 250 mm of irrigation water are supplied, that two-thirds of this water is transpired by the plants and that water use efficiency is 3 mg dry weight/gm of water, then about 2.0 tons/hectare/year rubber would be produced of which about two-thirds would be recoverable by harvesting the above-ground biomass. However, this is too optimistic of an estimate, as it uses a water use efficiency achieved by some plants in semi-arid regions (14), but of doubtful applicability to managed stands of guayule. Furthermore, this level of irrigation approaches that of some crop plants (e.g., cotton). Thus, at present at most one ton of rubber/acre/year could be predicted and even then competition with conventional agriculture may be severe. Assuming relatively high yields, the Firestone Co. estimated guayule rubber production costs at between $0.64-0.84/kg for agricultural production plus $0.45-0.61/kg for processing (25). This compares unfavorably with the current price of Hevea rubber of less than $1/kg. But this price is expected to escalate very significantly by 1985 when a natural rubber shortage is anticipated. Thus, in the intervening years a potentially profitable research and development opportunity exists, both in the processing and production of this plant. It should be emphasized in this connection that increasing yields (tons/hectare/year) should not be the main objective of such research but rather increased water efficiency and the ability to use land and water resources that cannot be presently used for food-feed production. Of course, very low yields will result in too large a harvesting effort required and long transportation distances to a processing plant. However, the current emphasis on high yields obscures the fact that water, not land, or sunlight, is the limiting factor and that very high yields do not allow optimal water use efficiency.

Recent legislation introduced in the U.S., the "Native Latex

Commercialization Act of 1978" (HR 12559), would appropriate
$60,000,000 for development of a guayule and related rubber contain-
ing North American plants industry. The Mexican government sponsors
guayule research, development and demonstration at Saltillo, as discussed above. The Goodyear Tire and Rubber Company is exploring
guayule cultivation in Arizona, and a number of other privately financed activities are underway. More research is required of other
rubber containing plants whose cultivation may be more appropriate
in other semi-arid regions outside the North American continent.
Natural rubber production could thus become an important future industry in semi-arid regions, a view now being adopted by important
segments of the chemicals-energy industry (25).

HYDROCARBON PLANTS: POTENTIALS AND LIMITATIONS

The above description of the potential of guayule and other
natural rubber plants of semi-arid environments has parallels in other
hydrocarbon-containing plants. However, the level of applicable,
specific knowledge is generally much lower. As perceived potential
is often inversely related to knowledge, it is not suprising that
such hydrocarbon plants have attracted considerable interest. The
popularization of the idea of growing such hydrocarbon plants for
chemicals owes much to the recent pioneering work of Melvin Calvin
(28). Recently, however, the concept of hydrocarbon plants has been
challenged (29). An examination of the issues is appropriate here.

A key part of the concept is that the more highly reduced the
final biomass product, the higher the "energy" yield will be in terms
of CO_2 fixed, which, of course, is proportional to water utilization
(everything else being equal). Thus, on a dry weight basis, a plant
would produce about twice as much carbohydrate than hydrocarbon per
weight of CO_2 used and water transpired. However, on an energy content basis, the hydrocarbon products will exceed that of the carbohydrates. This, however, assumes that all the carbon dioxide released during the biosynthesis of the hydrocarbon compounds from
sugars will be internally recycled and not released to the atmosphere.
Although reasonable, this assumption remains to be proven. Furthermore, the biosynthesis of reduced compounds will require a larger
solar energy input than carbohydrate manufacture, because of inef-
ficiencies in the metabolic pathways involved and because these
compounds are synthesized as end or "secondary" byproducts of the
plant differentiation process when metabolic efficiencies may be
expected to be low (29). However, in the desert environment where
sunlight is not the limiting factor, this argument is not a deciding
one. Until the increased water use efficiency argument can be
strengthened with experimental evidence, the major factor in favor
of the hydrocarbon plant concept is the generally higher value of
reduced organic compounds when compared to lignocellulosic biomass.

Several arguments may be brought up against the concept of hydrocarbon "farming." One is the fertilizer requirement of such plantations. If nitrogen contents were similar to whole corn (maize) plants at 1.5% of dry weight, fertilizer needs would be high. However, seed protein production is not the objective of such activities and nitrogen content would be significantly lower, possibly as low as corn residues (stalks) at 0.5% N. Although some fertilizers would be required for crop production under semi-arid conditions, it is not apparent that this would be a limiting factor. Indeed, a number of leguminous (and nonleguminous) annuals or perennial nitrogen-fixing plants native to desert areas could be considered to supply required soil fertility, either in crop rotations by intercropping, or even as hydrocarbon plants. Unfortunately too little is known about semi-desert nitrogen-fixing plants to evaluate their potential or water use efficiency. (Nitrogen-fixing crop plants require more water than non-fixing crops).

A more compelling argument against hydrocarbon plants is the processing problems associated with extracting plant hydrocarbons and converting them to useful products. Unlike seed oils, the hydrocarbons considered here would be distributed throughout the plant, requiring total harvest, breakdown of cellular structures and separation from lignocellulosic components. Thus it may be more appropriate to "harvest the greatest weight of carbohydrate and process that to more exotic substances" rather than have to deal with low concentrations of low molecular weight hydrocarbons in plants (28). This argument neglects, however, the fundamental difficulties of converting lignocellulosics to any type of valuable, or even useful, chemical. Although pyrolysis-gasification processes are reasonably well developed, the oils produced are of very low quality (even when using hydrogenation processes) and the synthesis gas produced could at best (and not yet economically) be upgraded to methanol (1). Without minimizing the processing problems that must still be resolved, it does appear that these will be less difficult than thermochemical conversions, for which much more experience exists. Indeed, it is the high specificity of plant biosynthetic pathways that must be considered the major attraction of hydrocarbon plants. Among the several thousand plants known to make oils (29) there are a myriad of organic chemicals that would tax the greatest ability of modern organic chemistry to produce. With the tools of modern plant genetics, without even calling on genetic engineering, it should be possible to create real "photosynthesis factories" by accentuating specific biosynthetic pathways and products, increasing their extractability, and maximizing both their relative content and total plant yield.

Unfortunately it has been the plant yield issue that has most clouded the arguments that can be made in favor of the hydrocarbon plant concept. Calvin's claim of "10 barrels of oil per acre per

year" (28) was based on measurements of single plants from small
irrigated test plots using an annual, Euphorbia lathyris, which contained 10% by weight (8% assumed to be extractable) of low molecular
weight (<10^5) pholyisoprenes. If the reference to oil is assumed
to be petroleum equivalent, this would represent at least 50 million
BTU of oil or (at 17,000 BTU/lb (28)) 3,000 lbs of extractable oil
and 18.5 short tons/acre/year of total biomass. Assuming an average
BTU content for biomass of 8,500 BTU/lbs, then this is equivalent to
20 tons/acre/year dry weight (44 metric tons/hectare/year) of normal
biomass. As has been pointed out by Loomis (29), this exceeds the
world record production for any C_3 crop plant (30). The implicit
suggestion that such yields are achievable in semi-arid "land that is
nonproductive" (28) is obviously not reasonable. Even for the irri-
gated-fertilized plots the calculations that reached the "10 barrels
of oil per acre per year" conclusion are based on single plants subject of edge effects and extrapolations from low to high planting
densities. The arguments made in the case of guayule apply here also.
Yields are best measured in terms of water use efficiency based on
relative amounts of water transpired, annual rainfall and supplemental irrigation. In case of the latter, irrigation water quality as
well as quantity are important parameters. The eventual yields will
be quite variable of course. However, on the average, in semi-desert
environments, dry weight yields one-tenth of those proposed by Calvin
would still greatly exceed those of native populations. Thus, even
one "barrel of oil" per acre/year would be a major achievement.

At any rate, the arguments over yields obscure the fact that
economics will be the deciding factor for which yields are only one,
and not even the deciding, factor. The key will be the production
of chemicals of higher value than the mere "barrel of oil" (fuel)
value of the biomass. This will subsidize the fuels deriveable from
the residues. A great deal of research and development will be required. The present selection of research subjects (e.g. Euphorbia)
should be considered only as a starting point. For example, it is
likely that perennials are, in mose cases, preferable over annuals as
they would have more favorable shoot/root ratios and production can
be accumulated over several years. At any rate, present information
does not support the view that "cultivating hydrocarbon plants on
(non-productive) lands can be started almost immediately, even without genetic improvements of the plants" (28). This is not even true
for guayule for which considerable experience and a well developed
market exist, and high product values (exceeding $1.50/kg) are fore-
cast. Only by immediately expanding research and development programs, will the guayule rubber market potential forecast for 1985
and beyond be met. For other hydrocarbon plants considerably longer
term research and development is required. Multiplant systems, rather
than monocultures, need to be examined.

MICROALGAE BIOMASS SYSTEMS

As discussed before, microalgae biomass systems are of potential application to semi-arid and saline environments. Microalgae are adaptable, both as a class and even as specific species, to wide variations in water quality, including salt concentration and types, and organic pollution. Microalgae are completely submerged, thus water use is strictly a function of free water evaporation at the specific locality (with, perhaps a 10-25% increase due to effect of mixing and increasing light absorption by the algae). Water use efficiency is, thus, solely a function of productivity (at any given site) with no tradeoffs between CO_2 utilization and water loss. The completely hydraulic nature of the microalgae production system is an advantage as it allows ready control and adjustment of large growing areas. The costs of the required ponding systems, including the necessary baffles, piping, paddle wheels, etc., are, however, considerable, requiring a minimum investment of $10,000/hectare of growth area, making favorable assumptions (31). The most expensive, and up to now most difficult, part of a microalgae cultivation system is, however, the harvesting of these microscopic plants. Their rapid growth rates and optimally low standing biomass (<100 g/m^2) require constant harvesting which means a one hundred-fold concentration (200-500 mg/l to 20-50 g/l) of the algae. At present the only potentially low cost harvesting processes are screening (applicable to large colonial or filamentous microalgae) or bioflocculation (which requires establishment of a self-flocculating culture) (7,32). Another difficulty is the supply of carbon dioxide to the ponds. This still requires development of a low cost process. The sources of the CO_2 is, of course, also an issue. All of these difficulties may be overcome, both through research and development and/or increased cost of production. The achievable yield (tons/hectare/year) thus becomes critical in this context (unlike that of higher desert plants where higher yields mean more transpiratory water loss), as most costs, including water, are fixed. A recent review of both worldwide experience in ourdoor microalgae mass cultures concluded that yields of 15-25 g/m^2/day are now commonly being achieved in a number of places for relatively prolonged periods of time, and short term yields exceeding 30 gm/m^2/day are attained often (34). Although not sustainable for a full year of operations, such achieved yields incidate a considerable production potential. Thus, a yearly average of 12.5 g/m^2/day (almost 45 tons/hectare/year), used as a basis for most recent analysis and calculations of microalgae biomass production costs (7,31), must be considered conservative. Indeed, yields of up to 100 tons/hectare/year are not unreasonable long term goals for controlled microalgae biomass production at favored locations.

Although in theory algae growth kinetics and yields may be estimated almost from first principles, in practice there is a lack of specific information which does not allow narrowing the gap between present achievements and future potentials. Specifically, cell and culture maintenance requirements (decay rates, predation rates, respiration, photorespiration, photoinhibition, etc.), light saturation levels, and cell concentrations and growth rates all enter into the final consideration of achievable productivities (33,34). The higher the light saturation levels and the lower the maintenance requirements, the higher the potential productivity of an algal culture. High saturating light levels are exhibited by marine flagellates and nitrogen-starved algae cultures, however, high productivities have not been reported for either case, probably because controlled mass cultivation has not been attempted or optimized. For both cases, as is discussed below, significant opportunities exist for production of high value chemicals from microalgae. Two other aspects must, however, be discussed first: the requirements for algae culture and population control, and the bioengineering and economic aspects of microalgae production.

The cultivation of plants for specific purposes (whether for food, feed, fiber, chemicals or fuels) requires the establishment of specific plant populations or species selected for their favorable properties. Microalgae culture requires similar specificity and selectivity of cultivation; allowing any algal "weed" to proliferate just won't do. Although in some cases, particularly municipal wastewater treatment, a large variety and variation of algal populations may be allowable, this will not be the case for algal cultivation for chemicals or other high value products (feeds). At any rate, harvesting technology alone requires some control over algal species (31,32). The proposition that it is possible, on a large scale in outdoor ponds, to control algal species is still controversial. The fast growth rates of microalgae and zooplankton allow very rapid changes in pond populations. Obviously considerable knowledge will be required about the growth responses to changing environmental conditions of the algae being cultivated and their common grazers or weeds. Constant information about pond conditions will need to be collected and this information must be translated into feedback to the pond operational variables controlled by the operator: detention time, mixing spped and schedule, depth, nutrient levels and pH, cell recycle and inoculation. The latter aspect is often neglected, however, operation of an algal culture on a semicontinuous basis with occasional restarting of the culture may be advantageous. Although sufficient work has not yet been carried out on those aspects, there are no theoretical grounds to dismiss algae species control as unfeasible. Halfhearted attempts at species control ending in failures cannot be used as evidence that algal species control is not possible. After all, weeds are also a major problem in higher plant cultivation.

Other arguments against the possibility of practical or at least low cost and high yield cultivation of microalgae are the assertions that yields are somehow inversely proportional to the scale of the system (34) and that net energy production is either not feasible or very doubtful (35). The first misconception is based on the fact that most microalgae production systems are small, and that the few large ones were either poorly operated with low productivity, or were very expensive, or both. In particular, the scale-up of ponds without provision for suitable mixing systems (36) can be a problem leading to such conclusions. Although scale-up problems are likely for any technology, it is not apparent why the simple high-rate pond design introduced by Oswald (37) cannot be easily scaled up. Although some uncertainties exist about maximum allowable channel widths and lengths, these are the only real unknowns, and minor ones at that. Actually the greatest problems are related to soil permeability and consistency; water loss must be minimized and silt suspension prevented. These issues are very site specific. Where soil conditions are unfavorable, a plastic liner must be considered. Microalgae cultivation is inherently a low energy process. Mixing power has been repeatedly calculated to be a small fraction (less than 10%) of biomass energy outputs as long as mixing speeds do not exceed 10 cm/sec and the pond surface is not very rough. Pond carbonation (CO_2 transfer from gas to liquid) power requirements are at present unknown but do not need to be high in theory. The only major power consuming process is harvesting. If microstraining or bioflocculation processes can be applied or developed this bottleneck of microalgae biomass production would be eliminated (31,32).

The economics of microalgae biomass production are still uncertain. Minimal capital costs for the algae growth pond - including grading, earthworks, baffles, paddlewheels, harvesting (settling) ponds and piping - may be estimated (1978 dollars) at about $15,000/hectare (31,32). If a plastic liner is required it would double capital costs. Power requirements for mixing, harvesting, and carbon dioxide supply, as well as fertilizer and labor costs would result in yearly operational costs of $1,500-3,000/hectare, depending on assumptions. Thus microalgae biomass production is, at present, a capital intensive and expensive process when compared to any conventional agriculture. These estimates assume low land costs ($1,000/hectare) and essentially no water or CO_2 costs. Even with such favorable assumptions microalgae biomass production costs must be currently estimated at from $100/ton to $200/ton. Although it may be possible to reduce this somewhat in the future, for the present microalgae cultivation will be restricted to wastewater treatment systems, where allowable costs are very high, or high value specialty foods and feeds (38). In addition, microalgae have significant potential in production of chemicals under saline-desert conditions.

One process already at the pilot plant stage involves the flagellate Dunaliella, isolated from highly saline environments such as the Dead Sea (39). These algae produce high concentrations of glycerol, up to 50% of dry weight, when exposed to high salt concen-

trations, exceeding those of saltwater (40,41). The glycerol is produced to allow the cell to resist the high osmotic pressures that
result from the exclusion of salt from its cytoplasm. The commercial
value of glycerol is about $1.00/kg, making it an attractive product.
At present not much information has been released about yields and
production technologies for these microalgae. The very high salinity
at which these algae are to be cultivated suggest a decrease in yields
over similar algae grown at lower salt concentrations, although they
may still be in the 40-50 ton/hectare/year range. It may be advantageous to go to a two (or more) stage process in which the algae
are first grown at a relatively low salt concentration (seawater) and
then harvested and transfered to a high salt concentration to maximize glycerol production. This would, however, involve multiple harvesting, which is still a problem. Although these algae exhibit
phototaxis (they swim to the light), this property has not yet been
successfully exploited as a harvesting tool. Centrifugation, presently used, is too expensive unless the algae are preconcentrated
by about ten-fold. This problem appears, however, readily solvable
and should not provide a major obstacle. The need to use very high
strength brines for <u>Dunaliella</u> production will restrict this system
to places where these occur naturally (such as desert salt pans) or
are manmade, such as solar salt evaporation systems. Indeed, there
is a potential for evaporative disposal of saline waters in large
solar ponds combined with algae production.

It is well known that microalgae contain a number of lipids,
fats and other reduced compounds which are of potential commercial
interest as a source of vegetable oils (42). Considerable interest
has recently been generated in the production of microalgae for their
lipids, an interest that goes back to the early days of microalgae
cultivation at the pilot plant scale (43). The concentration of such
lipids can exceed 50% of dry weight, depending on the microoaglae
species as well as the particular growth conditions. The condition
most strongly and universally favoring lipid accumulation is nitrogen
limitation or starvation. Unfortunately, nitrogen limitation conditions are not conducive to high productivities, which are achievable
only under conditions of light limitation.   However, nitrogen limitation has the potential of increasing total productivity by increasing the light saturation level of a culture, thereby allowing better
light utilization. This potential, however, is difficult to realize
because of overcompensation by the regulatory processes in nitrogen
starved algae which causes not only a drastic reduction in pigmenta-
tion but also a considerable loss of the enzymes involved in CO_2
fixation. Thus, the physiology of nitrogen starvation and its control requires more study. At present, it appears that a suitable
lipid production system would involve a multistage process in which
growth, nitrogen limitation, and nitrogen starvation (lipid production) would take place in successive ponds. Intermediate harvest
would not be required but depth, detention times, and other pond

variables would be adjusted. A related example of such a system, involving biophotolysis (the production of hydrogen and oxygen from water), has been recently operated successfully (44,45). This system involves nitrogen starved blue-green algae that, under an inert atmosphere, produce hydrogen. Continuous metabolic activity (hydrogen production) for over one month has been achieved with non-growing cultures, indicating that such systems are inherently stable. Obviously much more work is required in the development of these concepts.

One aspect of microalgae technology that has been neglected is the processing of the biomass produced. The drying processes employed are energy intensive, and either destructive or too gentle to break cell walls. Wet processing, either by solvent extraction or with chemical-biological means, needs greater attention. As with other plant materials from semi-arid regions, producing the biomass is only the first step. Processing technology is the key to actually obtaining a valuable product.

There are considerable water resources that could potentially be applied to microalgae production. For example, in California about 100,000 hectares of algae ponds could be operated with saline agricultural drainage water produced by irrigation. This drainage water presents a difficult disposal problem that may be solved by combining evaporation ponds with algal production (46). In other areas it may be possible to use saline groundwaters or local runoff for algae production. Under some circumstances plastic covered ponds may be considered, however this would greatly increase production costs. In conclusion, microalgae biomass systems will be restricted to special applications (wastewater treatment) or production of high value products. For the latter, an essentially flat site located near a waste CO_2 source and a saline water source are required. Climatically, microalgae production will be restricted to regions with high year-round sunshine. The proposition that microalgae could replace staple crops is not defensible at present.

ARID-SALINE PLANT RESOURCES IN PERSPECTIVE

The general concepts discussed above, do not exhaust the possibilities of this subject. Indeed, the separation made between chemical-fuels production and food-feed-fiber production is artificial; the final systems that are developed are semi-arid zone resource utilization are likely to incorporate and integrate all these possibilities. A case in point is that of mesquite (Prosopis juliflora) This valuable tree of arid zones of North and South America, now introduced in many other areas around the world, is a leguminous, nitrogen-fixing plant of significant potential for cattle food and fuel wood production. Its extremely long taproot allows it to reach water resources beyond the reach of other plants. Its protein rich

(12%) pods also contain up to 30% carbohydrate and, after grinding, can be used as cattle feed. An orchard of such plants in an area receiving 250-500 mm rainfall/year may produce up to 4 tons/hectare/year of such ponds without supplemental irrigation or fertilization after establishment of the stand.(47) It would, however, utilize groundwater originating outside of the orchard. The use of mesquite as a fuel wood is well established. Another example of a mixed system is that of the jojoba plant (Simmondsia chimensis) whose seed oil is extremely valuable and can replace sperm whale oil which will no longer be available after the imminent extinction of its source (48). Fiber plants also can, and are, being produced in the desert environments. Indeed, cotton is a crop requiring relatively little water and thus is suitable for many areas in semi-arid zones. Yucca and agave are being harvested from natural stands, both for their fiber and food content (agave being used in alcohol production). Residues from such plants could be sun dried and used as fuels. In addition, there may be some agronomic value to halophytes which accumulate salts in the reclamation of croplands inhibited by salt accumulation. The plants could be harvested and used as fuels with the ash (salt) residue removed to areas where leaching would not take place. In areas accessible to ocean waters the production of small seaweeds in coastal farms or on-shore ponds, would be feasible (49,50). Such plants are sources of valuable chemicals.

The resources of arid regions are, however, limited and their ecologies are fragile. The possibilities for production of chemicals with either novel or well established plant systems will be constrained by the realities of the harsh environments under consideration. Grandiose schemes to make the deserts bloom with water imported from far away, by covering them with plastic, or by mining subsurface groundwater, are not realistic in the context of energy limitations, engineering realities, and historical perspective. For example, in the southwestern U.S., groundwater irrigation is rapidly exhausting the resource; in some areas such agriculture already has ceased.

"New water" importation schemes from as far away as Canada are likely to require more energy (and money) than they will produce. A case in point is the Colorado River water desalinization project at the US-Mexican border. This project is designed to amerliorate the salinity problems engendered by agricultural operations further north, whose outputs do nut justify the money and energy required by the desalinization plant. The economics and engineering realities of greenhouse agriculture are often neglected by proponents who see this as a panacea to the problems of water limitation in arid zones. A ten-fold (at minimum) to hundred-fold (at present) increase in capital and operational costs (including energy use) cannot be justified on the basis of a possible two or three-fold increase in productivity. Even water conservation may be illusory as cooling will be required in most cases. Certainly greenhouse agriculture will continue to ex-

pand in size and importance. However, it will be restricted to very
high value produce in specific areas; it cannot supply staple foods
or be considered for the types of chemical production discussed above.
Appropriate technologies need to be developed which are adapted to
the environment in which they will be used and which can sustain a
viable human ecology for indefinite periods. In this quest, mankind's
greatest assets are the biological systems, evolved over billions of
years, and which we now have the power to control or destroy. Plants
can manufacture all the foods we need, they can provide us with fuels
and an astonishing variety of chemicals, and they can also purify our
environment. Their exploitation will be more difficult than mining
the earth for depletable resources but more rewarding in the end.

ACKNOWLEDGMENTS

This paper was prepared in part during work for the California
Energy Commission. I wish to thank Mr. Robert Hodam for this oppor-
tunity as well as Dr. O. Zaborsky for his encouragement and to Dr.
R.S. Loomis for his detailed and thoughtful comments.

REFERENCES

1. Benemann, J.R. 1978, Biofuels, A Survey, Electric Power Research
    Institute, ER-746-SR, Palo Alto, California.

2. Alich, J.A. and R.E. Inman. 1974, Effective Utilization of Solar
    Energy to Produce Clean Fuels. Final Report, Stanford Research
    Institute, Menlo Park, California.

3. Fraser, M. et al. 1977, The Photosynthesis Energy Factory: Analy-
    sis, Synthesis and Demonstration. Intertechnology/Solar Corp.,
    Warrenton, Virginia.

4. Inman, R.E. 1977, Silviculture Biomass Farms; Vol. I, Summary,
    Mitre Corp., McLean, Virginia.

5. Mariani, E.O. et al. 1978, The Eucalyptus Fuel Plantation as a
    New Source of Energy, Marelco Inc., Beverly Hills, California.

6. Oswald, W.J. and C.G. Goleuke. 1960, "Biological Transformations
    of Solar Energy", Adv. Appl. Microbiol. 2: 223-262.

7. Benemann, J.R. et al., 1978, 2nd Ann. Symp. Fuels From Biomass,
    "Fuels from Microalgae Biomass", in press.

8. Wilcox, H.A. 1976, The U.S. Navy's Ocean Farm Project, Code 8046,
    Naval Ocean Systems Center, San Diego, California.

9. North, W.J. 1976, Ocean Food and Energy Farm Project, Subtasks
    1 and 2: Biological Studies of M. pyrifera Growth in Upwelled
    Oceanic Waters, U.S. Energy Research and Development Admin.

10. Flowers, A. and A.J. Bryce. 1977, Energy Conversion from Marine
    Biomass, American Gas Association, Washington, DC.,

11. Wolverton, B.C. et al. 1975, Bioconversion of Water Hyacinths
    Into Methane Gas, Part I, NASA Tech. Memo TM-X72725.

12. Lipinsky, E.S. et al. 1977, Systems Study of Fuels From Sugarcane,
    Sweet Sorghum, and Sugar Beets; Vol. I, Comprehensive Evaluation,
    Battelle, Columbus Laboratories.

13. Benson, W.R. 1977, "Biomass Potential From Agricultural Produc-
    tion", in Conf. Proc., Biomass--A Cash Crop for the Future? Mid-
    west Research Institute and Battelle, Columbus Laboratories,
    Kansas City, Missouri.

14. Fisher, R.A. and N.C. Turner. 1978, "Plant Productivity in the
    Arid and Semiarid Zones," Ann. Rev. Plant Physiol., 29: 277-317.

15. Mudie, P.J. 1974, "The potential Economic Uses of Halophytes" in
    R.J. Reinold and W.H. Queens, eds. Ecology of Halophytes, Academic
    Press, New York., pp. 565-597.

16. National Academy of Sciences. 1974, More Water for Arid Lands,
    Washington, DC.

17. Solbrig, O.T. and G.H. Orians. 1977, "The Adaptive Characteristics
    of Desert Plants," Amer. Sci., 65: 412-421.

18. Berry,J.A. 1975, "Adaptation of Photosynthetic Processes to
    Stree," Science, 188: 644-650.

19. Schulze, E.D. and L. Kappen. 1975, "Primary Production of Deserts"
    in Photosynthesis and Productivity in Different Environments,
    J.P. Cooper, ed. Cambridge Univ. Press, Cambridge.

20. Bjorkman, O. et al. 1972, "Photosynthetic Adaptation to High
    Temperatures: A field Study of Death Valley, Calif." Science,
    1975: 786-789.

21. Boyko, H. 1975, "Saltwater Agriculture." Sci. Ameri., 102: 88-96.

22. Epstein, E. and J.D. Norlyn. 1977, "Seawater Based Crop Produc-
    tion: A Feasibility Study," Science, 97: 249-251.

23. National Academy of Sciences. 1977, Guayule. An Alternative
    Source of Natural Rubber, Washington, DC.

24. McGinnies, W.G. and E.F. Haase. 1975, Guayule: A Rubber Producing Shrub for Arid and Semiarid Regions, Office of Arid Land Studies, Univ. of Arizona, Tucson.

25. McGinnies, W.G. and E.F. Haase, eds. 1975, An International Conference on the Utilization of Guayule, Office of Arid Land Studies, Univ. of Arizona, Tucson.

26. Anderson, E.V. 1978, "Economics Improving for Guayule Rubber," Chem. Eng. News, Aug. 28, pp 10-11.

27. Reid, M.S. 1978, Jet Propulsion Laboratory, Pasadena, California, personal communication.

28. Calvin, M. 1978, "Green Factories", Chem. Eng. News, March 20, pp. 30-36.

29. Loomis, R.S. 1978, "Agriculture", presented at CHEMRAW Symp. on Organic Raw Materials, Intl. Union of Pure and Appl. Chem., Toronto, CANADA, in press.

30. Loomis, R.W., and P.A. Gerakis. 1978, "Productivity and Agricultural Ecosystems", in Photosynthesis and Productivity in Different Environments, J.P. Cooper, ed. Cambridge Univ. Press, Cambridge.

31. Benemann, J.R. et al. 1978, Cost Analysis of Microalgae Biomass Systems, U.S. Dept. of Energy, HCP/TI 605-01. 94 p.

32. Benemann, J.R., et al. 1978, Large-Scale Freshwater Microalgal Biomass Production for Fuel and Fertilizer, Final Report, Sanitary Engineering Research Laboratory, Univ. of California, Berkeley.

33. Oswald, W.J. 1977, "Determinants of Feasibility in Bioconversion of Solar Energy, in Research in Photobiology, A. Castellani, ed., Plenum Publishing, New York.

34. Goldman, J.C. 1978, Fuels for Solar Energy: Photosynthetic Systems State of the Art and Potential for Energy Production, USDOE, in Press.

35. Goldman, J.C. and J.H Ryther. 1977, "Mass Production of Algae: Bioengineering Aspects", in Biological Solar Energy Conversion, A. Mitsui, et al., eds., Academic Press, New York.

36. D'Elia, C.F. et al., 1977, "Productivity and Nitrogen balance in Large-Scale Phytoplankton Cultures", Wat. Res., 11: 1031-1040.

37. Oswald, W.J. 1963, "High-Rate Pond in Waste Disposal", in Dev. in Indux. Microbiol. pp. 112-119.

38. Oswald, W.J. and J.R. Benemann. 1978, "Biological Waste Treatment at Elevated Temperatures and Salinities", in this volume.

39. Ben-Amotz, A. 1978, Kors Industries Ltd., Israel, Private Communication.

40. Ben-Amotz, A. and M. Avron. 1973, "The Role of Glycerol in the Osmotic Regulation of the Halophilic Alga, Dunaliella parva", Plant Pysiol. 51: 875-878.

41. Borowitzka, L.J. and A.D. Borwn. 1974, "The Salt Relations of Marine and Halophilic Species of the Unicellular Green Alga, Dunaliella", Arch. Microbiology, 96: 37-52

42. Dubinsky, Z. et al. 1978, "Potential of Large-Scale Algal Culture for Biomass and Lipid Production in Arid Lands", Bioeng. Biotech., in press.

43. Burlew, T.S. ed. 1953, Algae Culture: From Laboratory to Pilot Plant, Carnegie Institute of Washington publication, Wash., DC.

44. Benemann, J.R. et al. 1977, Solar Energy Conversion with Hydrogen-Producing Algae, Final Report, Sanitary Engineering Research Laboratory, Univ. of Calif., Berkeley.

45. Hallenbeck, P.C. et al. 1978, "Solar Energy Conversion with Hydrogen-Producing Cultures of the Blue-Green Alga, Anabaena Cylindrica," Bioeng. Biotech., in press.

46. Benemann, J.R. and W.J. Oswald. 1977, "Algae Farms on Marginal Cropland", presented at Energy Farms Workshop, Biomass Alternative Implementation Div., Clif. Energy Resources Conservation and Development Commission, Sacramento, California,

47. Felker, P. and G. Weiner. 1977, "Potential Use of Mesquite as a Low Energy, Water and Machinery Requiring Food Source", unpublished manuscript, Univ. of California, Riverside.

48. National Academy of Sciences. 1975, Products from Jojoba: A Promising New Crop for Arid Lands, Washington, DC.

49. Ryther, J.H. et al. 1977, "Cultivation of Seaweeds as a Biomass Source for Energy", in press.

50. Neushul, M. 1977, "The domestication of the Giant Kelp, Macrocystis, As a Marine Plant Biomass Producer", in The Marine Plant Biomass of the Pacific Northwest Coast, R.W. Krauss, ed., Oregon State University Press.

ENZYME TECHNOLOGY: POTENTIAL BENEFITS OF BIOSALINE ORGANISMS

Robert W. Coughlin and Oskar R. Zaborsky

Department of Chemical Engineering
University of Connecticut
Storrs, CT   06028
        and
National Science Foundation
1800 G Street, N.W.
Washington, DC   20550

INTRODUCTION

A biosaline environment exists in many parts of the world where enzyme technology has not yet begun to flourish. Accordingly, there is little information available on the effects of the environmental conditions of high temperature and salinity on specific enzymatic processes of industrial importance or on the specific production of the animals, higher plants or microorganisms which presently serve as sources of enzymes. Up to now, it seems that industrialists have not sought and perhaps have avoided the practice of enzyme technology under biosaline conditions.

In view of the foregoing, it seems most appropriate for us to focus on some basic concepts and to review several important industrial uses of enzymes. We shall also mention new uses now in the development stage and give a brief description of industrial methods for producing enzymes. In the discussion of the known effects of temperature and salinity on microorganisms and enzymes, we shall focus on the special advantages and disadvantages that might be anticipated for enzyme technologies which could develop within the biosaline environment. We shall also raise issues and questions on which we may not have firm answers at this time. This is done in order to stimulate thought and further research as well as to relate our chapter to others in this book.

APPLICATIONS OF ENZYMES

All known enzymes are proteins that catalyze the chemical reactions in living cells. However, unlike the usual industrial catalysts (transition metals, supported oxides or salts of metals, and
inorganic acids), enzymes are extremely specific regarding the re-
actants and products and they are highly active at room temperature.

Enzymes can exist in a variety of forms which are useful for
industrial conversion processes. Of course, enzymes exist in viable,
intact cells of all living matter, and, in particular, enzymes in
microbial cells offer a very versatile reactor system through fermentation. This is a time-tested approach and one which has been
used to produce a host of products such as antibiotics, organic
acids, solvents, vitamins, amino acids and even the enzymes themselves. A more recent approach is the use of tissue cultures employing viable cell lines. Alternatively, particular enzymes in
intact but non-viable cells can be employed for chemical processing.
In this mode, cells containing the enzymes of interest are first
"killed" by chemical or physical treatments and then used as biochemical reactors. The non-viable cells can be further transformed
into immobilized counterparts, a topic which will be described
shortly.

Certain enzymes exist in integral bodies or organells in cells,
and once isolated, these can be used as enzyme reactors. Two common
examples or organelles are chloroplasts and mitochondria. Although
no commercial process is as yet based on the use of organelles,
these enzyme-containing bodies would most likely be employed in an
immobilized form. Finally, enzymes can exist in a cell-free environment. The advantages of this form are that the catalysts of interest can be used in a concentrated and more controlled manner, devoid of any possibly interfering organic cellular constituents.
The use of cell-free enzymes can afford higher yields and purity of
the product caused by simpler isolations. Reactions are also easier
to control than in fermentations, and there is essentially no waste
disposal problem. Cell-free enzymes have been used commercially in
both solution and immobilized forms.

The two major and most often stated advantages of enzymes over
conventional catalysts for industrial processing are high specificity and enormous catalytic power. Indeed, enzymes exhibit a very
high degree of specificity in their action. Some enzymes appear to
be extremely specific, while others show an activity toward a number
of compounds some of which do not appear to have any physiological
or apparent chemical resemblance to each other. On the contrary,
the usual lack of specificity of conventional catalysts has assured
a continued and ever-increasing interest in enzymes. For a given
reaction, there is often no alternative but to use an enzyme. It
is also important to note that the more specific a transformation

is, the less will be the formation of by-products and the lower
should be the costs of product separation and environmental clean-
up of effluent waste streams. In contrast, most conventional catalysts pose a difficult problem not only with regard to non-selective
product formation but also with effluent pollution.

The catalytic power of enzymes even when contained in intact
cells is remarkable. Although substantial progress is being made
on the synthesis of enzyme analogs and conventional metal catalysts,
no synthetic catalyst has yet been made which has the enormous power
of an enzyme such as catalase, carbonic anhydrase or urease. The
turnover number of these enzymes (the number of substrate molecules
transformed to product per second under saturating conditions) is
10^7, 10^5, 10^4, respectively.

Enzymes are also a renewable material. Although at present
enzymes can be obtained economically only from living sources, enzymes could possibly be synthesized from constituent amino acids in
the future. Also, effective enzyme analogs could be synthesized by
either chemical or enzymatic means.

The use of enzymes as industrial catalysts does have its associated problems, These are availability and purity, stability toward various inactivations, cofactor requirements and operational
limitations.

The availability of a desired enzyme at a reasonable cost and
purity is still a concern. This is especially so when dealing with
a newly discovered enzyme or one which is intracellular. As mentioned previously, most enzymes of commercial interest are obtained
presently by fermentation using suitable microorganisms. The
microbes need to be grown, harvested, and disrupted (for an intracellular enzyme), and the desired enzyme isolated through a host of
methodologies by trial-and-error. All of these steps consume energy,
materials, labor, and time and often make the cost quite high. Yet
as with other catalysts, production improvements can lower the cost
of enzymes.

Another disadvantage of enzymes is the loss of catalytic activity under storage and process conditions. Certain enzymes are
extremely labile and loss of activity can be caused by metal ion
inhibition or inactivation, chemical modifications (e.g., the oxidation of sulfhydryl groups by molecular oxygen) or by denaturation,
a process in which the spatial arrangement of the polypeptide chains
within the molecule is changed from that typical of the native protein to a more disordered arrangement. However, not all enzymes
are labile, and in reality a spectrum of stability exists even when
considering one parameter. The stability of an enzyme is a paramount consideration in its use, but surprisingly very little con-

scientious work has been conducted in this area (1, 2). At the
moment, the three approaches to enhance the stability of enzymes--
chemical, physical, and genetic--all seem to hold equal promise.
The chemical approach seeks to chemically modify an enzyme to a
superior state. The physical approach places an enzyme in its more
environmentally acceptable state. The genetic approach seeks to
create new proteins of superior chemical or physical properties while
retaining the same catalytic function.

A particularly important disadvantage of some enzymatic reactions is the need to regenerate coenzymes (3,4). Many enzymes,
especially those that are involved in synthetic reactions, require
a coenzyme molecule for full expression of catalysis, and regeneration is necessary if the envisioned process is to be commercially
viable. The regeneration of coenzymes in intact cellular systems
has been reported not to be a problem because coenzymes are regenerated in situ by associated enzymatic reactions. However, it remains
to be seen whether coenzyme regeneration is a problem with immobilized cells when employed in a lengthy continuous operation.

An additional problem associated with the use of enzymes is the
operational limitations associated with a soluble catalyst. Although
not all enzymes are soluble and many are particulate and attached
to water-insoluble cellular particles, most enzymes of present commercial significance are soluble. Furthermore, those enzymes which
are not soluble are often made so by the use of detergents or organic solvents in order to sever them from their natural support or
contaminating constituents. As with any soluble catalyst, operational limitations include mechanical loss, non-reusability (often
caused by a severe product isolation step), contamination of the
product and limited reactor design. With enzymes, reactions are
more or less restricted to an aqueous environment. Yet, this is
the exact domain in which conventional catalysts and organic procedures have their greatest limitation.

It is beyond the scope of this chapter to review extensively
the applications of enzymes. These biological catalysts have been
used for centuries, before recorded history and before their chemical nature was known. However, it is appropriate to mention some
of the more important uses.

The most widely used enzymes are the hydrolases--enzymes that
depolymerize natural polymers in the presence of water. These
include the amylases which hydrolyze polysaccharides and the pro-
teases which degrade proteins. Other enzymes of industrial importance are glucose oxidase, invertase, pectinase and glucose
isomerase. Table 1 provides an overview of the U.S. market for
enzymes (1, 2).

TABLE 1

ESTIMATED ENZYME MARKET BY MAJOR CLASSES

$Millions	1975	1980
AMYLASES	$12.50	$14.20
Glucoamylase	2.10	2.60
Fungal	2.40	3.10
Bacterial	8.00	8.50
PROTEASES	24.36	31.61
Fungal	0.82	0.91
Bacterial	1.00	1.30
Detergent	1.00	3.50
Pancreatin	0.80	0.80
Rennins		
Animal plus microbial	12.00	15.00
Pepsin	3.64	5.00
Papain	5.10	5.10
OTHER ENZYMES	14.25	20.80
Glucose oxidase	0.50	0.90
Cellulase	0.25	0.40
Invertase	0.20	0.20
Glucose isomerase	3.00	6.00
Pectinases	3.00	3.50
Medical, diagnostic, research, anti-inflammatory, and diagnostic aids	7.30	9.80
TOTAL	$51.11	$66.61

(Skinner, Ref. 5)

A major use of enzymes is for the conversion of starch to sugars according to the reaction given in Equation (1).

$$\text{starch} \xrightarrow{\text{acid or}} \text{dextrins} \xrightarrow{\text{glucoamylase}}$$
$$\text{glucose}$$
$$\text{glucose} \xrightarrow{\text{isomerase}} \text{fructose} \tag{1}$$

In this process, starch is first liquified to dextrins, either by acid hydrolysis or by an α-amylase (also called the liquefying enzyme). This enzyme attacks the α-1,4-glucan linkages in polysaccharides containing three or more glucose units linked by α-1,4 bonds but it does not hydrolyse 1,6 bonds. Glucoamylase, also known as amyloglucosidase, hydrolyzes α-1,4; α-1,6; and α-1,3 linkages and therefore acts on a wide variety of carbohydrates including starch, amylose, dextrins and glycogen. Amyloglucosidase represents one of the most important bulk enzymes in commercial use as essentially all glucose and glucose syrups are now prepared enzymatically. Glucose isomerase transforms glucose to its keto isomer fructose which is ca. 1.6 times sweeter. Whereas α-amylase and amyloglucosidase are used in soluble form and consequently lost from the reaction process, glucose isomerase is used in an immobilized form. Figure 1 is a schematic flow diagram of the process for the enzymatic isomerization of glucose (7). It should be mentioned that high fructose corn syrup (which contains about 50% of glucose and 42% fructose) now captures a sizeable portion of the multibillion lb/year U.S. industrial sweetener market. Table 2 is a summary of U.S. industrial activity using this enzyme (8).

Rennin is a proteolytic enzyme which finds wide use in clotting milk for cheese-manufacture. Its major reaction is the hydrolysis of kappa-casein which in turn causes coagulation of the milk protein upon application of heat in the presence of Ca^{2+} ion.

Other large-volume enzymes include pepsin, papain and pectinase. Pepsin finds use as a digestive aid, in the hydrolysis of soybean protein, as an ingredient in chewing gum and piglet food and as a partial replacement for rennin as a milk-clotting agent. Papain's major use is in tenderizing meat but it also is applied to hydrolyze the proteinaceous haze which forms during beer manufacture. Pectinase is used almost exclusively to clarify fruit juice by hydrolyzing colloidal pectin suspended therein; the commercial enzyme contain both polygalacturonase which hydrolyzes the α-1, 4-D-galacturonide links in pectic acid and pectinesterase which demethylates pectin to pectic acid.

IMMOBILIZED ENZYMES

The problem of operational limitations associated with enzymes or any soluble catalyst has received considerable attention during

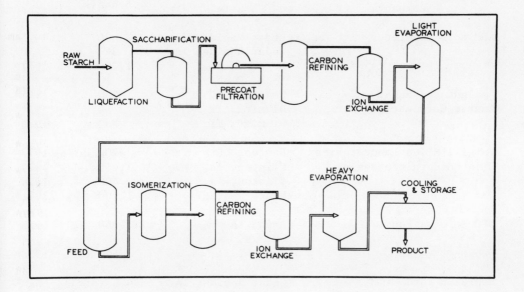

FIGURE 1

Flow diagram of the Clinton Corn Processing Company's glucose
to fructose isomerization processing using an immobilized glucose
isomerase (Mermelstein, ref 7).

TABLE 2

| U.S. High-fructose-corn-syrup capacity. | | Plant rating million lb/yr | |
Company	Site	(wet basis)	Onstream
ADM Corn Sweetener	Cedar Rapids, IA	690	1975
	Cedar Rapids, IA	690	1978*
	Decatur, IL	1,200	1977*
Amalgamated Sugar	Deactur, AL	350	1977*
Amstar Corporation	Dimmitt, TX	50	1974
	Dimmitt, TX	250	1976*
Cargill/Miles Laboratories	Dayton, OH	>200	1977*
Clinton Corn Processing	Clinton, IA	1,250	1972
	Clinton, IA	400	1976
	Montezuma, NY	1,000	1977*
CPC International	Argo, IL	240	1976
Hubinger	Keokuk, IA	500	1978*
Penick & Ford	Cedar Rapids, IA	3-400 (proj)	NA
A.E. Staley Mfg.	Morrisville, PA	600	1972
	Decatur, IL	600	1975
	Lafayette, IN	500	1977*

*Estimated time

(Rosenzweig, Ref. 8)

the past decade and can be circumvented by immobilization.

The term immobilization refers to the physical confinement or localization of enzyme molecules during a continuous catalytic process (9). This is largely an operational definition intended to encompass a host of varied methods that have been used to achieve this objective. Classically, the term immobilization has been used to describe the process of transforming a water-soluble enzyme into a water-insoluble conjugate—the immobilized enzyme. However, not all methods of immobilization involve the preparation of water-insoluble conjugates. Several methods consist of simply restricting the movement of enzyme molecules to a microspace, but the entrapped molecules still have considerable translational and rotational freedom and often retain their inherent solution characteristics. Insolubilization is only one means of immobilization.

There are three major reasons for immobilizing enzymes. Immobilized enzymes offer a considerable operational advantage over freely mobile enzymes. They also may exhibit selectively altered chemical or physical properties, and they may serve as model systems for natural in vivo membrane-bound enzymes. General operational advantages of immobilized enzymes are reusability, possibility of batch or continuous operational modes, rapid termination of a reaction, controlled product formation, greater variety of engineering designs for continuous processes, and greater efficiency in consecutive multi-step reactions. The properties of an enzyme that can be affected by immobilization, especially by chemical methods, are activity, pH-activity behavior, binding characteristics, maximum velocity and stability.

There are several ways of classifying the diverse array of methods for immobilizing enzymes. One is based on the nature of the interaction responsible for the immobilization, that is, whether the immobilization is caused by chemical bond formation or physical noncovalent forces. Chemical methods of immobilization involve the formation of at least one covalent bond or partially covalent bond between groups of an enzyme and a functionalized water-insoluble support or between two or more enzyme molecules. In reality, more than one covalent bond between the reacting components is usually formed. Thus, the immobilization of an enzyme by coupling it to a preformed polymer, by treating it with bifunctional reagents or by incorporating it into a growing polymer chain are all chemical methods. Until recently chemical methods have resulted in irreversible coupling. Once the original enzyme was attached to a polymer or another enzyme molecule, it could not be regenerated. However, this irreversibility is not an inherent feature of chemical methods.

Physical methods include procedures that involve localizing the enzyme in any manner which is not dependent on covalent bond

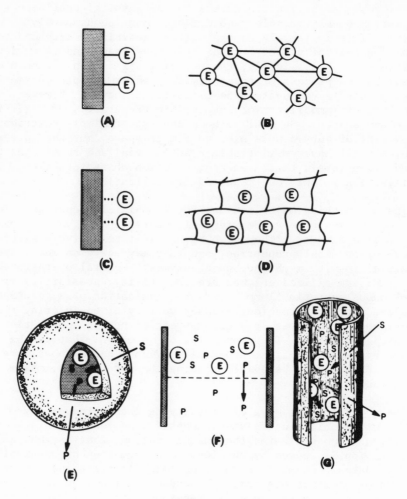

FIGURE 2

Schematic representations of immobilized enzyme systems. Letters
E, S, and P represent enzyme, substrate and product molecules, respectively. A, covalently bonded enzyme-polymer conjugate; B, covalently bonded intermolecularly crosslinked enzyme conjugate; C,
adsorbed enzyme-polymer conjugate; D, polymer lattice-entrapped enzyme conjugate; E, microencapsulated enzyme; F, ultrafiltration cell-
contained enzyme; G, hollow fiber-contained enzyme.

formation. In this group, the immobilization is dependent on the
operation of physical forces--for example, electrostatic interactions, entrapment of enzymes within microcompartments or the containment of the catalysts in prefabricated membrane-containing devices.

Figure 2 diagrams the most commonly employed immobilized enzyme systems. The covalent attachment of water-soluble enzyme
molecules via nonessential amino acid residues to water-insoluble,
functionlized supports is the most prevalent method used (Fig. 2A).
The normal procedure for producing these enzyme conjugates consists
of contacting an enzyme solution with a reactive water-insoluble
polymer. A host of different supports, organic and inorganic, using
different functional groups and reacting with various amino acid
residues of an enzyme can be employed. Commonly used supports include polymers of acrylamide, methacrylic acid, maleic anhydride
and styrene, polypeptides, polysaccharides (especially agarose,
cellulose and dextran) and glass. Immobilization should involve
only functional groups of an enzyme that are not essential for
catalysis. Polymer-bound enzyme derivatives can be filtered or
centrifuged and resuspended. This easy removal of a catalyst
permits selective transformations to be carried out and prevents
contamination of the product. Various forms of the support material
can be employed such as sheet, powder, or fiber. Many of the reactive water-soluble matrices and even enzyme-polymer conjugates are
commerically available.

The preparation of water-insoluble enzyme derivatives using low
molecular weight multifunctional reagents involves covalent bond
formation between molecules of the enzyme and the reagent to give
intermolecularly cross-linked water-insoluble species (Fig. 2B).
This method normally consists of adding the multifunctional reagent
to a solution of the enzyme under conditions that give the desired
water-insoluble derivative.

Systems C to G are classified as physical methods of immobilization. The preparation of water-insoluble enzyme conjugates by adsorption (Fig. 2C) consists of contacting the aqueous solution of
an enzyme with an adsorbent with suitable surface characteristics.
Adsorbents that can be used include anion and cation exchange resins,
carbon, clays, modified celluloses, and glass.

Enzyme can be immobilized by entrapment within the interstitial
space of crosslinked water-insoluble polymers (Fig. 2D). The method
involves the formation of a highly crosslinked network of polymer in
the presence of an enzyme. Enzyme molecules are physically entrapped
within the polymer lattice and cannot permeate out of the gel matrix;
however, appropriately sized substrate and product molecules can
transfer across and within this network to ensure a continuous conversion. The most commonly employed crosslinked polymer for enzyme

entrapment is polyacrylamide, but silicone rubber, starch, and silica
gel have been used.

Enzymes can also be immobilized by entrapment within semipermeable microcapsules whose mean diameters normally range from 5 to 300
microns. Permanent-membrane microcapsules (Fig. 2E) are formed by
interfacial polymerization or by coacervation or preformed polymers.
Coacervation is the phenomenon of phase separation in polymer solutions, and the formation of a microcapsule is dependent on the lower
solubility of the polymer (e.g., collodion or polystyrene) at the
interface between the aqueous microdroplet and the nonaqueous, poly-
mer-containing solution. Interfacial polymerization is the synthesis of a water-insoluble copolymer (e.g., nylon 6,10) at the interface between an aqueous microdroplet and a nonaqueous solution. One
reactant is partially soluble in both the aqueous and organic phase
(hexamethylenediamine) and the other reactant (the second component
of the copolymer, sebacoyl chloride) is soluble only in the organic
phase. The partition coefficient of the water-soluble reactant between the aqueous and organic phases primarily determines the properties of the membrane produced. In both processes for producing
microcapsules, an aqueous solution of the enzyme is first emulsified
in an organic solvent with an organic-soluble surfactant. The principle of operation using microencapsulated enzymes is based on the
perm-selectivity of the membrane. Enzyme molecules, being larger
than the mean pore diameter of the spherical membrane within which
they are entrapped, cannot diffuse through the membrane into the
external solution. On the other hand, substrate molecules, whose
size does not exceed the diameter of a pore, can readily diffuse
through the membrane and are transformed to product by the entrapped
catalyst. The products of the reaction then diffuse through the
membrane to the exterior phase. Enzymes can be immobilized also
within semipermeable non-permanent microcapsules or "liquid" membranes.

Various devices that contain a prefabricated semipermeable membrane can be used for immobilizing enzymes. These are available
commerically and are of a cell form, employing a flat disk-like
membrane (Fig. 2F), or of a cartridge or beaker configuration, employing bundles of hollow fibers (Fig. 2G) whose walls are the membranes. These devices are containers for the localization of an
enzyme in the same way that a beaker contains a solution. Yet,
because of the presence of the semipermeable membrane, they permit
a continuous operational mode. Immobilization consists of putting
the protein solution into the proper membrane-containing cavity
of the apparatus. The successful operation of this type of enzyme
reactor is dependent on the permeability characteristics of the
membrane employed; the membrane should retain the enzyme completely
but allow free passage of the substrate and/or product.

The flat membrane system (i.e., an ultrafiltration cell) is
especially suited for immobilizing soluble enzymes that act on
high molecular weight water-soluble or water-insoluble substrates.
The use of a soluble enzyme permits the necessary intimate contact
of catalyst with substrate in order to achieve an efficient conversion of macromolecules. Hollow fibers offer an extremely large surface area to volume ratio. In all membrane systems, the degree of
product distribution from enzyme degradation of high molecular weight
substrates can be controlled by the cutoff limit of the membrane
employed.

Figure 2 represents only the basic methods for immobilizing
enzymes, and many reported techniques involve variations and/or
combinations of those illustrated. An obvious and often asked question is, "What is the best method?" The answer to this question can
only be given after the characteristics of the particular enzyme are
known and the intended application has been stated because the choice
of the method is largely dictated by these two considerations. Also,
whether a process should be operated with soluble or immobilized enzymes will be dictated by economics. Important factors are cost of
enzyme, cost of support, cost of immobilization procedure, lifetime
on stream of the immobilized enzyme, regenerability of enzyme, and
reuseability of the support. Also entering such a choice between
the use of immobilized and soluble enzyme are technological considerations such as whether the reactor containing the immobilized
enzyme might be blocked by microbial growth, whether the pressures
necessary to pump through a bed of immobilized enzyme will be too
large, or whether the presence of the enzyme in the final product
is permissible.

The first industrial publicized application of an immobilized
enzyme was for the separation of L-amino acids from D,L racemic
mixtures of amino acids such as methionine, phenylalanine, tryptophan and valine using aminoacylase. After the racemic mixture of
amino acids is acylated, the enzyme very specifically hydrolyzes
only the acyl-L-amino acid, thereby producing a mixture of the L-
amino acid and the acyl-D-amino acid (Equation 2).

$$D,L - \underset{\underset{NHCOR^1}{|}}{R}CHCOOH + H_2O \xrightarrow{\text{aminoacylase}} L-\underset{\underset{NH_2}{|}}{R}CHCOOH + D-\underset{\underset{NHCOR^1}{|}}{R}CHCOOH$$

$$(2)$$

The mixture of products is easily separated by a solubility difference and the acyl-D-amino acid racemized chemically to produce a
D,L mixture which is returned to the enzyme reactor. This process,
which was pioneered by the Tanabe Seiyaku Co. of Japan, employs the
enzyme immobilized by ionic bonding to a DEAE-Sephadex support (10).

Related processes developed by the same company utilize immobilized whole cells containing the enzyme of interest. These include addition of ammonia to fumaric acid to give L-aspartic acid, production of malic acid from fumaric acid and water, the production of L-citrulline (urocanic acid), a sunscreening agent, and the production of 6-aminopenicillanic acid, a precursor of semisynthetic penicillins.

The use of immobilized glucose isomerase has already been mentioned and various supports are possible and are being used. In 1977, about 2.5 billion pounds of high fructose syrup were produced by this process—the largest volume of any material produced by immobilized enzyme technology. The enzymatic production of fructose and other chemicals was recently discussed by Zaborsky (11). Common configurations for enzyme reactors are given in Figure 3.

Immobilized enzymes are also finding growing uses in analytical applications. Various instruments are now available which can assay for substrates such as glucose which produce H_2O_2 by an enzymatic reaction (Equation 3).

$$\text{glucose} + O_2 + H_2O \xrightarrow{\text{glucose oxidase}} \text{gluconic acid} + H_2O_2$$

(3)

The hydrogen peroxide is detected electrochemically with an electrode placed in close proximity to a membrane-containing immobilized glucose oxidase. Another assay technique involves the enzymatic production of an ion (e.g., NH_4^+ from urea by the enzyme urease) at an electrode specific for that ion. The ion selective electrode is usually surrounded by a porous sheet of the immobilized enzyme.

EMERGING ENZYME PROCESSES

For illustrative purposes, only a few of the many potential processes will be mentioned.

Many organisms possess the enzyme systems necessary to recycle back to glucose the more than 100 billion tons of cellulose produced on earth each year by plants. Although this can also be accomplished by acid hydrolysis, there is hope that eventually it will be more economical to carry out the process under mild conditions using cellulase enzymes obtained from microorganisms. The glucose produced from cellulose could be used as a raw material to grow single-cell protein for nutritive purposes, or it could be fermented to ethanol for use as fuel, or an intermediate for other chemicals. Many problems remain, however, regarding cellulase enzyme production, purification and isolation as well as reactor design (12).

In recent years Coughlin, Charles and co-workers (13) have developed and brought to the pilot-plant stage a process using

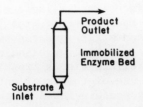

PACKED BED REACTOR

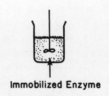

BATCH REACTOR

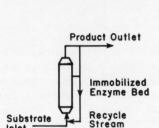

RECYCLE REACTOR

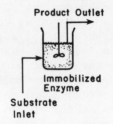

**CONTINUOUS FLOW
STIRRED TANK REACTOR**

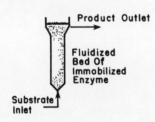

FLUIDIZED BED REACTOR

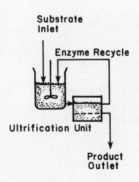

**ULTRAFILTRATION COMBINED WITH
CONTINUOUS FLOW STIRRED TANK
REACTOR**

TUBE WALL REACTOR

FIGURE 3

Reactor processing schemes for enzyme reuse

lactase immobilized on porous alumina to hydrolyze the disaccharide lactose to its constituent monosaccharides glucose and galactose. When applied to whey, a by-product of cheese manufacturing, the 5% lactose in the liquid is converted, thereby conferring greater sweetness and making the whey more attractive and valuable for blending into foods. This process offers the hope of eliminating a troublesome water-polluting waste and making better use of the excellent nutritive value of the extremely high-quality protein found in cheese whey. As part of the work on this process, an economic analysis was carried out and a typical result is exemplified by Figure 4 which shows the cost of hydrolysis as a function of lifetime of the immobilized enzyme and purchase price of the native enzyme.

Xylitol is a natural carbohydrate currently obtained commercially by the hydrogenation of xylose, which in turn is obtained from the hydrolysis of xylan-containing materials. Xylitol is as sweet as sucrose and has a cool taste because of its higher heat of solution. In addition, tests have shown that xylitol causes a large reduction in the formation of dental cavities and plaque. The production of xylitol from hemicellulose of birch wood via an acid hydrolysis/hydrogenation sequence has been explored and commercialized. Although xylitol can be produced in this fashion, a considerable cost of the final process is borne by the high purification costs of separating xylose from the crude hydrolyzate or xylitol from the hydrogenated mixture. The possibility of using enzymes for hydrolyzing xylans is enticing in that the high specificity of enzymes would circumvent the inherent high separation costs.

SOURCES OF ENZYMES

Although the enzyme ribonuclease has been synthesized from the basic chemical building blocks, this technique is very expensive and the practical sources of enzymes remain living organisms, especially the microbes. As illustrations, the sources of many of the enzymes mentioned previously are given in Table 3. The advantages of microbes as sources of enzymes are many. These are large-scale production, ease of extraction, predictable output and activity, and the ability to control and optimize enzyme productivity by the regulation of pH, temperature and composition of the environment for microbial growth. That fermentative production of enzymes is a rapidly developing scientific technique is supported by a statement published as late as 1960 (14) that enzymes from plants and animals were more important than those from microbial sources. The situation is just the opposite today in the 1970's.

Important considerations in fermentative production of enzymes are high levels of sterility, high aeration rates, proper microbial strain and metabolic behavior, control and manipulation of growth rate, nature and composition of the medium, environmental conditions

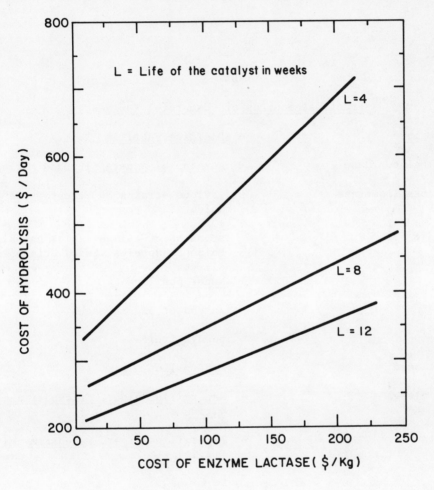

FIGURE 4

Cost of hydrolysis of lactose as a function of the cost of the
enzyme and the life of the immobilized enzyme in the reactor.

TABLE 3

Practical Sources of Commercial Enzymes

α-amylase   -   Bacillus subtilis

amyloglucosidase   -   strains of Aspergillus and Rhizopus

glucose isomerase   -   various strains of Streptomyces
                                   molds

rennin   -   stomachs of calves, lambs and kids
                                   Rhizopus species, Endothecia para-
                                   siticus
                                   Mucor pusillus

pepsin   -   mucosa of pig stomachs

papain   -   papaya fruit

pectinase   -   Aspergillus niger

cellulase   -   Aspergillus niger, Trichoderma
                                   viride

lactase   -   Aspergillus niger, Saccharomyces
                                   fragillis

of pH, temperature, composition, dissolved oxygen and surface active agents, stage of the growth cycle for enzyme expression, regulatory mechanisms such as induction, feedback repression, and catabolic repression and gene dosage. Growth in submerged culture now appears to be widely practiced but semi-solid culture is still used in many instances.

It should be stressed that parameters such as temperature, pH and various compositions might assume one value for optimum growth of a given enzyme-producing microbe but quite different values for optimum production of a particular enzyme. Neither does maximum excretion of enzymes necessarily occur during the growth phase. This behavior is related to the operative regulatory mechanisms in the particular microbe.

The large-scale extraction and purification of enzymes from microbial cells is still a developing technology. The same techniques are also applicable, more or less, to enzyme production from the tissues of animals and higher plants. Microbial enzymes are generally thought of as extracellular or intracellular. The latter class requires that the cells be broken apart, after which the purification procedures are about the same for both classes. Common procedures for releasing enzymes from cells include extraction by alkali, lysozyme, detergents, cold shock, osmotic shock, sonication, freezing and thawing, and application of shearing stress. Purification often requires removal of nucleic acids and concentration by operations such as precipitation, ultrafiltration, freeze-drying, gel chromatograhy, ion-exchange chromatography, affinity chromatography, adsorption and electrophoresis.

OPPORTUNITIES-IMPLICATIONS OF SALINITY AND TEMPERATURE

How can the unique aspects of marine, saline and arid land biomass systems contribute advantageously to enzyme technology? This question will be considered from the standpoint of the enzymes themselves and also with respect to the needed sources.

The catalytic activity of an enzyme is intimately associated with the tertiary structure of the protein. This structure can be strongly influenced by temperature as well as by high salt concentrations. On the one hand, proteins are often selectively precipitated by increasing the salt concentration (especially with NH_4Cl), and they can even retain their native conformation in this process. In fact, enzymes can exhibit catalytic activity in the crystalline state (15). On the other hand, high concentrations of NaCl, urea and other compounds are known to denature proteins. With regard to temperature, most, but not all, proteins become increasingly unstable above 40-50°C. and begin to denature with a concomitant loss of catalytic activity. However, both of these generalizations on enzymes with temperature and salt solutions are based on the easily isolated and studied enzymes obtained from easily grown mesophilic microorganisms and not from thermophilic or halophilic organisms.

The rate of growth of microbes as a function of temperature is exemplified by Figure 5 but this behavior can vary greatly (16). Some bacteria isolated from hot springs can grow at temperatures exceeding 95°C. The growth temperature ranges of several procaryotes are shown in Figure 6. Many specific proteins from thermophilic bacteria are considerably more heat stable than their homologes from mesophilic bacteria, and nearly all proteins of a thermophilic bacterium remain in the native state after a heat treatment that denatures virtually all the proteins of a related mesophile. Thus we can expect that enzymes produced by thermophiles are capable of operation at higher temperatures and, on this basis, would be of greater industrial value in certain situations. The recent publication by Brock (17) provides an excellent discussion of thermophilic microorganisms and some aspects of the associated enzymes. Thermophilic proteins and enzymes have also been discussed by Singleton and Amelunxen (18) and by Doig (19).

The reason for the enhanced stability of enzymes obtained from thermophilic microbes in comparison to enzymes from mesophilic organisms is still uncertain and is probably due to a multiplicity of associated causes. Frequently mentioned reasons are an enhanced binding of bivalent ions between side chains in the protein, enhanced hydrophobic interactions, or increased hydrogen bonding. However, comparative examinations of enzymes from thermophilic and mesophilic microorganisms have, to date, revealed only very minor differences in their physico-chemical properties and the cause of the enhanced stability must be due to subtle changes. At the present time, it is perhaps best not to generalize too much about the exact nature of the phenomenon without additional firm data. Otherwise, similar misleading generalizations may be made about this class of enzymes as was the case with enzymes and thermal stability.

Bacteria also vary greatly in their response to salinity. Some are able to grow in very dilute solutions and some (the osmophiles or halophiles) in solutions of very high sodium chloride concentrations (see J. Lanyi chapter). Some examples of bacteria are given in Table 4. Most enzymes become inactive in high NaCl concentrations but, for extreme halophiles, a high NaCl concentration is essential to maintain both the stability and the catalytic activity of enzymes, as shown in Figure 7.

The foregoing discussion immediately suggests that enzymes produced by microbes adapted to hot, saline environments would be expected to possess unique and potentially valuable characteristics. Many of the microorganisms which flourish in these environments have been little studied and essentially no concerted effort has been

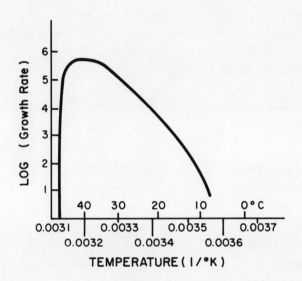

FIGURE 5

Arrhenlus plot of the relationship between growth rate and
temperature for E. coli. (Stanier et al., Ref. 16).

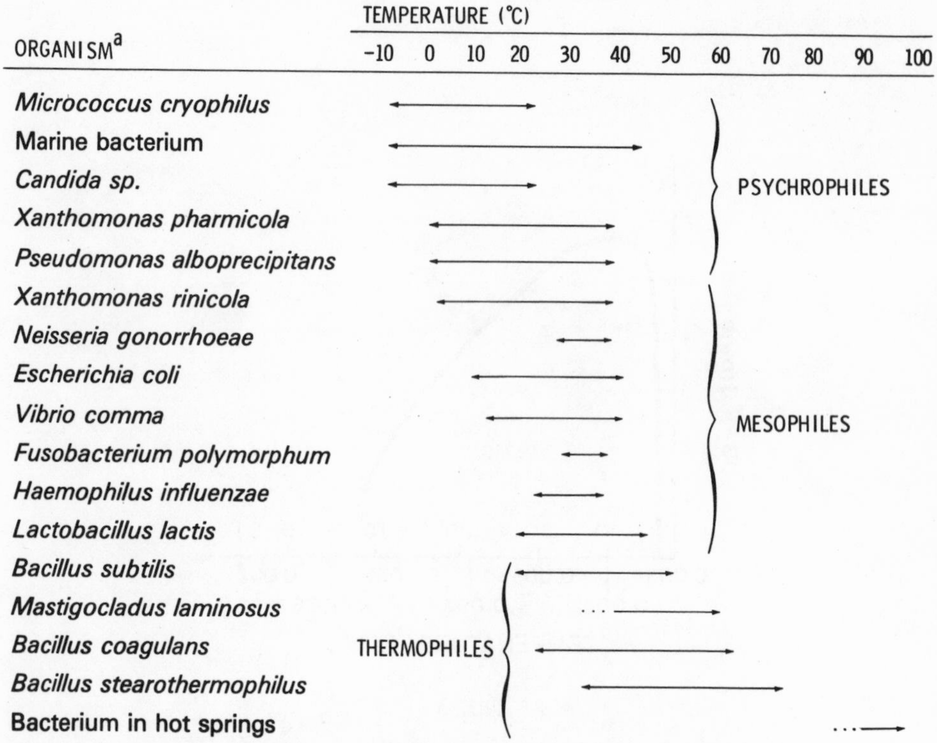

TEMPERATURE (°C)

ORGANISM[a]	-10	0	10	20	30	40	50	60	70	80	90	100

Micrococcus cryophilus

Marine bacterium

Candida sp.

PSYCHROPHILES

Xanthomonas pharmicola

Pseudomonas alboprecipitans

Xanthomonas rinicola

Neisseria gonorrhoeae

Escherichia coli

Vibrio comma

MESOPHILES

Fusobacterium polymorphum

Haemophilus influenzae

Lactobacillus lactis

Bacillus subtilis

Mastigocladus laminosus

Bacillus coagulans   THERMOPHILES

Bacillus stearothermophilus

Bacterium in hot springs

[a]LINES TERMINATING IN SINGLE ARROWS INDICATE ESTABLISHED TEMPERATURE LIMITS OF GROWTH
FOR AT LEAST ONE STRAIN OF THE INDICATED SPECIES; VARIATIONS EXIST AMONG DIFFERENT
STRAINS OF SOME SPECIES. DOUBLE-HEADED ARROWS INDICATE THAT THE ACTUAL TEMPERATURE
LIMIT LIES BETWEEN THE ARROW POINTS. SOLID LINES TERMINATING IN DOTTED LINES INDICATE
THAT THE MINIMUM GROWTH TEMPERATURE IS NOT ESTABLISHED.

FIGURE 6

Temperature range of growth of several procaryotes.

TABLE 4

Osmotic Tolerance of Certain Bacteria

Physiological Class	Representative Organisms	Approximate Range of NaCl Concentration Tolerated for Growth (%, g/100ml)
Nonhalophiles	Spirillum serpens	0.0–1
	Escherichia coli	0.0–4
Marine Forms	Alteromonas haloplanktis	0.2–5
	Pseudomonas marine	0.1–5
Moderate halophiles	Micrococcus halodenitrificans	2.3–20.5
	Vibrio costicolus	2.3–20.5
	Pediococcus halophilus	0.0–20
Extreme halophiles	Halobacterium salinarium	12–36 (saturated)
	Sarcina morrhuae	5–36 (saturated)

(Stanier et al., Ref. 16)

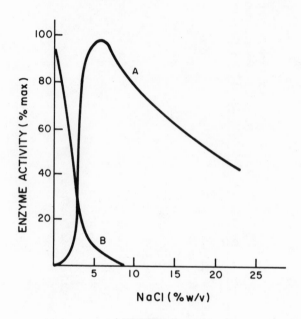

FIGURE 7

Effect of NaCl on the activity of the enzyme, malic dehydrogenase, from an extreme halophile (a) and from liver (b). Like most
enzymes, the enzyme from liver becomes inactive in high concentrations of NaCl. The enzyme from the extreme halophile requires NaCl
for activity. (Stanier et al., Ref. 16).

made to develop fermentation processes based on them--processes which might produce new enzymes, drugs and other useful chemicals as well as valuable thermo- and halotolerant homologes of already known and useful enzymes.

Another characteristic of thermo- and halophilic microbes which might be of industrial value is their potential ability to grow and produce useful products in growth media which might not require sterilization. High salinity and temperature might be sufficient to prevent the interfering simultaneous growth of contaminating microbes. Yet another feature of importance could be the potentially greater response of osmophiles to osmotic shock as a means of extracting internal enzymes. The storage stability of such enzymes would also be superior. Additionally, the use of thermophilic microbes in regions of the world having high temperatures (e.g., the Middle East) would preclude the extensive use of cooling water to maintain the delicate and necessary conditions for fermentation and other biotransformations (20).

By analogy with microbes, the higher plants adapted to hot, saline environments might be expected to synthesize enzymes of correspondingly higher thermo- and halotolerance. However, the expectation of enhanced salt tolerance in enzymes isolated from salt-tolerant plants may not be realized. Initial studies have shown that the enzymes from salt-tolerant and salt-sensitive plants appear to have similar physico-chemical characteristics (see W. Rains chapter). Yet halophytes and other plants associated with dry climates are known to have enhanced levels of amino acids and organic acids such as malic acid (21, 22). Thus, higher plants from a biosaline environment might be interesting sources of chemicals, drugs, and enzymes. The complex organic macromolecules which these plants produce might also display enhanced properties and stabilities.

To conclude, we strongly urge that more research be conducted on microorganisms and higher plants able to live and/or adapt to elevated temperatures and saline conditions. Most important, a thorough examination of the associated enzymes is mandatory if these are ever to be used industrially.

REFERENCES

1. A. Wiseman, Process Biochem. 8, 14 (1973).

2. O. R. Zaborsky, in Enzyme Engineering (E.K. Pye and L.B. Wingard, Jr., eds.) Vol. 2,115, Plenum, New York (1974).

3. W. H. Baricos, R.P. Chambers and W. Cohen, Enzyme Technology Digest 4, 39 (1975).

4. G.M. Whitesides, in Techniques of Chemistry, Applications of
    Biochemical Systems in Organic Chemistry, Part II, (J.B. Jones,
    C.J. Sih, D. Perlman, eds.), 901, Wiley, New York (1976).

5. K.J. Skinner, Chem. Eng. News 53 (Aug. 18), 22 (1975).

6. B. Wolnak & Associates, Present and Future Technological and
    Commerical Status of Enzymes NTIS: PB-219 636 U.S. Dept. of
    Commerce (1972).

7. W.H. Mermelstein, Food Technology, June, 20 (1975).

8. M.D. Rosenzweig, Chem. Eng. (Sept. 27), 54 (1976).

9. O.R. Zaborsky, Immobilized Enzymes, CRC Press, West Palm Beach,
    FL (1973).

10. I.Chibata and T. Tosa, in Advances in Applied Microbiology (D.
    Perlman, ed.), 22, 1, Academic Press, New York (1977).

11. O.Zaborsky, World Conference on Future Sources of Organic Raw
    Materials, Toronto, Canada, 1978. Pergamon Press, in press.

12. C.R. Wilke, ed., Cellulose as a Chemical and Energy Resource,
    Biotechnol. Bioeng. Symposium No. 5, John Wiley, New York (1975).

13. E.K. Paruchuri, M. Charles, K. Julkowski and R.W. Coughlin,
    AIChE Symposium Series No. 1972, Vol. 74, 40 (1978).

14. E.J. Beckhorn, Production of Industrial Enzymes, Wallterstein
    Laboratories'Communications, 201 (1960).

15. F.A. Quiocho and F.M. Richards, Proc. Natl. Acad. Sci., U.S., 52,
    833 (1964).

16. R.Y. Stanier, E.A. Adelberg and J.L. Ingraham, The Microbial
    World, Prentice-Hall, Englewood Cliffs, NJ (1976).

17. T.D. Brock, Thermophilic Microorganisms and Life at High Temp-
    eratures, Springer-Verlag, New York (1978).

18. R. Singleton, Jr., and R.E. Amelunxen, Bact. Rev. 37, 320 (1973).

19. A.R. Doig, Jr., in Enzyme Engineering (E.K. Pye and L.B. Wingard,
    Jr., eds.), Vol. 2, 17, Plenum, New York (1974).

20. G. Hamer, in Microbial Conversion Systems for Food and Fodder
    Production and Waste Management, Proceedings of the Regional
    Seminar, Kuwait Institute for Scientific Research (KISR),
    November 12-17, 1977 (T.G. Overmire, ed.) 109, KISR (1978).

21. M.C. Williams, Plant Physiol. 35, 500 (1960).

22. C.B. Osmond, Aust. J. Biol. 20, 575 (1967).

POTENTIAL FOR GENETIC ENGINEERING FOR SALT TOLERANCE

J. Mielenz, K. Andersen, R. Tait, and R.C. Valentine

Plant Growth Laboratory
University of California
Davis, CA 95616

All living processes are ultimately coded for by heredity material, usually DNA. Salt tolerance is no exception, and it is reasonable to speak of "salt tolerance genes" codings for this property in both simple as well as complex organisms. Genetic engineering of salt tolerance is the application of genetic techniques for harnessing and manipulating the salt tolerance genes for production of food, fiber, oil and energy. For example, it is clear from the studies of Epstein and co-workers and Rains, reported in this volume, that higher plants have evolved salt tolerance genes allowing growth at different levels of salt in the environment.   It is also well known that a variety of microorganisms have evolved salt tolerance genes. The potential for using techniques of recombinant DNA for genetic engineering of salt tolerance genes in bacteria is discussed here with the objective of increasing the utility and productivity of these organisms in a salty world. A brief discussion of various natural sources for isolating salt tolerance genes of microorganisms will be followed by a description of the current state of the art of "genetic engineering technology" as it might be applied to salt tolerance with the concluding sections dealing with symbiotic N_2 fixation. These organisms share the outstanding property of harnessing solar energy as a power source for biological nitrogen fixation.

Obviously, the major goal of a theoretical article of this type is to stimulate the interests of others to bring their expertise to bear on an area which is ripe for development.

BACTERIA AS SOURCES OF SALT TOLERANCE GENES

The relations of microorganisms to salt have recently been reviewed by Brown (1) (see also the paper by Lanyi, this volume), and by Measures (2). One can distinguish between two distinct ways that microorganisms have used to solve the problem of growing in the presence of varying concentrations of NaCl: 1) the salt is excluded from the cell so that the intracellular NaCl and KCl concentrations are relatively low and constant, and specific intracellular organic solutes are produced to balance the osmotic pressure of the outside medium; 2) salt is not excluded from the cell. The sum of the intracellular NaCl and KCl concentrations are approximately equal to the outside concentrations. The intracellular concentration of K^+ is in both cases usually larger than that of Na^+.

A. <u>Extremely halophilic bacteria</u> use mechanism 2) above (see Table 1). These bacteria not only grow at very high salt concentrations, they also require a high NaCl concentration for survival (3). They have apparently undergone the most elaborate evolutionary adaptation of any organism to a life in environments with extremely high salt concentrations. Perhaps the most remarkable adaptation involves the enzymatic machinery of the cell whose activity requires high concentrations of salt (see Lanyi, this volume). While it has not been formally proven by genetic experimentation, it seems clear that the activity and stability of a variety of enzymes of <u>Halobacterium</u> spp. to a high salt environment reside in their unique primary structure of amino acids whose genetic information resides in the genes themselves. Undoubtedly, a multitude of such genes are involved, making it unreasonable to speak of a particular set of salt tolerance genes or operons for <u>Halobacterium</u> spp. Therefore the gene pool of <u>Halobacterium</u> spp. may not represent a suitable source for the salt tolerance genes which when transferred to ordinary bacteria will make them salt tolerant. However, the presence of so many unique genes coding for extremely salt tolerant enzymes makes <u>Halobacterium</u> spp. a potential resource of stress resistant enzymes for use by enzyme engineers who seek to adapt enzyme technology to a wider scope of uses involving high salt reaction conditions.

B. <u>Moderately Halophilic bacteria</u> have a lower requirement for growth than the extreme halophiles, and have been less extensively studied. Some have high intracellular salt concentrations (Table 1), while others have significantly lower concentrations than in the surrounding medium (5).

C. <u>Non-Halophilic bacteria</u> require less than 2% NaCl for growth (3); many seem to have no NaCl requirement. However, some species are very adaptable and can tolerate high salt concentrations (Table 1); although their intracellular enzymes are quite sensitive to salt. One example is <u>Achromobacter</u> sp. 2 which can grow in the

TABLE I

Salt tolerance and osmoregulation in various organisms.

Organism	NaCl Tolerance (M)	Intracellular solute contributing to osmotic adjustment	Reference
Non-halophilic bacteria			
Escherichia coli	0-1.4	glutamate α-aminobutyrate proline	(2)
Klebsiella aerogenes	0-1.7	glutamate proline	(2)
Pseudomonas aeruginosa	0-1.7	glutamate	(2)
Staphylococcus aureus	0-3.5	proline	(2)
Moderately halophilic bacteria			
Vibrio costicola	0.2-4	KCl, NaCl	(4)
Extremely halophilic bacteria			
Halobacterium spp.	2.6-saturated	KCl, NaCl	(1)
Halotolerant algae			
Dunaliella spp.	0.2-saturated	glycerol	(1)
Halotolerant yeasts			
Saccharomyces rouxii	0-3	polyols	(1)
Halotolerant plants			
Triglochin marifina	0-0.6	proline	(9)
Hordeum vulgare	0-0.6	proline	(9)

presence of up to about 3.5M NaCl (6). Very little is known about
the mechanism by which this organism adapts to high salt concentration. Genetic systems consisting of generalized transducing phages
and DNA transformation have been developed in this organism (7,8).
It is, therefore, an interesting organism for study of the biochemical genetics of high salt tolerance. Several non-halophilic bacteria
seem to respond to increasing NaCl concentration according to mechanism 1) discussed above. Evidence has been presented that amino
acids can serve an osmoregulatory function, reaching very high pool-
concentrations (Table 1). Several gram-positive bacteria accumulate
proline, while the glutamate-pool increases in several gram-negative
bacteria (2). It is interesting to note that some of the biochemi-
cally and genetically best characterized bacteria, like Escherichia
coli and Klebsiella aerogenes, can grow at NaCl concentrations
above sea water concentration (2). The activity of enzymes involved
in glutamate synthesis, especially glutamate dehydrogenase, in these
organisms are sensitive to the K^+ concentration. According to
Measures (2), the cells first respond to osmotic stress by loosing
water, thereby increasing the intracellular K^+ concentration, which
then leads to increased glutamate production. These bacteria may,
therefore, serve as useful model-organisms for understanding the biochemical mechanisms of low to moderate salt tolerance, and for identi-
fying the genes involved.

Several eucaryotic organisms can also tolerate high NaCl concentrations, and some have been reported to respond by accumulating high
pool-concentrations of glycerol or other polyols (Table 1). Evidence
has been presented that some salt tolerant plants respond to salt by
increasing their proline or betaine pools (Table 1; for review, see
9). Betaine or related compounds may also play a role in osmoregulation in a halotolerant bacterium (10). Studies of the mechanism of
salt tolerance in bacteria may, therefore, provide a basis for
understanding, and ultimately increasing the salt tolerance of important crop plants.

Many of the microorganisms responding to salt according to mechanism 1) discussed above can adapt to a remarkably wide range of salt
concentrations. One reason for this may be that the solutes used for
osmoregulation are less inhibitory to the intracellular enzymes than
KCl and NaCl. One could imagine that the simplest genetic event
leading toward salt tolerance might be the genetic derepression of a
secondary metabolic pathway to the status of a primary pathway whose
new function is to provide osmotic balance to the cell. As we have
seen, a whole series of metabolites can apparently serve this purpose
in different organisms. The most effective compounds are expected
to be those that 1) have a high potential for maintaining osmotic
balance, i.e., have low molecular weights; 2) have a chemical configuration limiting their futile escape into the environment; 3) are
metabolically least expensive to synthesize; and 4) are nontoxic to

other vital cellular processes. If this notion of multiple mechanisms of salt tolerance could be experimentally substantiated it would
be very encouraging for genetic engineers because it would mean that
a number of independent genes (operons) could be isolated each
leading to some degreee of salt tolerance. The methods available for
isolating salt tolerance genes are described in the next section.

RECOMBINANT DNA TECHNOLOGY

It seems appropriate at this point to summarize some recent
advances and implications of recombinant DNA technology as a prerequisite to the discussion of potential application for salt tolerance. Assuming that a source of salt tolerance genes has been identified in bacteria (see above section), the next step is to utilize
current recombinant DNA technolgoy for isolation (cloning) and propagation (amplification) of the DNA.

In recent years, techniques have been developed that allow the
in vitro manipulation of genetic information and the construction of
recombinant DNA molecules from a wide group of living organisms (11).

The use of restriction endonucleases has become a fundamental
part of the analysis and restructuring of DNA (12). Restriction
endonucleases are enzymes that recognize and cleave specific base
sequences in DNA. Enzymes are designated by an abbreviation of the
name of the organism from which the enzyme was isolated. The enzyme
EcoRI, for example, was isolated in Escherichia coli, while HaeIII
was isolated in Hemophilus aegypticus. Many restriction enzymes,
such as EcoRI, cleave DNA in a manner that generates short, non-co-
valently complementary termini that can be reannealed after digestion
to generate hybrid DNA molecules. The nicks in the hybrid molecules
can be sealed by the enzyme DNA ligase to regenerate the functionally
intact double stranded DNA molecule. Other enzymes such as HaeIII,
cleave in a manner that leaves no overlapping termini. In a process
known as "blunt end ligation," T-4 DNA ligase can join these blunt
ends to one another, generating covalently intact double stranded DNA
molecules. The abilities to digest DNA in a site specific manner and
to repair double stranded breaks in DNA are the basis for the in vitro
manipulation of genetic information.

While the use of restriction enzymes allows the isolation of
specific DNA fragments, a particular gene may represent on the order
of 0.1% of the cellular genome. In order to obtain sufficient quantities of the gene to allow its analysis, one must be able to replicate the purified DNA fragment containing the gene. In some instances, it may be desirable to introduce a specific gene into bacteria in order to alter the bacterial phenotype. Replication of
specific DNA fragments can be accomplished by the use of bacterial

plasmids (12). Plasmids are small extrachromosomal elements that are able to replicate in bacteria. Their presence can be detected by a variety of phenotypic markers, such as the generation of anti-biotic resistance. A plasmid can be cleaved with a restriction en-zyme and a DNA fragment can be ligated into the cleavage site. When this is done in a manner that does not disrupt genes necessary for the replication of the plasmid or the expression of phenotypic mark-ers, the recombinant plasmid can be inserted into bacteria. The plasmid will replicate and generate many copies of itself. When large amounts of the cloned DNA fragments are desired for in vitro analysis, digestion of the recombinant plasmid with the same restriction enzyme that was used during its construction will release the DNA fragment. This DNA fragment can be purified from the plasmid DNA and used in in vitro studies of gene function. In some instance, genes present on the DNA fragment inserted into the plasmid may be functionally ex-pressed, allowing alteration of the bacterial phenotype. Ratzkin and Carbon (13) have been able to complement leu⁻ and his⁻ auxotrophs of E. coli using cloned fragments of yeast DNA.

The utility of these techniques in the in vitro manipulation of bacterial phenotypes was demonstrated by the construction of a bac-terial plasmid that produces a synthetic precursor to the human hor-mone somatostatin. Somatostatin is a tetradecapeptide of known amino acid sequence. A short piece of DNA containing the genetic informa-tion for the somatostatin was chemically synthesized. This DNA frag-ment was inserted into a plasmid containing the control elements of the lactose utilization operon and the gene for β-galactosidase in such a manner that transcription and translation results in the syn-thesis of a fused gene product consisting of a portion of β-galacto-sidase fused to somatostatin. This synthetic precursor to somato-statin is biologically inactive, but treatment of the protein with cyanogen bromide cleaves at a methionine residue, releasing somato-statin from the β-galactosidase fragment. In the presence of an in-ducer of the lac operon, bacteria containing this plasmid synthesize an inactive synthetic precursor to the hormone somatostatin. Treat-ment of bacterial extracts with cyanogen bromide results in the re-lease of biologically active somatostatin.

GENETIC ENGINEERING OF SALT TOLERANT LEGUMES

The article by Epstein in this volume describes the dramatic progress being made in the breeding of salt tolerant crops, particu-larly the cereal grains. Today there is considerable worldwide in-terest in leguminous crops such as soybean, alfalfa, clover and peanuts because these crops have evolved the capacity to synthesize their own supply of available nitrogen using the process of biologi-cal nitrogen fixation. This is a case of symbiosis in which benefi-cial bacteria called Rhizobium spp. utilize the energy rich products

derived from solar energy conversion (photosynthesis) to power nitrogen fixation, a process which leads to the cleavage of the very inert
triple bond of gaseous nitrogen. Thus nitrogen is made available to
the plant utilizing radiant energy in contrast to the cereal grains
which must be supplied with heavy doses of commercial fertilizer made
from non-renewable energy resources. The study of salt tolerant legumes involves an additional challenge compared to salt tolerance in
cereals such as barley and wheat. That is, one must take into account both the plant as well as its crucial microsymbiont. In the
following paragraphs, some recent advances in genetic engineering of
Rhizobium spp. (root nodule bacteria) are discussed with an eye toward developing salt tolerant strains suitable to be paired with salt
tolerant legumes. The study of genetic engineering of root nodule
bacteria which we have undertaken is not expressly directed toward
salt tolerance genes but rather the broader goal of key symbiotic
genes affecting energy efficiency, host specificity, and stress tolerance. Salt tolerance in these bacteria is of considerable agronomic
importance with most research to date focusing on isolation of naturally occurring strains capable of nodulation in saline soils (for review see 15). In short, it is clear that some strains of Rhizobium
spp. possess some degree of salt tolerance, genes of importance in
the context of the "gene bank" for Rhizobium spp. as described next.

The symbiotic bacterium R. japonicum was chosen for genetic engineering because this is the most important N_2 fixing organism in
U.S. agriculture today. Approximately 30 million acres of soybean
are inoculated annually with this organism whereas the remaining 20
million acres of this essential crop rely on the natural rhizobial
population in the soil.

The approach was to create a "gene bank" of R. japonicum in
Escherichia coli (E. coli) using recombinant DNA technology. Random
fragments of the chromosome of R. japonicum were linked to a plasmid
vector and introduced into E. coli by transformation. By random
selection of enough clones of E. coli (> 3,000), the total chromosome
of R. japonicum should be represented in the collection.

A gene bank of R. japonicum has been constructed using the restriction enzyme HindIII. The details of this procedure involved
isolation of DNA from R. japonicum 3IIb110 (110) and the plasmid
cloning vehicle, pBR322. These DNAs are digested by restriction enzyme HindIII. HindIII cleaves at staggered sites on the two strands
of DNA, thus generating small cohesive single strand ends. These
ends can anneal with any other HindIII generated ends whether on the
same or different molecules. The cloning vehicle was digested with
HindIII and then treated with bacterial alkaline phosphatase (BAP)
to eliminate self-linkage (16). This allows the plasmid to be linked
exclusively to the rhizobial DNA. HindIII, BAP digested plasmid vehicle, pBR322, was mixed with HindIII digested R. japonicum DNA.

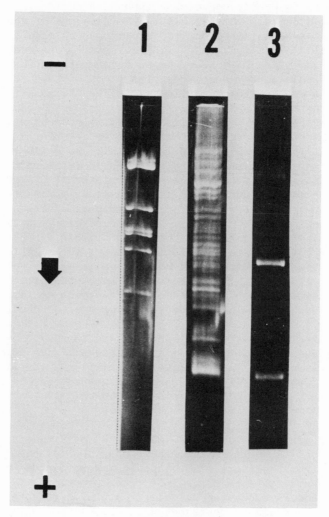

FIGURE 1

Plasmid gene bank of Rhizobium japonicum 3I1b110. The figure shows bands of DNA separated by electrophoresis in agarose gels. Separation is on the basis of size with the largest fragments of DNA at the top. Well 1 and 3:Control DNAs with bacteriophage lambda DNA after digestion with EcoRI (well 1) and cloning plasmid pBR322 without inserted DNA showing two circular forms (well 3). Well 2 contains the total plasmid bank with DNA of Rhizobium japonicum 3I1b110 linked to the cloning plasmid pBR322. The large number of bands of DNA in well 2 indicate a wide variety of different sized fragments of rhizobial DNA were linked to the plasmid vehicle.

The small cohesive ends were allowed to anneal followed by covalent linkage of the plasmid to the rhizobial DNA. Part of this DNA was introduced into E. coli C600RM by transformation. Since the HindIII site on pBR322 is in the promotor region of the genes for tetracycline resistance, insertion of foreign DNA at that site should reduce or eliminate tetracycline resistance. Analysis of the transformed E. coli for the two drug resistance phenotypes carried on pBR322, ampicillin (amp^r) and tetracycline (tet^s) showed that 99% of the cells were amp^r tet^s, with transformation frequencies of 5 x 10^3 and 2 x 10^1 per microgram DNA for amp^r and tet^s, respectively. Therefore, the BAP treatment allowed the creation of a population of E. coli containing inserted DNA (tet^s) in a single step with no involved enrichment procedure. The ampicillin resistant E. coli were kept and constituted the HindIII "gene bank" for R. japonicum.

The size of the inserted DNA will determine how useful a given gene bank will be for genetic or structural analysis. If any one size of fragments predominate in the gene bank, the cloned fragment collection would not be representative of the total rhizobial DNA. The problem is realistic since smaller fragments will be ligated into the vehicle and transformed into E. coli at a higher frequency. Therefore, the size of inserted DNA in the gene bank was determined. A portion of the transformed E. coli was grown with ampicillin and the plasmid content was amplified by chloramphenicol. The plasmid was isolated and analyzed on agarose gels where the largest DNA is found at the top of the gel. Figure 1 (well 2) shows the total plasmid bank after separation on 0.7% agarose gels. The majority of the plasmids contain recombinant DNA as judged by increased size of the plasmids compared to the cloning vehicle alone (well 3). Further analysis of the recombinant plasmids showed the cloned DNA pieces range in size from about 15 x 10^6 to less than 1 x 10^6 daltons with a good distribution of all sizes of fragments. Therefore, it appears the gene bank contains cloned DNA-pieces of sizes representative of the total chromosome of R. japonicum.

In summary, a functional gene bank of Rhizobium spp. should include genes coding for salt tolerance as well as genes permitting growth under a variety of environmental stresses.

ACKNOWLEDGMENT

We gratefully acknowledge the National Science Foundation for their continued support of our studies of biological nitrogen fixation. This report was prepared with the support of the National Science Foundation (Grant Number AER 77-07301). Any opinions, findings, conclusions, or recommendations expressed in this publication are those of the authors and do not necessarily reflect the views of the National Science Foundation. Dr. J. Mielenz is the recipient of a research fellowship from the Continental Grain Company. We especially thank Dr. Charles Hershberger in whose laboratory at the Eli Lilly Corporation the experiments on the rhizobial gene bank were initiated.

REFERENCES

1. Brown, A.D.   (1976) Bacteriol. Reviews 40, 803–846.

2. Measures, J.C.   (1975) Nature 257, 398–400.

3. Larsen, H. (1962) in "The Bacteria" (Gunsalus, I.C. and Stanier, R.Y., eds.) vol. 4, pp. 297–342, Academic Press: New York.

4. Shindler, D.B., Wydro, R.M. and Kushner, D.J. (1977) J. Bacteriol. 130, 698–703.

5. Matheson, A.T., Sprott, G.D., McDonald, I.J. and Tessier, H. (1976) Can. J. Microbiol. 22, 780–786.

6. Thomson, J.A., Woods, D.R. and Welton, R.L. (1972) J. Gen. Microbiol. 70, 315–319.

7. Woods, D.R. and Thomson, J.A. (1975) J. Gen. Microbiol. 88, 86–92.

8. Thomson, J.A. and Woods, D.R. (1973) J. Gen. Microbiol. 74, 71–76.

9. Flowers, T.J., Troke, P.F. and Yeo, A.R. (1977) Ann. Rev. Plant Physiol. 28, 89–121.

10. Shkedz–Finkler, C. and Avi–Dor, Y. (1975) Biochem. J. 150, 219–226.

11. Nathans, D. and Smith, H.O. (1975) Ann. Rev. Bio. 44, 273–293.

12. Roberts, R.J. (1976) Critical Rev. Biochem. 4, 123–164.

13. Ratzkin, B. and Carbon, J. (1977) Proc. Nat. Acad. Sci. 75, 487–491.

14. Itakura, K., Hirose, T., Crea, R., Riggs, A.D., Heyneker, H.L., Bolivar, F. and Boyer, H.W. (1977) Science, 198, 1056–1063.

15. Vincent, J.M. (1974) in "The Biology of Nitrogen Fixation" (A. Quispel, ed.), North Holland: Amsterdam, pp. 277–367.

16. Ullrich, W., Shine, J., Chirgwin, J., Pictet, R., Tischer, E., Rutter, W.J. and Goodman, H.M. (1977) Science 196, 1313–1319.

ECOLOGICAL CONSIDERATION OF BIOSALINE RESOURCE UTILIZATION

John L. Gallagher[1]

University of Georgia Marine Institute

Sapelo Island, Georgia

Ecological impact evaluation should be concurrent with techno-logical development and economic assessment of the feasibility of exploiting a new biosaline resource. The ecological questions focus on both the internal and external functions of the system. The internal functions are processes such as primary productivity, secondary productivity, resource storage, decomposition and mineral cycling. Among the factors which determine the degreee to which the natural state of these processes will be altered are: the type of resource being exploited, the buffering capacity of the ecosystem, the nature of the expolitation technology and the seasonal timing of the activity.

First, let us consider the case of relatively light impact on the internal functioning of a natural saline wetlands. Spartina patens grows relatively high in the intertidal zone in marshes along the Atlantic and Gulf coasts. It has been harvested mechanically for a number of years (Hitchcock, 1972). This activity obviously removes large portions of the primary production for other than its natural use. Since the energy fixed in these plants is the base of the food web, the storage of energy in the system as detritus will be reduced. Likewise the nutrient pool in the macrophytes will also be shunted away from its usual pathway. Animals dependent on that energy and nutrient source will be reduced if the energy, nutrient or shelter provided by the macrophytes are limiting secondary production. It cannot automatically be assumed that the removal of Spartina patens will be detrimental however, since other factors may be limiting the secondary productivity.

---

[1]Present Address: U.S.E.P.A., Corvallis, Oregon.

Algal growth on the soil surface may be enhanced because macrophyte shading is reduced. Diverting the primary production of the
Spartina from the marsh detritus food web to a more direct link with
man could increase the activity of the algal producers. Although
this has not been shown for Spartina patens marshes Sullivan and
Daiber (1975) demonstrated a similar effect in a Spartina alterniflora marsh in Delaware. An increase in secondary production of
animals feeding on algae rather than macrophyte detritus would likely
result from the clipping. Snails such as, Melampus bidentatus, which
may eat epiphytic algae on plant stubble and the crab Uca pugilator
which forages on the soil surface in low density vegetation areas
(Kraeuter and Wolf, 1974) are likely favored. Carrying the scenario
further, these possible secondary productivity increases coupled
with the loss in protective cover may result in increased predation
of marsh invertebrates. Mink and racoons feed on these invertebrates
during low tide (Shanholtzer, 1974) and fish, such as, drum feed in
the marsh when tidal water inundates the vegetation (Odum, 1966).
Thus, although standing crops may be reduced, secondary productivity
may be higher. There are few data sets which document the impact
of forage removal either mechanically or by grazing animals. In the
latter type of situation Reimold, et al. (1975) reported reduced
primary production for natural food webs and smaller macroinverte-
brate populations in a southeastern salt marsh. Trampling by grazing
animals (donkeys, cattle, horses and pigs) was apparently one of
the major factors in reducing invertebrates. In a review of grazing
on wetland meadows of the World Reimold (1975) cites only 23 references on the use of saline marshlands. Of these only a few report
on the quantitative impact of the activity on the internal functions
of these systems. There is an obvious need for information on the
impact of low level biosaline resource utilization on natural systems
ranging from coastal areas to those in arid regions such as those
discussed by McGinnies in another chapter of this volume. The impact
of such use is relatively slight as long as man acts as a gatherer
of already existing food, fiber and fuel in a controlled fashion.
When he becomes a farmer of marshes, seagrass and algal beds or dune
and desert areas the changes are more significant. The practice of
irrigating coastal deserts with saline water to grow salt resistant
varieties of traditional crops as reported by Epstein et al. in this
volume or newly domesticated halophytes as proposed by Mudie (1974)
and Somers (1975) would totally alter the ecosystem. As these practices develop in our coastal areas, the growing of exotic species
of salt resistant plants adjacent to natural saline ecosystems presents the potential problems of a cultivar becoming wild.There are
all too many examples of cases where an exotic was introduced and
displaced more valuable native species. Care must be taken to minimize such risks with developments proposed in this volume for the
use of saline resources.

Another alteration of the natural system which has involved an
even more basic saline resource has been the conversion of coastal
saline soils to glycophyte agricultural use. In the last century

large acreages of southeast Atlantic coast brackish marshlands were
diked and used for rice culture. Under present economic conditions
maintenance of the structures has not been feasible. A number of
acres of old rice fields south of Darien, Georgia on the Altamaha
River Estuary are shown in Figure 1. In most of the area the dikes
are breached and the marshland has returned to at least a semi-
natural state. Some areas near the center of the photograph have
intact dikes and are managed for waterfowl habitat by the State
Department of Natural Resources.

In the past 50 years similar diking has taken place in many
of the west coast marshes. Some of the acreage is in intensive row
crop cultivation but much is in low intensity agricultural uses such
as, pasture and native grass hayfields. These uses of saline resources alter the natural internal functioning of the systems much
more than the earlier examples. However, some processes such as
denitrification may occur very much the same in anaerobic soils
under rice culture as in brackish tidal marshes.

Do we as a society wish to preserve a certain percentage of the
potentially convertable biosaline resources in the natural state so
they can carry out their natural internal functions? The judgments
are essentially political once the ecological and socio-economic
factors have been evaluated.

As long as we are dealing with the internal functions of systems, the decisions although complex and difficult are relatively
straight forward. They effect only the area used and the geographical extent of the impact is known. The most challenging ecological
considerations arise when the external functioning of these systems
is examined.

When the impact of various kinds of biosaline resource exploitation on internal functions was discussed the effects were about the
same whether we were concerned about terrestrial (coastal deserts,
dune complexes or other uplands) or aquatic (marshland, algal beds,
seagrass beds, mangrove forests) systems. When we turn our attention to the external functions, that is, the interactions of a system
with an adjacent system, the ecological implications of exploitation
are much more wideranging with aquatic systems. These systems are
bathed in a high density fluid which can float particles and in which
all substances are at least slightly soluble. In the case of most
saline aquatic systems tides and tidal currents further increase
the connectivity between adjacent ecosystems. The interactions between five coastal wetlands are shown in Figure 2. Harvesting mangrove swamp forests for fuel or timber, for example, would impact
several adjacent areas through either the role detritus plays or
because of the habitat changes of aquatic organisms during various
stages of their complex life cycles. In some ways the inter-

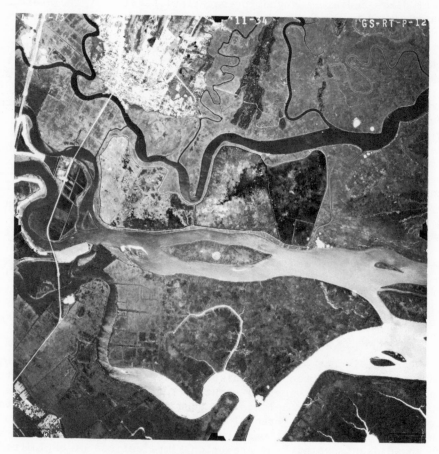

FIGURE 1

Aerial view of abandoned rice fields south of Darien, Georgia.

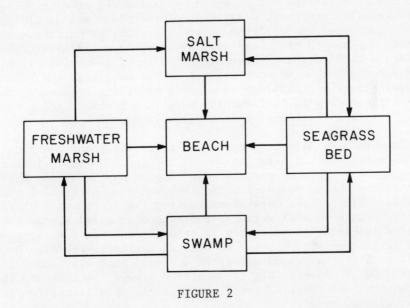

FIGURE 2

Connectivity of tidal wetland ecosystems (from Gallagher, in press).

active nature of adjacent aquatic ecosystems works against projects
which are designed to exploit a particular resource. On the other
hand, the effect of resource utilization on the internal functions
may be reduced by compensatory interaction with adjacent systems.
In the earlier example of harvesting forage from intertidal wetlands
it was evident that nutrient cycles would be disrupted by placing
a large man-made leak in the system. In practice the loss of nutrient resulting from biomass removal may be partially replaced by the
input of nutrients from incoming tidal water. Obviously, the more
regularly flooded wetland would receive the most potential supplement. In fact, the ability of the marsh to absorb the nutrient is
a key factor. The response of 16 marshes along the Atlantic coast
to nitrogen fertilizer was reported by Gallagher, et al (1977).
Under the pulse treatment used, most of the plant stands did not
exhibit increased productivity. The importing of nutrients represents an example of energy subsidy which Odum (1974) attributes
to tidal wetland systems under his category of a subsidized solar
powered ecosystem. The more natural energy subsidies can be used
in developing the technology to exploit biosaline resources, the
greater the saving in fuel subsidy which has become a major ecological cost in conventional modern agriculture.

Thus far, we have been concerned primarily about the impact of
biosaline resource utilization on the chemical and biological aspects
of the system. In some cases a major impact may be on the physical
component of the surrounding area. For example in Figure 1 the
original diking of the marshlands removed the storm water storage
capacity of the coastal wetlands during medium sized storms and
resulted in higher flood waters in other areas. In designing such
dikes for future exploitation of coastal wetlands, the dike height
should be such that flooding of the enclosed land will occur prior
to flooding of adjacent highly developed areas. In that way the
altered area can function for flood control, as well as, for growing
saline adapted crops.

The development of heretofore under-utilized resources seems
inevitable in view of pressures for increased food, fiber and fuel
production. The only questions relate to how to integrate ecological requirements, socio-economic considerations, and agricultural
technology into programs which will benefit mankind in the long
term.

A series of questions, such as, those listed below should be
compiled during the early stages of developing a biosaline resource.

1. What are the federal, state, and local regulations concern-
    ing the proposed natural resource to be altered?

2. What is the value of the area as a physical buffer between
    the forces of nature and man-made ecosystems?

3. What are the food web relationships within the ecosystem before and after it is altered?

4. What are the interactions (nutrients, sediments, energy flow) between this ecosystem and surrounding ecosystems before and after the resource is developed?

5. What opportunities are there for mitigating the adverse affects of the development by enhancing similar natural systems in the area?

6. What are the social and aesthetic consequences of the resource development?

Ideally these questions should be quantitatively answered, but realistically with our present state of knowledge, most answers will be qualitative and many will be speculation.

REFERENCES

Gallagher, J.L., F.G. Plumley and P.L. Wolf, 1977. Underground biomass dynamics and substrate selective properties of Atlantic coastal salt marsh plants. Technical Report D-77-28. U.S. Army Engineer Waterways Experiment Station, Vicksburg, MS.

Gallagher, J.L., In Press. Estuarine angiosperms productivity and initial photosynthate dispersion in the ecosystem. In Wiley, M. (ed.) Estuarine Interactions. Academic Press Inc.

Hitchcock, S.W., 1972. Fragile nurseries of the sea. National Geographic 141: 729-65.

Kraeuter, J.N. and P.L. Wolf, 1974. The relationship of marine macroinvertebrates to salt marsh plants, p. 449-462. In Reimold, R.J. and W.H. Queen (eds.) Ecology of Halophytes. Academic Press, Inc. 605 pp.

Mudie, P.J., 1974. The potential economic uses of halophytes, p. 565-597. In Reimold, R.J. and W.H. Queen (eds.) Ecology of Halophytes. Academic Press, Inc. 605 pp.

Odum, E.P., 1966. Regenerative systems: discussion (A.H. Brown, discussion leader). In Human Ecology in Space Flight (D.H. Calloway, ed.). The N.Y. Acad. Sci. Interdisciplinary Communications Program, New York, pp. 82-119.

Odum, E.P., 1974. Halophytes, energetics and ecosystems, p. 599-602. In Reimold, R.J. and W.H. Queen (eds.) Ecology of Halophytes. Academic Press, Inc. 605 pp.

Reimold, R.J., 1976. Grazing in weland meadows, p. 219-225. In
  Wiley, M. (ed.) Estuarine Processes I. Academic Press, Inc. 541 pp.

Shanholtzer, G.F., 1974. Relationship of vertebrates to salt marsh
  plants, p. 462-472. In Reimold, R.J. and W.H. Queen (eds.) Ecology
  of Halophytes. Academic Press, Inc. 605 pp.

Somers, G.F. (ed.), 1975. Seed-bearing halophytes as food plants.
  Proceedings of a conference. Univ. of Delaware, Newark, p. 15-17.

Sullivan, M.J. and F.C. Daiber, 1975. Light, nitrogen and phosphorus
  limitation of edaphic algae in a Delaware salt marsh. J. Exp.
  Mar. Biol. Ecol. 18: 79-88.